The
MODERN
MYTHS

The

MODERN
MYTHS

Adventures in the Machinery

OF THE

Popular Imagination

———

PHILIP BALL

THE UNIVERSITY OF CHICAGO PRESS
CHICAGO & LONDON

The University of Chicago Press, Chicago 60637
The University of Chicago Press, Ltd., London
Published 2021
Printed in the United States of America

30 29 28 27 26 25 24 23 22 21   2 3 4 5

ISBN-13: 978-0-226-71926-9 (cloth)
ISBN-13: 978-0-226-77421-3 (e-book)
DOI: https://doi.org/10.7208/chicago/9780226774213.001.0001

Library of Congress Cataloging-in-Publication Data

Names: Ball, Philip, 1962– author.
Title: The modern myths : adventures in the machinery of the popular
    imagination / Philip Ball.
Description: Chicago : University of Chicago Press, 2021. | Includes
    bibliographical references and index.
Identifiers: LCCN 2020044847 | ISBN 9780226719269 (cloth) |
    ISBN 9780226774213 (ebook)
Subjects: LCSH: Literature and myth. | Myth in mass media.
Classification: LCC PN56.M94 B35 2021 | DDC 809/.915—dc23
LC record available at https://lccn.loc.gov/2020044847

♾ This paper meets the requirements of ANSI/NISO Z39.48-1992
(Permanence of Paper).

Myths often do a lot of theoretical good, while they are still new.

GILBERT RYLE, *The Concept of Mind*

A whole volume could well be written on the myths of modern
man, on the mythologies camouflaged in the plays that he
enjoys, in the books that he reads. The cinema, that "dream
factory," takes over and employs countless mythical motifs.

MIRCEA ELIADE, *The Sacred and the Profane*

CONTENTS

✿

Chapter 4

UNCHAINING THE BEAST
The Strange Case of Dr. Jekyll and Mr. Hyde (1886)

[130]

✿

Chapter 5

THE BODY AND THE BLOOD
Dracula (1897)

[165]

✿

Chapter 6

WHO SHALL DWELL IN THESE WORLDS
The War of the Worlds (1897)

[223]

✿

Chapter 7

REASON WEARS A DEERSTALKER
The Sherlock Holmes stories (1887–1927)

[275]

✿

Chapter 8

I AM THE LAW
Batman (1939–)

[311]

✿

Chapter 9

MYTHS IN THE MAKING,
MYTHS TO COME

[351]

✿

Chapter 10

THE MYTHIC MODE

[370]

Chapter 1

HOW CAN
A MYTH
BE MODERN?

We can start almost anywhere, and there's no virtue in being high-brow about it. So why not with the 2004 movie *Van Helsing*, starring Hugh Jackman as the famous vampire-hunter? Aside from his name and calling, this youthful, dark-locked, brawny action hero has nothing in common with Dracula's venerable nemesis in Bram Stoker's novel. It's not only the bloodsucking count he's pursuing but a legion of monsters that includes Frankenstein's creature and the brutish Mr. Hyde. We know we are in safe hands here, for Jackman and his screen lover, Kate Beckinsale, are genre stalwarts, much as Christopher Lee, Peter Cushing, Boris Karloff, and Bela Lugosi were in their day. The texture of the movie is familiar too: comic-strip gothic, lit by moonlight and bristling with razor-fanged CGI beasts, the framing and aesthetic echoing the graphic novels of Frank Miller and Alan Moore. The imagery is iconic: here is Dracula's castle, much as it was in the 1931 Lugosi movie directed by Tod Browning. There is Frankenstein's electrified laboratory, full of sparks and shadows, where Karloff's creature rose up in the same year. And here comes the flat-headed monster himself, his patchwork skull apt to fly open, in slapstick fashion, to reveal a sparking brain. We forgive Van

Helsing for becoming a werewolf and killing his paramour; heroes these days are prone to such things.

Van Helsing is a stupendously silly homage, about as scary and unsettling as a soap opera, and I rather enjoyed it.

There doesn't, though, appear to be much we can learn about our modern myths from this sort of good-natured romp, with its relentless computer-game dynamic, cheap sentimentalism, and makeshift, Frankensteinian mosaic of motifs. Surely it does for these myths only what Ray Harryhausen did for the classical myths of Greece, turning them into a parade of cinematic effects with not a care for coherence or poetry. (I don't mean that in a bad way.)

Still, you know what I'm talking about, don't you? You know what *Van Helsing*, in its clumsy, carefree fashion, is up to. You know that the seam it is exploiting, the language it is using, is indeed that of myth.

***

I hope I am not being condescending when I suspect that few among *Van Helsing*'s target audience—hormonal adolescents eager for violent action, sexy vampires, and spectacle—will have read Stoker's *Dracula*, Mary Shelley's *Frankenstein*, or Robert Louis Stevenson's *The Strange Case of Dr. Jekyll and Mr. Hyde*. These are stories that everyone knows without having to go to that trouble. They have seeped into our consciousness, replete with emblematic visuals, before we reach adulthood. I met Robinson Crusoe in a black-and-white French television series from the 1960s, in which nothing much seemed to happen save for the discovery of that momentous footprint in the sand; like many of my generation, I can hum the theme tune today. Frankenstein arrived as a glow-in-the-dark model kit of the Karloff incarnation, arms outstretched to claim his next victim. (It was not actually "Frankenstein," of course, but his monster, although the box didn't tell me that.) The stingray alien craft of George Pal's 1953 movie *The War of the Worlds* were ominous enough to distract us from the wires from which they dangled. Dracula—well, he was Christopher Lee, everyone knew that.

This cultural osmosis is how we learn our modern myths. For myths are what these stories are, and to suggest (as some purists do) that the Hammer films or Hollywood adaptations traduced the "real" story is to miss their point. In this book I propose that the Western world has, over the past three centuries or so, produced narratives that have as authentic a claim to mythic status as the psychological dramas of Oedipus, Medea, Narcissus, and Midas and the ancient universal myths of creation, flood, redemption, and heroism.

Myths have no authors, although they must have an origin. They escape those origins (and their originators) not simply because they are constantly retold with an accumulation of mutations, appendages, and misconceptions. Rather, their creators have given body to stories for which retellings are deemed *necessary*. What is it in these special tales that compels us so compulsively to return to them? Each age finds different answers, and this too is in the nature of myth.

Why are we still making myths? Why do we need *new* myths? And what sort of stories attain this status?

In posing these questions and seeking answers, I shall need to make some bold proposals about the nature of storytelling, the condition of modernity, and the categories of literature. I don't claim that any of these suggestions is new in itself, but the notion of a *modern myth* can give them some focus and unity. We have been skating around that concept for many years now, and I can't help wondering if some of the reticence to acknowledge and accept it stems from puzzlement, and perhaps too a sense of unease, that *Van Helsing* is a part of the story. Not just that movie, but also the likes of *I Was a Teenage Werewolf* and *Zombie Apocalypse*, as well as children's literature and detective pulp fiction, not to mention queer theory, alien abduction fantasies, video games, body horror, and artificial intelligence.

In short, there are a great many academic silos, cultural prejudices, and intellectual exclusion zones trammeling an exploration of our mythopoeic impulse. Even in 2019, for example, a celebrated literary novelist dipping his toe into a robot narrative could suppose that *real* science fiction deals in "travelling at 10 times the speed of light in anti-gravity boots," as opposed to "looking at the human dilemmas of being close up." It is precisely because our modern myths go everywhere that they earn that label, and for this same reason we fail to see (or resist seeing) them for what they are. As classical myths did for the cultures that conceived them, modern myths help us to frame and come to terms with the conditions of our existence.

Evidently, this is not all about literary books. Myths are promiscuous; they were postmodern before the concept existed, infiltrating and being shaped by popular culture. To discern their content, we need to look at comic books and B-movies as well as at Romantic poetry and German Expressionist cinema. We need to peruse the scientific literature, books of psychoanalysis, and made-for-television melodramas. Myths are not choosy about where they inhabit, and I am not going to be choosy about where to find them.

The idea of a modern myth, admits literary critic Chris Baldick,

> simply should not exist, according to the most influential accounts of what a 'myth' is . . . the consensus in discussion of myths is that they are defined by their exclusive anteriority to literate and especially to modern culture. 'Myth' . . . is a lost world, to which modern writers may distantly and ironically allude, but in which they can no longer directly participate.

Like Baldick, I think this traditional view is entirely the wrong way to understand what a myth is. I will try to explain why.

The word *myth* is bandied about with dreadful abandon, as if it doesn't much matter what it means. Often now it is used to stigmatize a widely held misconception: the myth that the moon landings were staged in a Hollywood studio, or that Nelson Mandela died in the 1980s, or that eating carrots improves your eyesight. It can mark a clumsy attempt to disguise a franchise as an epic: *Star Wars* is mythic, right? (We'll see about that.) Or perhaps a story becomes a myth just by being much retold?

Experts, I fear, aren't much help here. You can collect academic definitions for as long as your patience lasts. "The word *myth*," as Northrop Frye rightly says, "is used in such a bewildering variety of contexts that anyone talking about it has to say first of all what his chosen context is." Folklorist Liz Locke put it more bluntly in 1998: "such a state of semantic disarray and/or ambiguity is truly extraordinary."

Finnish folklorist Lauri Honko nonetheless gives a definition that kind of *sounds* like what you'd expect from an expert: a myth is

> a story of the gods, a religious account of the beginning of the world,
> the creation, fundamental events, the exemplary deeds of the gods as
> a result of which the world, nature and culture were created together
> with all parts thereof and given their order, which still obtains. A myth
> expresses and confirms society's religious values and norms, it provides
> a pattern of behavior to be imitated, testifies to the efficacy of ritual
> with its practical ends and establishes the sanctity of cult.

This is a fair appraisal of how myth has often been regarded by anthropologists. But it is fraught with dangers and traps. Like the word "anthropology" itself, it seems to offer an invitation to make myth something "other": something belonging to cultures not our own, and most probably to ones that even in the circles of liberal academics retain an air of the "primitive." Gods, creation, ritual, cult: these are surely notions that we in the developed world have left behind and only pick up again with an air of irony. Our "gods" are not real beings or agencies but metaphorical cravings ("he worships money") or celebrities (rock gods and sex goddesses). Our rituals, not invested with any spiritual content (except in churches, mosques, and temples attended by the devout), are empty or, at best, time-honored habits we indulge for the social sanction they offer—marriages and funerals, say. Our cults are brainwashing sects isolated from regular society. And so likewise, our "myths" are things that many people believe to be true but that aren't really—or, as "urban myths," oft-told tales that likely never happened.

Our popular narrative, then, is that we shed mythology in its traditional sense, probably during the process that began in the Enlightenment, in the course of which the world became "disenchanted" by the advance of science, and that has led since to a secular society on which the old deities have lost their grip. We grew out of gods and myths because we acquired reason and science.

This picture is tenacious, and I suspect it accounts for much of the resistance to the notion (and there is *a lot* of resistance, believe me) that anything created in modern times might deserve to be called a "myth." To accept that we have never relinquished myths and mythmaking might seem to be an admission that we are not quite modern and rational. But all I am asking, with the concept of myth I use in this book, is that we accept that we have not resolved all the dilemmas of human existence, all the questions about our origins or our nature—and that, indeed, modernity has created a few more of them.

One objection to the idea of a modern myth is that, to qualify as myth, a story must contain elements and characters that someone somewhere believes literally existed or happened. Surely myths can't emerge from works of fiction! The anthropologist Bronislaw Malinowski asserted as much, saying of myth that "it is not of the nature of fiction, such as we read today in a novel, but it is a living reality, believed to have once happened in primeval times, and continuing ever since to influence the world and human destinies."

But this is simply the grand narrative with which Malinowski and his generation framed their study of the myths of "primitive" cultures. It allows us to insist (as they wished to) that we advanced societies have no myth left except religion (and even that is no longer believed in quite the same way as it was a couple of centuries ago).* As Baldick puts it, in

---

* Malinowski's distinction becomes (inadvertently) more fruitful once we recognize the *desire to believe*—and to cultivate belief—that is evident in modern myth. Many modern myths, from *Robinson Crusoe* onward, have been presented as "true" accounts, for example, by means of epistolary and diaristic devices. The most avid fans of Sherlock Holmes talk of him as a real person and furnish him with biographies and obituaries. As for *The War of the Worlds*, the literal belief in the story when it was broadcast on US radio in 1938 is notorious. Malinowski would have done better to describe part of the work of myth as blurring what is real and what is fictional.

this view "myth is the quickest way out of the twentieth [and now the twenty-first] century."

Even in its own terms, however, Malinowski's definition is tendentious. Did the author(s) we know as Homer believe he was merely writing history, right down to, say, Athena's interventions in the Trojan war? To assert this would be to neglect the long and continuing scholarly debate about what Homer was really up to—was he, for instance, a skeptic, or a religious reformer? Worse, it would neglect the even longer and profound debate about *what storytelling is up to*. It might be unwise to attach any contemporary label to Homer, but one that fits him more comfortably than most is to say he was a poet, and that he used poetic imagination to articulate his myths. Stories like his relate something deemed culturally important and in an important sense "true"—but not as a documentary account of events. Plato admitted as much in the fifth century BCE; are we then to suppose that Greek myth was already "dead" to him?

To ask if ancient people "believed" their mythical stories is to ask a valid but extremely complex question. It is much the same as asking if Christian theologians, past and present, "believe" the Bible. Yes, they generally do—but that belief is complicated, multifaceted, and contentious, and to imagine it amounts to a literal conviction that all the events and peoples described in the holy book occurred as written is to misunderstand the function of religion itself. What's more, while we can adduce a range of interpretations about these beliefs today, it is not clear we can ever truly decide how these correspond (or whether they even need to correspond) to the convictions of the people who created the original text.

Many early anthropologists and scholars of myth (and some still today) thought that myths must be "sacred": they must have the aspect of religious belief, and perhaps have been used in ritual. This was the position taken by Edward Burnett Tylor, one of the nineteenth-century founders of cultural anthropology, and also by James Frazer in his seminal early work on comparative myth, *The Golden Bough* (1890). For Tylor, Frazer, and Malinowski, myth was a prescientific way of understanding the world, and thereby of trying to control it. The French anthropologist Claude Lévi-Strauss argued that in fact myth was a kind of primitive science: logical and concrete in its own terms. Philosopher Karl Popper believed that science arose out of efforts to assess the validity of myth through empirical, rational investigation of its effectiveness. The evaluations of all these commentators stemmed largely from a focus on *creation* myths, which are the easiest to map onto questions about how the physical world is constituted and governed.

This view of myth as primarily religious and at any rate prescientific is certainly a convenient way of keeping myth at arm's length from today's secular, technologically sophisticated society. But to insist that a religious role is a necessary, defining feature of myth would be an arbitrary stricture that tells us little about the social work myths did. It is a view that mistakes function for process—as if to say that the function of a church service is to enable hymns to be sung. It doesn't even fit with the way (as far as we can say anything at all about this) myth functioned in ancient times. As classicist William Hansen has said, there is "no real evidence that I can see that the Greeks regarded myths as 'sacred stories,' unless you take 'sacred' in a very watered-down sense."

It's undeniable that religion and myth are tightly entwined, however—not least because both address cosmic questions about *meaning*, which lie outside the domain of scientific cosmology. For religious scholar Mary Mills, myth offers a "cosmic framework . . . which indicates where cosmic power resides, what it is called, and so how it can be used." To which one might reply: sure, sometimes. The world emerging from the body of the giant Ymir, maybe. Theseus's struggles with the Minotaur or Medea, not so much.

The temptation is to suppose that modern myths are today's replacement for religion. That would be to fall for another a category error, however, for religion might be regarded as a particular social and institutional embodiment of myth—and not vice versa. Religion explores some of the same questions—about life and death, meaning, suffering and fate, origins—and so it is not surprising that we will find our modern, secular myths still returning to religious themes, questions, and experiences. Institutional religion tends, however, to crystallize from this exploration codes of conduct, values, and norms. At its worst, it seeks to escape from the ambiguities of myth by imposing a false resolution to irresolvable questions. At its best, it leaves those doors open.

This brings us to the tidy function ascribed to myth by Honko's definition: to express and confirm society's religious values and norms and provide a pattern of behavior to be imitated. This is the sober, objective rationalist's view of myth as normative narrative. But whose behavior exactly is to be imitated among the Titans and Olympians? Cronus the father-castrator? Zeus the sex-pest and rapist? Dionysius the libertine? Heracles with his boneheaded pragmatism? Self-absorbed Narcissus?*

No, myth is not tidy. Myth is the *opposite* of tidy! Was Gong Gong, who caused the Great Flood of China's formation myth, a wicked demon king who broke one of the Pillars of Heaven and made a hole through which the waters poured (as some versions suggest)? Or was he an opportunist human king who simply tried to exploit the catastrophe? Or was he an inept engineer employed by the Emperor Yao to drain the waters? Or was he in fact the father of the engineer-hero Yü who finally resolved the mess? Take your pick. This, I think, is the best answer we can give to such questions based on the conflicting versions of myths that have survived (while no doubt many others have been lost): *it depends on the story you want to tell.* Gong Gong could represent many influences and agencies, contingent on what the myth-teller wanted the people to believe about flood control and state authority. A myth is typically not a story but an evolving web of many stories—interweaving, interacting, contradicting each other. Boil it down and the story falls away: there are no characters (which is to say, individuals with histories and psychologies), no location, no denouement—and no unique "meaning." You're left with a rugged, elemental, irreducible kernel charged with the magical power of *generating versions of the story.*

---

* Mythologist William Doty suggests that myths involving trickster figures, of which Dionysius might be considered an example, "often convey social norms precisely by describing behaviors taught to be just the opposite of proper living." But if that's so, they are filled with the same ambivalence we will find in modern myths—for such tricksters are almost always not simply wicked but attractive and seductive.

A function of China's flood myth (for example) is thus not simply to give citizens a quasi-religious account of how their nation began, but also—and more importantly—to help them deal with the ever-present fear of massive flooding. It sanctifies the authority of someone who can cope with such a natural disaster, and even proposes a theoretical basis for flood control: carve out channels to let the waters flow, don't dam them up. But more than that: it creates *a vehicle for thinking about the problem*, without offering either a definitive version of events or an unambiguous solution. When a myth becomes a dogmatic social and ethical code, it has become prescriptive religion or something like it: we have arrived at Daoism or Confucianism, say. When a myth permits of a resolution—when it indeed succeeds in what Lévi-Strauss describes as its aim, of providing "a logical model capable of overcoming a contradiction" (though how could it, if the contradiction is real?)—it has become a moral fable. But while it retains ambiguity and contradiction, it stays a myth. It acts not as a cultural code but as what cultural historians Amanda Rees and Iwan Rhys Morus call (apropos the kind of modern myths discussed in this book) a "cultural resource."

It's unfair to judge the early anthropologists too harshly for the limited and inadequate picture of myth they presented, which was of course a product of its time. They deserve credit for taking seriously the function it serves; as Malinowski said, myth "is not an idle rhapsody, not an aimless outpouring of vain imaginings, but a hard-working, extremely important cultural force." Yet it wasn't until around the middle of the twentieth century that such scholars began to recognize the obvious corollary: that the work demanded of this cultural force does not simply vanish with modernity, and that there is nothing essentially ancient about myth. Philosopher Ernst Cassirer was perfectly happy with the idea that myths are still being produced, especially in the political sphere—he regarded Nazism as an instance. "In all critical moments of man's social life," he wrote, "the rational forces that resist the rise of the old mythical conceptions are no longer sure of themselves. In these moments the time for myth has come again."

Still, for Cassirer, mythmaking remained an atavistic throwback, a strategy necessary only in "desperate situations." It took a theologian to dismantle that idea: the German Rudolf Bultmann, who sought to "de-mythologize" the Christian New Testament. To seek historical verifica-tion of the scriptures or reconciliation of their account of cosmogenesis with science is not, he argued, a meaningful pursuit. Rather, their mes-sage is ethical and philosophical. It's the same with all myth, he said: it expresses timeless human experience:

> The real purpose of myth is not to present an objective picture of the world as it is, but to express man's understanding of himself in the world in which he lives. Myth should be interpreted not cosmologi-cally, but anthropologically, or better still, existentially.*

If this is indeed at least a part of what myth is about, then mythmaking can cease only when "the world in which we live" has ceased to change, and when we have solved all of our problems and resolved all of our anx-ieties. That, I suspect we can agree, will not be any time soon.

The idea that myths confirm social norms and values—that they tell us how to behave—is nevertheless especially persistent. What is the myth of Oedipus if not a warning against incest? Yet, on closer inspection, that reasonable-sounding supposition dissolves. Taboos against incest are pretty much universal,† and for good biological reasons: inbreeding promotes genetic dysfunctions, a fact obvious enough from experience long before Darwinism suggested a means of rationalizing it. And for that reason, too, no society really needs a cautionary tale. To suggest that the myth of Oedipus has this utilitarian function is ludicrous—we all know from the outset what a harmful and terrible thing the hapless

---

* Praise for Bultmann must be qualified, however, for he believed that myths have a single meaning, which it is the scholar's task to identify. His critic Karl Jaspers was more astute here, acknowledging that the interpretation of a myth may adapt to different circumstances.

† They have been sometimes set aside, however, when they conflict with notions of hierarchy-preservation, leading to inbreeding in royal and noble lineages from the ancient Egyptians to the Hapsburg monarchs.

Oedipus is doing, and it's obvious he'd never contemplate it if he knew the truth himself. No, the Oedipus myth exists because, even knowing already that incest is bad, still we worry that it might happen. We call the protosexual attraction of a child to his or her parent "oedipal" today not because it looks a bit like what happened in the Oedipus story, but because the Oedipus story encodes the fact that *this is what happens to us.* That's why the story exists: to create a narrative space for thinking about the unsettling fact that a young man awakening to adulthood may start to look at his mother in a different way (and the mother, in turn, might do the same). To be the object of your child's first sexual impulses is unsettling for a parent, but also very common. Such disquieting but widespread, if not universal, experiences create a demand for myth.

We should be wary, though, of concluding that a myth lays hold to a specific dilemma and toys with it. Myths are certainly "about" things, but they are rarely about a single thing. The Oedipus myth explores what we now call oedipal anxieties, but bound up with them are questions about how individuals and societies respond to taboo-breaking, and where responsibility for it lies. These issues change with the times. Did the ancient Greeks also consider the story to be primarily about father-son rivalry and fears around incest, or were they perhaps more impressed with the way it explored the arbitrariness of fate? An openness to multiple readings is a hallmark of modern myth, and surely of classical myth too.

Another line of resistance to the concept of modern myth takes the form of evasion: So, do you mean stories from modern times with a whiff of the supernatural? Say, the "Angel of Mons," glimpsed by troops in World War I? But we have a perfectly good word for such stories already in anthropology: folklore. Now, it's true that there is plenty of academic discussion about where to fix the boundaries between folklore and myth, but this is beside the point. One thing we can say is that modern myths have little, if any, use for the supernatural. On the contrary, they are keen to advertise their modernity. What's more, they are not synonymous with isolated archetypes, beliefs, or apparitions. They are *stories.* That is what *mythos* means.

This is vital to everything that follows, for it is the narrative nature of myths that allow them to do their cultural work. As professor of religion and scholar of myth Sarah Iles Johnston says in the course of outlining her own definition of Greek myth, there are "certain things stories do that simpler statements cannot." They help us to refine our skills, our thoughts, our conception of the world, Johnston says, by allowing us vicariously and without actual hazard to place ourselves in new situations, to consider the responses of the characters, "weighing their choices and considering whether we would do the same under similar circumstances." Ideas formulated as stories are congruent with our experience of the world—it is not simply for the purposes of entertainment but more for the purposes of aiding cognition that they take this form.

So then, what kind of stories serve this role? Fantastical ones often work best: in myth, says Lévi-Strauss, "everything becomes possible." Such stories, says Johnston, "can coax us to look beyond the witnesses of our five senses and imagine that another reality exists, in addition to the reality that we experience every day." The very departure from realism that has led mythical narrative to be derided and belittled in the modern literature of the fantastic, in Gothic, horror, and science-fiction novels, is a key enabling feature of the work of myth.

And yet not everything possible becomes actual in myths. The limitless profusion of possibilities condenses into rather specific, if schematic, narratives. The question is what directs that emergence.

Mythologist Robert Segal simply says that a myth is a story about "something significant"—more precisely, that it "accomplishes something significant for adherents." This of course raises the tricky question of what makes someone an "adherent." As I've said, I don't think adherence need entail a belief in the literal truth of the story. I think we are all adherents of the modern myths I explore in this book, simply by virtue of the fact that we know the core stories and recognize them as being deeply embedded in our culture—and because we want to hear them, again and again, and to juggle with their implications.

The idea that modern myths exist should not really be controversial, for most of the stories I consider in this book are routinely referred to in these terms, without inciting protest. All I am doing, in a sense, is to take this designation seriously and to ask what it means—about these stories in particular and about our need for new myths in general.

Literary scholar Ian Watt was one of the first to argue the case for modern myths. In *Myths of Modern Individualism* he identifies four of them: *Don Quixote, Faust, Don Juan,* and *Robinson Crusoe.* All are good candidates, although one can quibble. (One can always quibble about such selections, and I anticipate and even hope you will do so with mine.) *Faust,* for example, has a history that goes back at least to the biblical story of Simon Magus, and was a morality tale until Mary Shelley spun it into something far more ambiguous and contemporary. And it's hard to see much of a cultural heritage to *Don Juan* after Byron had his way with it, partly because the early forms of the tale are so heavily invested in ideas of divine morality. (But I'll accept the 1966 film *Alfie* among its progeny.)

Where I would respectfully take greatest issue with Watt, however, is in his assertion that "unlike the myths of Greece, the new Romantic myths were conscious inventions; and they were the product of single individuals." On the contrary, I will insist that, to the extent that these stories were conscious inventions, they are not mythical; what seems a necessarily if not sufficient ingredient for myth is a lack of authorial control. What is more, while the founding texts of modern myths are indeed generally the work of individuals (although they need not be), the myth is not identical to its founding text. Myths are the work of a culture (as Virginia Woolf said of *Crusoe*)—but of course they have to start somewhere.

One more bone of contention (Watt's contribution is very important, so let's have it over with): Watt claims modern myths give us more detail than the ancient ones. With the original texts still in our possession (or at least in our archives), that is not surprising—but those details are, in general, *precisely what we discard* as story turns to myth. I summarize each of my modern myths in a paragraph at the start of their respective chapters, and I suspect most readers could produce something similar, regardless of whether they've read the books in question. Do you remember Crusoe's first shipwreck off the Norfolk coast, or the tragic fate of Safie in *Frankenstein*, or who mocked up the Hound of the Baskervilles with luminous paint? I thought not.

Rather than lay out a set of rigid qualifications, I want merely to suggest some general factors that our modern myths seem to share:

- They are stories that lend themselves to many reworkings, some barely recognizable as versions of the original form.
- They don't have morals. They explore human questions that are irresolvable, and they are *polysemic*: able to seed many different interpretations.
- They do, however, have significance—even if, as novelist Pat Barker says of the ancient myths,* "we don't agree on what that significance is."
- They are not consciously invented, merely crystallized—often unwittingly and messily, though sometimes with a degree of genius—by their first teller. As mythologist William Doty states, "Myths are not the creation of individual authors, but collective products elaborated over relatively long periods of time."
- They often burst forth from works of rather prosaic literary merit. If a story is too carefully crafted, the characters too finely drawn, it can't be altered without losing the essence, and so becomes a poor vehicle for myth.
- Their core narratives can be stated concisely—but not *too* concisely. Taken too far, the process of condensation leaves only a residue of banality: "we have a dark side," "we might invent something

---

* Barker, one of several writers commissioned to write "modern" versions of ancient myths, revisits *The Iliad* in *The Silence of the Girls*, reshaping the story (as do others engaged in the project) to suit the demands of the present. Barker, for example, offers a female and often feminist perspective, while Margaret Atwood and Jeanette Winterson let modern science infiltrate old narratives. We have been doing much the same for years, centuries even, with our modern myths, but without quite recognizing that is what we are up to.

dangerous," "we fear invasion." Only a few stories are myths, but there are no myths without stories.

⟶◆⟵

Well, OK then, you can have your modern myths, the skeptic might (at last) allow. But what does that mean, except that there are stories we keep returning to? Is *Pride and Prejudice* a modern myth? *A Christmas Carol*?

No and no—but that's just my opinion. I'd be happy for someone to make a case for one or both of these novels as myths. But you would need to do more than say simply that people still read them and base films, plays, and TV series on them. Do their messages and characters ever really change, even when we boldly set them in some time and place other than nineteenth-century England? Is a telling of *A Christmas Carol* ever anything other than the story of a miserly misanthrope turned kindly by seeing how lonely and wretched his behavior has made him? Is *Pride and Prejudice* ever about anything except how we can be too hasty, in affairs of the heart, to judge others and interpret their nature and intentions? I would contend that the goals, trajectories, and characters of these two popular stories are too fixed, too lacking in ambiguity, to make them mythical. But I'm open to persuasion (and perhaps even to *Persuasion*).

There are mythical and archetypal elements in Shakespeare too, but his works cannot engender their own myths, partly because if you lose Shakespeare's words then you lose their substance. Dickens's stories too would be slight fare without their finely wrought characters and dialogue. Fictions of this kind, with characters in whom we can believe and with whom we can sympathize, and with plots that intrigue, inspire, move, and astonish, can always hope to find an audience. But myths are something else, and serve a different function. They erect a rough-hewn framework on which to hang our anxieties, fears, and dreams. The timber must be rather crudely cut and loosely assembled so that the frame does not too tightly constrain what is suspended on it. As Lévi-Strauss said, a myth's true substance "does not lie in its style, its original music or its syntax, but in the *story* which it tells." Myths arise by accident, not design—but they become myths because they provide something our cultures need. The themes and meanings of modern myths are not necessarily located in the books that initiated them but are decided collectively and dynamically, constantly shifting with the times. We can agree on the plots, but not on the meanings.

This characteristic addresses yet another objection to designating anything as a modern myth: that it was *written by someone*. Myths are no more written by an individual than the First World War was caused by an assassination. The originating text—if there is one—supplies the material for a culture to work on. "The popular imagination," writes cultural critic James Twitchell, "is continually finishing these stories, plugging loopholes, locating gaps, reiterating important characteristics, and, most important, introducing new players, especially victims." The media that nurture and convey myths need not, he adds, be literary:

> The myth may appear in a literary text and it may appear on the movies
> or on television, but it may also be told in comics, on bubble gum
> cards, on dolls, on Halloween masks, in toys of all description, even in
> the names of breakfast cereals . . .

. . . to which we might now add: on the internet, in video games, in social media. What Roland Barthes calls "mythical speech" can show up in "photography, cinema, reporting, sport, shows, publicity." Myths themselves aren't written down; they live in the warp and woof of the human world.

And while we may make aesthetic judgments about these revisions and retellings and refractions, this need not reflect their importance for the evolving myth in question. "All versions of each text are valuable," says literary scholar Michael Preston, although some are better than others. The texts that seep most deeply into the cultural discourse—Preston cites *Robinson Crusoe*, *Gulliver's Travels*, and *Alice in Wonderland*— become malleable to the form required of them. "If one cannot simply re-read a text so that it accords with one's values, then it is adapted," Preston says.

Notions of purity or authenticity then become otiose, just as they do for classical myths. "The mythical value of the myth is preserved even through the worst translation," Lévi-Strauss asserts—which is what makes it the opposite of poetry (and undermines the idea that Shakespeare wrote myths). But we can go further than that: poor translation often forces us to see the gist even more clearly. As Twitchell says, "The more derivative and exploitative the version, the more revealing it may be."

The skeptic perhaps now grants, begrudgingly, that she can see what we're getting at. So, then, what are these modern myths? *The Lord of the Rings*? *Star Wars*? After all, wasn't Joseph Campbell, doyen of the Hero myth, consulted by George Lucas?

Sorry, but neither of these tales qualifies. (Nor does *Harry Potter*, I'm afraid.) For *a modern myth is not an old myth retold in modern times.* We can argue about how well *Star Wars* integrates Campbell's conception of the Hero's Journey, but it adds nothing substantial to it. (A more interesting question might be how satisfactorily the elements of classical myth can be blended with the demands of a modern movie franchise.) Certainly, these Manichean tales demonstrate that our appetite for the old myths is undiminished—but we knew that already, not least from that most characteristic modern reimagining of the Hero myth: the Western. But let's not mistake them for something new.

Lucas wanted quite explicitly to give *Star Wars* a mythic structure. And that is the whole point: he took this structure ready-made from Campbell and set it in space. New myths—genuine modern myths— have no such template. They aren't wrought with a mythopoeic aim in mind. I don't think anyone has ever sat down with the express intent of writing a modern myth and actually succeeded in doing so. I suspect no one ever will.

What is it, then, that distinguishes modern myths from the exploits of Thor or Theseus? It is this: these stories *could not have been told* in earlier times, because their themes did not yet exist. Modern myths explore dilemmas, obsessions, and anxieties specific to the condition of modernity.

It's this novelty that explains why modern myths are being created at all: because the modern world confronts us with questions and problems that have no precedent in antiquity. Modern industrialized cultures face challenges that our ancestors did not: in the search for meaning within an increasingly secular society; in the disintegration of close-knit community and family structures; in the opportunities and perils presented by science and technology. And so our new myths deal with issues of identity and status, individualism, isolation and alienation, power and impotence, technological transformation, invasion and annihilation. They speak of scientific discovery and spiritual ennui, sexual dysfunction and erotic displacement, dystopia and apocalypse. Modern myths do not feature kings and queens, dragons and heroes. They draw less distinction between hero and villain, human and monster. We ourselves play the roles of gods, and of course we are as vain, fallible, and compromised as the deities of Olympus and Asgard ever were. The evil forces, likewise, do not manifest as demons and malign deities but lurk inside us all.

Myth is where we go to work out our psychic quandaries: to explore questions that do not have definitive answers, to seek purpose and meaning in a world beyond our power to control or comprehend. By looking at the narratives that have become modern myths, we can examine the contemporary psyche and reveal some of the dilemmas and anxieties of our age: what we dream, what we fear. These stories provide a mental map of our dark thoughts. They are more honest than we dare to be.

Of the seven modern myths I propose here, only one arose from a story written by a woman. I don't pretend to feel comfortable about this, but it confronted me with a choice. I could have pursued gender balance by rooting around for female-authored tales for which I might then patch together a justification for inclusion. But that seemed to me not only unsatisfactory but dishonest. Myths mine our cultural unconscious for anxieties, prejudices, and obsessions, so if most of our modern ones have excluded (sometimes, as we'll see, comprehensively) a female perspective, then that is part of the material we need to work with. To pretend that the situation is otherwise would be to forgo the opportunity to ask what the male dominance of our modern mythopoeia tells us about our culture. I want instead to try to let voices that have been silenced in these stories speak out. They are already doing that. Many commentators, especially women, have laid bare the unsettling and damaging ways in which these modern myths exclude or traduce a female perspective.

Similar considerations apply to the fact that these are all Western myths written by white authors. I don't doubt for an instant that there are modern stories in non-Western cultures that can lay claim to mythical status, and I would love to hear what they are. And the more we can allow the voices of storysmiths of color to be heard, the less our mythic explorations of the dilemmas of modernity will be seen only through this narrow aperture. This is already happening, as we will see—but the gestation period of modern myths is typically longer than the timescale over which racial bias and discrimination has received serious challenge. Again, this is partly the whole point: for some of these myths Eurocentricity, racial anxiety, and xenophobia are at the core. No one should pretend that myths set out to satisfy a modern liberal ethic of inclusion and diversity. They often do the opposite.

My seven myths have, moreover, a specifically Anglophone, and indeed a British, bias. My own cultural blinkers might be partly to blame for that, but I believe it also reflects the historical circumstances: the British imperial hegemony throughout much of the period I am examining (as well as the increasing dominance of American culture during the twentieth century). I can think of a few potential modern myths in non-English languages (you're about to hear about one of them), and would be delighted to learn of others, but I genuinely struggle to identify any that have had quite the global pervasiveness of those I consider here. Once more, rather than pretend this were otherwise, it seems more honest and valuable to ask how the British character of much of the modern mythopoeia has determined its themes—themes, for example, of class and propriety, of empire and its vulnerabilities, of brutality, repression, and dominance. Sexual anxiety is universal, but Great Britain has cultivated a very particular variety of it.

One thing I don't apologize for is having made a list at all—for the simple reason that I make no claims for it. My choice of modern myths is neither exhaustive not definitive, and I very much hope that others will argue the case for candidates I've neglected. That is in fact my main objective: to promote further discussion. It should be fun, and hopefully interesting, but I think it is also important, and overdue.

————◆————

My modern myths start more or less where the modern novel is often said to start: with Daniel Defoe's *Robinson Crusoe*. But there is an earlier contender for that accolade, and arguably too for the status of first modern myth, albeit one of an unusually benign nature. So let it serve here as a brief and gentle introduction to the matter of what modern myths can be and where they can go.

Miguel Cervantes's *Don Quixote* (1605) is celebrated for its literary influence: pioneering, for example, in its use of dialogue, metanarrative, and irony. These attributes are not, however, what makes it a myth. The story is a quest narrative—but whereas quests in classical myths do not fail, that of Don Quixote is doomed from the outset. That's what gives the story the power to speak to modernity (even though, when it was written, the modern age itself was just emerging). We can't be sure things will work out, but with tragicomic resolve we will try to keep going anyhow. There's no right way to proceed, and *Don Quixote* offers our sympathies a very modern choice. We know that the Don's sidekick, Sancho Panza, has the more trustworthy perspective, but it is also the dullest, the most literal: how much finer the world appears through the delusions of Quixote! When the old man sees the error of his ways before he dies, we're not sure it's an improvement.

This was Cervantes's genius: to see that Don Quixote mustn't just be held up to ridicule. We're on his side, all the more so as the real world increasingly confounds and frustrates his dream. We must *want* to be seduced by that dream, just as the narrator is seduced by the impractical, romantic enthusiasms of Zorba the Greek in Nikos Kazantzakis's 1946 novel, which borrows more or less explicitly from Cervantes.

Like many modern myths, *Don Quixote* didn't come out of nowhere. Other satires on the chivalric literature preceded it—but Cervantes found the elusive formula that gives a tale currency beyond its immediate context. The old gentleman, Alonso Quixano, is a *hidalgo*, the lowest order of Spanish nobility, who is addicted to books of chivalry. He decides to become a knight errant himself, roaming the world on horseback in search or opportunities to perform noble deeds. He puts on his rusty ancestral armor, names his old nag of a horse Rocinante, and anoints himself Don Quixote de La Mancha—a knight of the poor rustic village, south of Madrid, in which he lives. His squire, Sancho Panza, is a local farmer. As a knight errant, he needs a lady whose heart he must win, and so decides that his great love is for a farm girl from a nearby village, who becomes "Dulcinea del Toboso."

The Don's exploits in the Spanish countryside are filled with the pathos of his fantasy. Some people humor him in his delusion, others mock him, and Sancho does his best to keep his master from disaster. Several of their adventures, especially the encounter with windmills that Don Quixote mistakes for giants, are universally known, episodes more or less liberated from their source text—a common signature of a mythic text. Thus we understand, however scant our knowledge of the novel, what is meant by "tilting at windmills" and by "quixotic" ventures. Edward Riley, perhaps the foremost twentieth-century authority on Cervantes, writes that

> the figures of Don Quixote and Sancho Panza have attained a life of their own quite independent of the book they first appeared in. They are visually known to countless people who have never read the novel. In other words, they have attained mythic status, such as is only now and then achieved by the creations of literary art (and not always the most prestigious of them). Were it not for this mythic, simplified, and easily visualized quixotism, it is unlikely that Cervantes' invention would have come to inspire the extraordinary number of adaptations and refashionings it has in almost every genre and medium—from opera to animated cartoon, from sculpture to computer game.

These many modern reworkings of the Quixote myth have a clear enough pedigree without feeling any need to ape Cervantes's story. The Don's delusions can easily be transformed into a form more relevant and contemporary than a love of chivalric romance, even if it is a love of drink (Zorba, and the priest in Graham Greene's 1982 *Monsignor Quixote*) or a simple (and perhaps misplaced) conviction that life will turn out fine in the end. In *The Pickwick Papers*, Dickens recounts the adventures of the roving old gent Samuel Pickwick and his astute servant and companion Sam Weller—and drafts a blueprint that transforms *Don Quixote* into the madcap "road trip" narrative, out of which emerges everything from Hunter S. Thompson's *Fear and Loathing in Las Vegas* to Neil Gaiman's mythological reboot *American Gods*, and innumerable buddy movies, among them, *Thelma and Louise* (1991) (not nearly as radical-feminist a rewriting as Kathy Acker's *Don Quixote* [1986]) and *Sideways* (2004). The lovably foolish, deluded old man who keeps getting into scrapes is the model for Toad of Toad Hall; the faithful servant who humors his master's unworldliness is Jeeves to the Don's Bertie Wooster. The set-up has infinite scope for variation, but at its core is a psychological truth. Don Quixote and Sancho Panza, said George Orwell, represent the familiar duality of human nature: "noble folly and base wisdom, [which] exist side by side in nearly every human being." If you look into your own mind, Orwell continues,

> which are you, Don Quixote or Sancho Panza? Almost certainly you
> are both. There is one part of you that wishes to be a hero or a saint,
> but another part of you is a little fat man who sees very clearly the
> advantages of staying alive with a whole skin.

And all this, says Ian Watt, is conveyed with the memorable visual emblems that are so often a requirement of the modern myth: "If we should ever see a stick and a ball advancing together side by side down a road, we would immediately recognize them as Don Quixote and Sancho Panza."

Don Quixote is a book unencumbered by expectations. In its nascent form the European novel can do anything, moving from the fantastical to the realistic, from allegory to farce and satire. The heroes can become famous for starring in the very novel that records their fame; there is an intertextuality to *Don Quixote* that, Riley points out, makes it "very congenial to the age of Borges and Eco, Barthes and Derrida."

This situation changes at the start of the nineteenth century. One could be a Jane Austen, writing minutely observed realist novels about manners, society, and personality. Or one could be a Mary Shelley, subsuming character and to some degree plot (certainly consistency of plot) into the fantastical outpourings of the subconscious—and thereby writing in the mythic mode. I'll return to this division at the end of the book. To take the mythic path was to risk being dismissed as a genre writer. On the face of it, you were then simply writing to entertain. But mythmakers were really up to something more dangerous.

That's one reason we should attend closely to our modern myths: what they are and what they signify. It's a peculiar thing that we have made such widespread use of them while remaining so ambivalent about their status. The closer to us these stories are historically, the less carefully we examine them. *Crusoe* and *Frankenstein* have benefited from many serious critical readings (even if many of these are relatively recent and had at first to justify themselves in the face of critical disdain). H. G. Wells and Arthur Conan Doyle still suffer from being labeled genre writers, while serious scholarship about comic-book narratives such as Batman is in its infancy (even though the form itself is a century old). This is not so much because modern myths take a long time to be recognized and assimilated (although that tends to be true too), but because we have been slow to throw off the shackles of literary prejudice. The result is that we fail to see the important psychic work these stories are doing. I'm put in mind of what William Doty says about Plato's determination to guard the morality of the republic's citizens against the corrupting influence of the artist's deceptions:

> Plato worried about the influence of old wives' tales in the education of the elite citizens of his idealized "republic." Out of the rulers' control, such materials—and the poets' elaborations of them—were to be excluded because they were unphilosophized and unrationalized.

I think Plato would exclude as well the unruly, uncouth, unconventional, undisciplined, uncanny. The modern myths I have chosen are all of those things, but perhaps more than anything else the force that drives them is *unconscious*. Plato could not keep these qualities out of his culture's stories, and neither—even if we wanted to—can we keep them from ours. We create our new myths because they do necessary work. It's worth getting to know them, and to understand what they are up to. Let's make a start.

❧

Chapter 2

JOHN BULL ON A BEACH

ROBINSON CRUSOE (1719)

A man is left stranded in a remote, inhospitable place, and by dint
of his ingenuity and perseverance forges a solitary life—
before perhaps discovering a faithful, subservient companion.
He is eventually rescued and returns to civilization.

═══════════════════════

Some years ago the *New Yorker* placed a ban on desert-island cartoons. The ragged, bearded fellow making some satirical observation from his little patch of sand with its lone palm tree had become a tired formula. It wasn't that he had run out of things to say, but that he'd become a cliché.

Yet he is still with us, reinvigorated with self-aware irony. The age of Wifi and social media offers new possibilities: the castaway forgets his password (was it "coconut" or "fish"?) or gets spammed by messages in bottles.

Why is this minimalist scenario so well suited for commenting on

"DAMN, I FORGOT TO UNSUBSCRIBE"

"SO WHEN WAS YOUR FIRST DESERT ISLAND CARTOON?"

"PASSWORD...PASSWORD...WAS IT FISH?...OR COCONUT..?"

Spam, online security, health and safety, Russellian self-reference, literary criticism . . . is there anything the desert-island cartoon can't embrace?

the idiosyncrasies of modern life? The castaway, faced with isolation and hopelessness, still clings to familiar petty, futile obsessions, routines, platitudes, and delusions. We can't go on, we go on, our Beckettian resolve both noble and pathetic.

Daniel Defoe's *Robinson Crusoe* (1719) was not the first story of shipwreck and marooning on a desert island, but it turned previous real-life accounts into a cultural emblem. Like the desert-island jokes it spawned, *Crusoe* reduced the world of its day to a single, stark schema. It couldn't have done so at a better time—or perhaps it is better to say, Crusoe arrived just when he was needed. The novel became a sensation, an Enlightenment trope, and a best seller.

Out of the hazards of maritime adventure it forged an enduring metaphor and myth that spoke to the emerging notions of individuality, autonomy, and self-determination. It became a representation of empire and colonialism, but also a fable about isolation, both personal and existential. It has endured as a children's story and an economic parable, and it established the island as a place of new possibilities. Lying beyond the bounds of established norms and rules, the island could be a utopia—or a place of nightmare.

Being marooned alone on some distant, isolated shore had been a hazard from the earliest days of seafaring. But before the eighteenth century it would not have been deemed a meaningful proposition that there was a story to be told about such a predicament. What story ever had but a single protagonist? Without human relationships, what was there to say?

The concept of the self-made man—without which there could have been no *Crusoe*—crystallized only in the early modern period. By the eighteenth century one's role and fate were not preordained by birth, stars, or gods; you could improve your lot by your own efforts. It's central to Crusoe's tale that he comes from a family made prosperous and respectable by trade. He was a representative of the book's target audience, enacting a fable of capitalist self-sufficiency.

A superficial reading makes *Robinson Crusoe* an opportunistic dramatization of several real-life accounts of sea adventures and castaways in Defoe's time. One widely read example was William Dampier's *A New Voyage Round the World* (1697), which recounted the English explorer's exploits aboard the ships of buccaneers and privateers, as a kind of semi-sanctioned pirate, off the coast of the New World. Dampier sailed around Cape Horn into the Pacific and traveled from Central America to Manila, China, and Australia on board the ship of the privateer Charles Swan. In the archipelago of the Juan Fernández Islands off Chile,

Dampier's crew found a "Ploskito Indian" who had been marooned for three years, surviving by hunting goats. Dampier was later stranded himself with two others on an island in the Indian Ocean, but he managed to reach Sumatra and ultimately found his way back to England, where he published his account to great acclaim.

So much so, indeed, that Dampier was given command of a warship of the Royal Navy and sent on a mission to New Holland in Australia. On the return journey he was forced to land his damaged ship at Ascension Island in the middle of the South Atlantic, where he and his crew were stuck for five weeks before being rescued by a vessel of the East India Company.

In 1703 Dampier was sent with a gunship to defend English interests against Spain and France during the War of the Spanish Succession. Rounding Cape Horn, he arrived again at the Juan Fernández Islands in 1704, where one of his officers, a Scot named Alexander Selkirk, complained that the *Cinque Ports*, a smaller ship accompanying Dampier's vessel, was not seaworthy. The commander of the *Cinque Ports*, one Thomas Stradling, was glad enough to grant the troublesome Selkirk's request and put him ashore on the island of Más a Tierra. (Selkirk was soon proved correct, when the *Cinque Ports* sank off the coast of what is now Colombia.)

How Selkirk thought he would survive alone on his island wasn't clear. But he did, for four years. He was finally rescued in 1709 by another English privateer, Woodes Rogers, who gave an account of Selkirk's solitary survival in his book *A Cruising Voyage Round the World* (1712). Rogers's book, too, was wildly popular, and Selkirk became a reluctant celebrity in English society.

Selkirk first appeared to his rescuers, Rogers wrote, as "a Man cloth'd in Goat-Skins, who look'd wilder than the first Owners of them." The account he gave of his time on the island bears all the ingredients of Crusoe's strategies:

> He had with him his Clothes and Bedding, with a Fire-lock, some
> Powder, Bullets, and Tobacco, a Hatchet, a Knife, a Kettle, a Bible,
> some practical Pieces, and his Mathematical Instruments and Books.

He diverted and provided for himself as well as he could; but for the
first eight months he had much ado to bear up against Melancholy,
and the Terror of being left alone in such a desolate place. He built
two Hutts with Piemento Trees, cover'd then with long Grass, and
lin'd them with the Skins of Goats, which he kill'd with his Gun as he
wanted, so long as his Powder lasted.

He survived by eating crawfish and goat, killing the latter by hand when
he ran out of gunpowder. He grew turnips from those that Dampier's
crew had left him, and found "Cabbage-Trees" and small black plums.
He got by without shoes when his wore out, and stitched crude clothes
from goatskins and a store of linen. He established a daily routine resembling Crusoe's: keeping a log of his hunts, taming goats, reading the
scriptures. By the time Rogers found him, however, he had almost for-
gotten how to speak: "we could scarce understand him, for he seem'd to
speak his words by halves."

The privateer captain saw a moral in this:

> We may perceive by this Story the Truth of the Maxim, That Neces-
> sity is the Mother of Invention, since he found means to supply his
> Wants in a very natural manner, so as to maintain his Life, tho not so
> conveniently, yet as effectually as we are able to do with the help of all
> our Arts and Society. It may likewise instruct us, how much a plain
> and temperate way of living conduces to the Health of the Body and
> the Vigour of the Mind, both which we are apt to destroy by Excess
> and Plenty.

Rogers admonishes himself for coming on here more like a "philosopher
and Divine than a Mariner." But he'd hit the target squarely: Selkirk's
tale was not just entertaining but exemplary.

While it seems all too obviously the template for *Crusoe*, we can't
assume that Rogers's account was Defoe's main source. Stories of daring
voyages, attended by storm, mutiny, disaster, pirate battles, and fantastical new lands, had become so popular by 1710 that the Earl of Shaftes-
bury denounced them as hackneyed, the equivalent of those old books

of chivalrous deeds that had bewitched Don Quixote. Selkirk's exploits were particularly apt to be recycled—they feature, for example, in Edward Cooke's *Voyage to the South Sea* (1712) and an article in the periodical *The Englishman* in 1713. But there were other models too, such as sea captain Robert Knox's account of a twenty-year captivity while on a voyage for the East India Company in his *Historical Relation of Ceylon* (1681). For an ambitious writer like Daniel Defoe, always looking out for the next wave to ride, the ubiquity of these tales was no deterrent. On the contrary, the public hunger for them presented an opportunity.

Defoe is slippery, there's no two ways about it. He was a social agitator and Dissenter who advocated revolutionary Puritanism. In 1703 he was sentenced to public pillory in the stocks for writing a leaflet criticizing the repression of Dissenters encouraged by the Tory party during the reign of the Catholic Queen Anne. The court condemned him as "a Seditious man and of a disordered mind, and a person of bad name, reputation and Conversation." And yet he claimed (although there is no real evidence to support it) that he acted as a secret agent of the Crown in the 1690s, and he later became a Tory apologist.

The Foes—Daniel added the grandiose "De" himself—were dissenting Presbyterians, a compromising situation that Defoe compounded with his wheeler-dealer activities. Although he acquired a tidy sum by marriage, he squandered it on speculative business projects such as his efforts to breed civet cats for the perfume ingredient they secrete. By 1692 he was in the debtors' prison in Newgate, bankrupt by an enormous sum, alleged by some to be a barely credible £17,000.

Despite his radical politics, Defoe would write on just about anything for anyone if the price was right. You'd be hard-pressed to find a more prolific writer in England in those times, although the quantity of his output can't be said to have done much for its quality: paid by the page, he was determined to fill it. Between around 1704 and 1713 he produced a periodical, *The Review*, three times a week, to which he was the sole contributor.

It wasn't just for need of cash that Defoe was prepared to overlook his political principles. His conviction for sedition drew an unusually harsh sentence, and it wasn't clear how long he could expect to languish in jail. So he petitioned the Tory politician Robert Harley, Speaker of the House of Commons, who got him released and pardoned. In return, Harley seemed to consider that Defoe was now his man, and Defoe was happy to oblige, writing essays supporting Harley's views where previously he had called the Tories "plunderers" and "betrayers of liberty." He traveled the length and breadth of the country gathering information that might help Harley consolidate power and influence, and defended his master against attacks from the Whigs.

He knew how incongruous, not to say hypocritical, this was. On occasion he even published anonymous articles criticizing the views he had expressed in *The Review*. Still, it was a profitable trade: Harley rewarded him well, and by the 1720s he was living in a large house in Stoke Newington, just north of London, with a retinue and carriage. Most of the books Defoe produced date from this period, from around 1715 until his death in 1731. They were not the kind of thing a gentleman would write: tales of disreputable characters like Moll Flanders ("Twelve Years a Whore, five times a Wife (whereof once to her own Brother) Twelve Years a Thief, Eight Years a Transported Felon in Virginia") and Colonel Jack ("Born a Gentleman, put 'Prentice to a Pick-Pocket, was Six and Twenty Years a Thief, and then Kidnapp'd to Virginia, Came back a Merchant; was Five times married to Four Whores"). But they sold, and it's not hard to see why.

We don't do Defoe any real injustice, then, by calling him a hack. In *Robinson Crusoe* his literary shortcomings are plain to see, plot being one of them. He is evidently making it up as he goes along—for example, he neglects to tell us until well after the event that Crusoe has saved a dog and two cats from the wreckage. "The stream of his writing merely flowed around any obstacle in its way," says critic James Sutherland, adding that "many of his best things came to him on the spur of the moment." That was probably as true in Defoe's life as it was in his writing.

Robinson Crusoe shows us the modern novel in infancy: a mirror for the issues of the day, but messy, undisciplined, and sometimes baffling. Character psychology is crude at best, and the plotting amounts to little more than stringing episodes together one after the other. We all know about the shipwreck, Crusoe's hardships on the island, and the appearance of Friday—but what surrounds these events often feels like mere baggage.

But Defoe didn't know how to write a novel, because no one did; some scholars identify him as its inventor. The result was often as makeshift as Crusoe's first shack, full of holes and not to be prodded too hard lest the entire edifice collapse.

One curious aspect of *Crusoe*'s genesis is that its author seemed to insist on its being taken as fact. In the manner of the day, it doesn't have a title so much as a précis:

> The Life and Strange Surprizing Adventures of Robinson Crusoe of York, Mariner, Who lived Eight and Twenty Years all alone in an un-inhabited Island on the Coast of America, near the Mouth of the Great River of Oroonoque; Having been cast on Shore by Shipwreck, wherein all the Men perished but himself. With An Account how he was at last as strangely deliver'd by Pyrates.

Defoe's name did not appear on the title page of the first edition, printed by W. Taylor "at the Ship in Pater-Noster Row." Rather, the book is "Written by Himself"—that is, Crusoe. Now, there is nothing so unusual in this at one level. The title page resembles those of *Moll Flanders* ("Written From her Memorandum") and *Colonel Jack*, and anonymous publication was common—that was how Samuel Richardson's *Pamela* first appeared in 1740, and Laurence Sterne's *Tristram Shandy* in 1759. (Henry Fielding, in contrast, put his name to the picaresque *Tom Jones* in 1749.) However, not only did this mode of presentation blur any distinction between fiction and memoir, but Defoe maintained the pretense for years, leaving it ambiguous whether he had concocted the tale or merely related it (he called it an "emblematick history"). When accused of lying or inventing the tale, Defoe defended himself with a hint

THE
LIFE
AND
STRANGE SURPRIZING
ADVENTURES
OF
ROBINSON CRUSOE,
Of *YORK*, MARINER:

Who lived Eight and Twenty Years,
all alone in an un-inhabited Ifland on the
Coaft of AMERICA, near the Mouth of
the Great River of OROONOQUE;

Having been caft on Shore by Shipwreck, where-
in all the Men perifhed but himfelf.

WITH

An Account how he was at laft as ftrangely deliver'd by PYRATES.

Written by Himfelf.

L O N D O N:
Printed for W. TAYLOR at the *Ship* in *Pater-Nofter-
Row.* MDCCXIX.

Title page of the 1719 first edition of Robinson Crusoe.

of the very modern discussion about whether *any* fiction is just "making things up"—or whether it serves a deeper truth. After all, how much of alleged reported fact is also a good part fiction?

Still, Defoe's motive for this ploy isn't obvious. Was he eager to create an illusion of authenticity as a sales gambit, thinking that readers would accept the book more readily if they thought they were reading about real events? At any rate, by presenting the story as a series of entries in a journal, he was using a form audiences would recognize from genuine memoirs. In its attention to minute detail, Defoe wrought a narrative that could plausibly pass as true; so much so, in fact, that in its time the book was not widely seen as "literary" at all.

But when you look closely, the structure of *Crusoe* is incoherent. The "Journal" section only begins many pages into the book, after we have heard about Crusoe's early life, his shipwreck, and his efforts to create a serviceable home inside a cave. This journal plods back over events related previously, but before long the daily entries peter out and Crusoe, feebly pleading a shortage of ink, falls back into a continuous narrative while pledging "to write down only the most remarkable Events of my Life." Even if he had no blueprint for a novel, Defoe seems to have lacked the will to impose a stable structure on his tale.

For the reader anticipating the familiar castaway narrative, Robinson Crusoe takes a long time to get himself marooned. He goes to sea against his father's wishes, only to be nearly drowned in a storm off the English coast. Too obstinate to return home after this scare, he signs up for a voyage to Guinea. Between the Canaries and the north African coast the ship is attacked and overwhelmed by Turks. Crusoe is taken prisoner and sold into slavery under the Moors of Morocco. He escapes by canoe with a young Moorish lad named Xury and is picked up by a Portuguese captain, who bears him across the Atlantic to the Brasils. At last, we think: the shipwreck is coming.

But not yet. With the money he left in England, Crusoe buys a sugar plantation and becomes a prosperous colonist. And there he might have remained and thrived, were he not tempted by a proposal from some other merchants and plantation owners of his acquaintance. They are short of slaves but can buy them only with the consent of the kings of

Portugal and Spain, and then at a high price. So they plan to sail secretly to Guinea to acquire slaves and divide them among themselves. Knowing that Crusoe is a canny trader, they ask if he might come with them to conduct the negotiations.

Crusoe is wealthy enough not to need this bounty, but his seafaring urge gets the better of him, and off he goes on a 120-ton ship. The slave-hunters are blown off course by a hurricane, and as they head up the South American coast they are caught in a second storm and driven onto a sandbank offshore from some unknown territory. They figure the stranded ship will be wrecked, and attempt to row to shore in a boat. But they are overturned by a wave, and Crusoe finds himself washed up on the sand, alone.

So at last begins the account of how Crusoe makes his island a home. He builds a sturdy palisade, at first fearing attack by wild animals; finding none on the island, he then worries (with more justification, it transpires) about visiting cannibals. He shoots goats, but has the foresight also to catch and domesticate them and raise a herd. He builds a canoe to circumnavigate the island but is almost carried out to sea in a current. Despite coming to feel comfortable in his island kingdom, he makes plans to escape. When he saves Friday from being slaughtered by cannibals, he enlists the man's help to construct a sailing boat.

But before they are ready to set out, Crusoe spots the sails of an English ship. In due course a party comes ashore, whereupon Crusoe and Friday discover that the crew has mutinied. They help the commander and those loyal to him overcome the mutineers, and (with remarkably little ceremony given that Crusoe has dwelt on the island for twenty-eight years) the castaway and his companions make for England. Crusoe finds that his plantations have flourished and that he is now a rich man.

Here the story ends—but the book does not. In a bizarre coda, Crusoe returns with Friday from having settled his estate in Lisbon, journeying by land with a group of other travelers across the Pyrenees. As they pass through Languedoc and Gascony, the party is attacked in the woods by wolves and a bear. This encounter is dragged out for several pages, including an almost slapstick account of how Friday taunts and shoots the bear. The whole escapade seems like an afterthought. Might Defoe have intended some allegory about the threats from Catholic powers (wolves) and the Russian tsar (the bear)? Or might he simply have read an account of an attack by wolves and bears on Languedoc villagers in the 1718 *Mist's Weekly Journal* (perhaps he even wrote it—he was a contributor) and decided to tack it on regardless of the narrative value?

Robinson Crusoe was a great success. The first print run of a thousand to fifteen hundred sold out quickly, even at the high price of five shillings—not bad when the entire reading public in England probably numbered in the tens of thousands. There were four more editions that same year, as well as serializations and (appropriately enough) pirate copies. Yet Defoe scarcely benefited financially. Writers tended to be paid in a lump sum, and Defoe's fee was likely to have been a modest fifty pounds or so. It's small wonder, then, that he was eager to capitalize on the book's success by publishing sequels. *The Farther Adventures of Robinson Crusoe* (1719) describes Crusoe's return to his island and travels elsewhere, while *Serious Reflections During the Life and Surprising Adventures of Robinson Crusoe* (1720) is a collection of essays musing on what the author has learned from his experiences.

Quite who "the author" was is another matter. Some of *Serious Reflections* was presented as Crusoe's work, some as Defoe's. Still, Defoe defended the pretense that Crusoe's tale is historically true and not just a "romance." In particular he refuted the charges of the writer Charles Gildon, who satirized the invention in his book *The Life and Strange Surprizing Adventures of Mr. D— de F—, of London, Hosier* (1719). In his riposte, Defoe makes a comparison to the fictional *Don Quixote*, reminding readers that that tale too was (allegedly) based on the real-life exploits of the Duke of Medina Sidonia. What was to be gained by maintaining the deception with such laborious effort is not clear.

Who is Crusoe anyway? If he was Defoe's alter ego then that does the writer few favors, for he is one of the least attractive heroes in adventure literature. It's not that he is unpleasant; after all, villainous protagonists have their own appeal. Rather, there is a moral and spiritual emptiness to the man. He discharges obligations dutifully and repays his debts when he can, but he forges no strong relationships that are not ultimately pragmatic: everyone must serve his own ends. He mouths pious formulas without evincing any deep engagement with his faith. He is a callow dullard with little capacity for self-reflection. James Joyce thought he had the measure of Crusoe ("the English Ulysses"), and it was a character that a twentieth-century Irishman knew all too well. He is, said Joyce,

> the true prototype of the British colonist, as Friday . . . is the symbol
> of the subject races. The whole Anglo-Saxon spirit is in Crusoe: the
> manly independence; the unconscious cruelty; the persistence; the
> slow yet efficient intelligence; the sexual apathy; the practical, well-
> balanced religiousness; the calculating taciturnity.

Crusoe is a bookkeeper and bean counter with not an ounce of artistry or poetry in his soul. All he does is utilitarian: making tools and furniture, baskets and clothes, obsessively inventorying his domain. Even once comfortably settled and supplied, he indulges no urge to express himself, nor gives any sign of having the imagination to do so. Despite his exotic escapade, Crusoe seemed to Virginia Woolf to personify the humdrum existence of the bourgeois middle class:

> Everything appears as it would appear to that naturally cautious,
> apprehensive, conventional and solidly matter-of-fact intelligence. He
> is incapable of enthusiasm. He has a slight natural distaste for the sub-
> limities of Nature. . . . Everything is capable of a rational explanation,
> he is sure, if only he had the time to attend to it.*

---

* You might infer from this that Woolf doesn't think much of *Robinson Crusoe*. On the contrary, she saw plenty to praise it its style, commending Defoe for using the small, telling observation in place of labored descriptive prose—as when, from Crusoe's discovery of a pair of mismatched shoes on the sand, we understand that all the rest of the shipwrecked crew are dead.

Robinson Crusoe, from the 1719 first edition.

Everything in the book, Woolf says, is seen through "shrewd, middle-class, unimaginative eyes."

Even in his introspection Crusoe thinks like an actuary. He weighs up the pros and cons of his situation "very impartially, like Debtor and Creditor," listing as if on a spreadsheet his bad and good fortune:

Evil	*Good*
I am cast upon a horrible desolate Island, void of all hope of Recovery.	But I am alive, and not drown'd as all my Ship'd Company was.

And so forth. True, God seems to have thrown him into a difficult place. But it could have been much worse. One hears the ancestral ghost of that most British of dispositions: "Mustn't grumble."

Charles Dickens called *Robinson Crusoe* "the only instance of an universally popular book that could make no one laugh and could make no one cry." It has no heart, he complained, no "tenderness and sentiment"—and Crusoe himself lacks the depth to be changed by what he has experienced: he's no different after twenty-eight years of a mostly solitary existence than he was before. Risking the classic confusion of author and character, Dickens (of all people!) suspected that Defoe was not dissimilar: "I have no doubt he was a precious dry disagreeable article himself"—a man devoid of humor, a defect that for Dickens places him beyond the pale.

Dickens's near-contemporary, the critic Leslie Stephen, regarded Crusoe's imperturbability as implausible: no one could be so resilient to such utter and protracted isolation. Wordsworth agreed: the "extraordinary energy and resource of the hero," he said, were "so far beyond what was natural to expect, or what would have been exhibited by the average of men." Robert Louis Stevenson's Ben Gunn in *Treasure Island* suggests what a pathetic state the marooned sailor might really soon fall into. Some went mad, like the Dutch seaman stranded on St. Helena who (the story went) dug up a dead colleague and set out to sea paddling in his coffin. As Aristotle said, "He who is unable to live in society, or who has no need because he is sufficient for himself, must be either a beast or a god." He can't be an accountant.

There seems no reason to think that Defoe intended his hero to appear so unsympathetic. On the contrary, his levelheaded pragmatism was probably meant to be admirable. He was the epitome of the stolid, contented bourgeoisie, the self-made (and self-satisfied) English middle class. This "middle state," his father tells Crusoe, is

> the most suited to human happiness, not exposed to the miseries and hardships, the labour and sufferings of the mechanic part of mankind, and not embarrass'd with the pride, luxury, ambition and envy of the upper part of mankind.

Perhaps we expect too much of our heroes these days; making them capable yet a little self-pitying and dim is no longer enough. But Crusoe's very blandness serves the myth: he is a cypher, a blank canvas, existing to tell the tale but too unremarkable to give it any particular flavor. He is a vehicle, indeed a mere chassis, ready to carry our own invention.

No one has ever suggested that *Robinson Crusoe* exhibits much in the way of literary style. (They'd have a job on their hands if they felt so inclined.) The early nineteenth-century essayist Charles Lamb was impressed by the way it managed to appeal both to the educated and the "servile" reader: in a fine example of appreciative condescension he said that the book "is an especial favourite with sea-faring men, poor boys, servant-maids," and that Defoe's novels in general "are capital kitchen-reading, while they are worthy, from their interest, to find a shelf in the libraries of the wealthiest and the most learned." He added pointedly, however, that *Crusoe* "is written in a phraseology peculiarly adapted to the lower condition of readers." When Lamb said that "it is like reading evidence in a court of Justice," he might not have meant it to sound quite as negative as it does now—but he's not far wrong.

It seems to be the style, as much as the content, that recommended the book to children, who would scarce be troubled by Crusoe's psychological vacuity and will see nothing amiss in his perfect lack of any hint of a sexual life. The only relationships he has for most of his island existence are with his pets—and which child will take exception to that?

In *Mythologies*, Roland Barthes perceives the childlike appeal of Crusoe's Island myth, suggesting the delight comes from the comforting finiteness, expressed also in the way children love huts, tents, and dens: "To enclose oneself and to settle, such is the dream of childhood." Of Jules Verne's Crusoe-inspired *The Mysterious Island*, Barthes writes:

> The man-child re-invents the world, fills it, closes it, shuts himself up
> in it, and crowns this encyclopaedic effort with the bourgeois posture
> of appropriation: slippers, pipe and fireside, while outside the storm,
> that is, the infinite, rages in vain.

For the early-twentieth century Freudian psychoanalyst Ernest Jones, "All aspects of the idea of an island Paradise are intimately connected with womb phantasies."

Children were considered a particularly suitable audience for what educators in the Enlightenment and Romantic era deemed to be the moral message of *Crusoe*. Jean-Jacques Rousseau said of it:

> Nothing upon earth can be so well calculated to inspire one with
> ardour in the execution of a plan approved by so great a genius. Might
> there not be found means to bring together so many lessons of instruc-
> tion that lie scattered in so many books; to apply them through a single
> object of a familiar and not uncommon nature, capable of engaging
> the imitation, as well as rousing and fixing the attention even at so ten-
> der an age? . . . I see thine expand already, thou ardent philosopher. But
> be not in pair; we have found such a situation. It is described, and, no
> disparagement to your talents, much better than you would describe
> it yourself, at least with more truth and simplicity. . . . It shall be the
> text to which all our discourses upon natural science shall serve as a
> commentary. It shall be the criterion of our taste and judgment; and,
> as long as these remain uncorrupted, the reading of it will always be
> agreeable to us. Well, then, what is this wonderful book? Is it Aristotle,
> Pliny, Buffon?—No: it is Robinson Crusoe.

Edgar Allan Poe felt the book was the ideal articulation of the castaway story: "The idea of man in a state of perfect isolation ... was never before so comprehensively carried out." Wilkie Collins loved it; his character Gabriel Betteredge in *The Moonstone* says, "I have worn out six stout *Robinson Crusoes* with hard work in my service," and its influence is not hard to spot in Collins's first (unpublished) novel, set among the "noble savages" of Tahiti.

Like Jonathan Swift's *Gulliver's Travels* (1726), which continues even today to be bowdlerized for young audiences, *Crusoe* is unlikely now to shake off its image as a children's story. A Victorian pantomime of *Crusoe* featured the famous clown Joseph Grimaldi, while Beatrix Potter's *The Tale of Little Pig Robinson* (1930) gave the pig from Edward Lear's "The Owl and the Pussycat" a backstory as a castaway who narrowly escaped becoming a feast for a crew of sailors.

Robinson Crusoe was among the obvious satirical targets of *Gulliver's Travels*, although of course Swift had wider political and intellectual targets in his sights. He did not like Defoe's Nonconformist tendencies and deplored *Crusoe's* bourgeois focus on the necessities of everyday life, which he felt would degrade literary sensibilities. In short, he found the book a bit commonplace. It's ironic, then, that *Gulliver* suffered the same fate as *Crusoe* in being recast as a children's book. Perhaps that was a way of rendering troublesome texts harmless.

There is certainly much more to it than a yarn. But what, exactly?

A great deal of the literary analysis *Crusoe* has attracted—and this applies equally to most of the other modern myths in this book—has held a spotlight on *what the author is trying to say*. This matters, but not as much as you might think. Often it's little more than speculation, since authors rarely left detailed accounts of their methods and meanings before the contemporary rounds of media interviews and literary festivals. What's more, the authorial intentions behind a mythopoeic story might have little to do with the features that give it this status. They might allude to issues of the day that seem to us now parochial, archaic, and arcane. Even more importantly, we will see time and again that the themes that have made these stories lenses for focusing the anxieties of successive eras tend to appear in the tales without the authors' conscious intent. The writers inadvertently penetrate deep psychic strata, and what comes up is beyond their control.

Defoe does seem to have intended that *Crusoe* contain political allegory. His hero returns to England in 1687, just before the Glorious Revolution of 1688 replaced the Stuart king James II with William, Prince of Orange—a maneuver designed to fend off the possibility of a Catholic Stuart succession. In the decades that followed, Britain was rife with religious tension. The Act of Settlement of 1701 excluded Catholics from the throne, but unease about religious tolerance was stoked by the British involvement in the War of the Spanish Succession, in which Britain's status as a colonial power was at stake.

The microcosmic society that Crusoe constructs on his island can be read as a miniature version of the sovereignty that, in Defoe's view, the British ought to enjoy. Once Crusoe and Friday are joined by Friday's father and a castaway Spaniard—both captives of the cannibals, who were going to feast on them—his kingdom is replete:

> My Island was now peopled, and I thought my self very rich in Subjects; and it was a merry Reflection, which I frequently made, how like a King I look'd. First of all, the whole Country was my own meer Property; so that I had an undoubted Right of Dominion. *2ndly*, My People were perfectly subjected: I was absolute lord and Law-giver; they all owed their lives to me, and were ready to lay down their Lives, *if there had been Occasion of it*, for me. It was remarkable too, we had but three Subjects, and they were of three different Religions. My Man *Friday* was a Protestant, his Father was a *Pagan* and a *Cannibal*, and the *Spaniard* was a Papist: however, I allow'd Liberty of Conscience throughout my Dominion.

This appeal to "liberty of conscience" would have been understood by all Defoe's readers as an allusion to religious tolerance. James II had claimed to extend such liberty to his subjects, while ensuring that Catholics were amply represented in civil and military offices. Yet Stuart England—of which Queen Anne's rule from 1702 to 1714 was the last gasp—had permitted little liberty to a Nonconformist like Defoe, as his spell in the stocks confirmed. *Crusoe* might, then, be seen as an allegory of exile to a place of greater religious liberty, where Defoe's own Nonconformist beliefs wouldn't be condemned. Crusoe shares those beliefs, saying that the doctrine of the salvation of Christ can be understood "without any Teacher or Instructor."

With its themes of punishment and repentance, sin and grace, *Crusoe* invites further religious interpretations. Is Crusoe condemned to his predicament for defying his father's wishes by going to sea in the first place? (If so, the moral is rather undermined by Crusoe doing so well in the end, profiting handsomely from his exploits in the New World.) In his delivery from the sea and his systematic enumeration of hardships, Crusoe is a kind of colonialist blend of Jonah and Job. In a particularly biblical image he sows seeds rescued from the ship; some fall inadvertently to the soil, sprouting and flourishing with a hint of divine providence. But what, really, has he done to deserve that?

In fact, if Defoe wants us to derive a Christian moral from Crusoe (as several of his early readers believed), the narrative shows how little grasp the author had on it. There's a kind of gloating glee in how he tots up the wealth Crusoe has accumulated during his absence, dissipating any sense that the castaway has sinned and been redeemed. Crusoe doesn't even have to reconcile himself with his father, who is long dead by the time he returns to England. As historian of literature Leo Damrosch says, "Far from punishing the prodigal Crusoe for disobedience, the novel seems to reward him for enduring a mysterious test."

Yet this ambiguity about the religious themes is a part of what makes *Robinson Crusoe* so modern. The novel pays lip service to Christian obligations while seeming to feel not especially bound by them. In the end, Crusoe is author of his own fate, and God can't do much about it. Harold Bloom identifies this as the defining feature of "all quest-romances of the post-Enlightenment": they are "quests to re-beget one's own self, to become one's own Great Original." To do, in other words, pretty much what Milton's Satan wished to do in *Paradise Lost*. And why not? God was losing his ability to intervene. "Why God no kill the Devil?" Friday reasonably asks Crusoe when informed of the Christian faith. Crusoe has no answer.

◆

Ah, Friday. All our modern problems with *Crusoe* start with the castaway's obedient cannibal. The morality of slavery was hotly contended even in Defoe's time, and one of the features that makes the book an uncomfortable staple of the children's canon now is Crusoe's unabashed ownership of sugar plantations in the West Indies, dependent on slave labor. Defoe perceived the injustice and inhumanity of slavery, which, after all, Crusoe himself suffers. But the author's position that the slave trade needed to be regulated, not abolished, is shared by his hero, who is content enough to profit from it on his own estate, and indeed to go off in search of more slaves. Some critics wonder if Crusoe's shipwreck is meant to be seen as retribution for undertaking that illicit voyage. But since after his travails he prospers so heartily, it's hard to sustain the idea.

Robinson Crusoe is not so much an allegory of imperialism as a simple illustration of it, wrought in miniature. With the pragmatic sangfroid of the British colonist, Crusoe gets on with whatever practical task is needed to subdue and domesticate his environment, turning it in short order into a home away from home. Left with nothing but a pocketknife and a pipe, says James Joyce (with considerable exaggeration), he becomes "an architect, a carpenter, a knife grinder, an astronomer, a baker, a shipwright, a potter, a saddler, a farmer, a tailor, an umbrella-maker, and a clergyman." What hope do the indigenous dwellers have against such stolid resolution? An assumption of natural privilege and cultural superiority pervades the story.

To modern eyes Crusoe treats the young Moor Xury, with whom he escapes from slavery, appallingly. Although the two were mutually dependent allies in their escape, Crusoe happily sells the lad into servitude to their Portuguese rescuers—on condition that Xury be freed after ten years if he converts to Christianity. Xury, Crusoe reassures us, was perfectly content with this arrangement. He then compounds the injustice by commenting, as his plantation in Brazil begins to thrive, "I had done wrong in parting with my Boy *Xury*" — not because there was any moral misdeed involved, but because he'd let go someone who could have acted as a voluntary slave for *him*.

When Crusoe encounters the "savage" Friday, the relationship is immediately one of master and servant, no questions asked; Friday seems to assume this is the natural order of things. Even that imposed pet name reminds Friday of his indebtedness to his master: it was on this day that his life was saved. It doesn't occur to Crusoe that Friday might already possess a name.* First, he says,

> I made him know his Name should be *Friday*, which was the Day I sav'd
> his Life; I call'd him so for the Memory of the Time; I likewise taught
> him to say *Master*, and then let him know, that was to be my Name.

Is it just misunderstanding that causes people today to believe Friday is called "Man Friday"? Or is that the convenient way we overlook the fact that all such references in *Crusoe* are to "my Man *Friday*" —which is a manner of saying, "my servant-slave *Friday*"? (For all this, Crusoe can't shake off residual fears about his servant: can he be trusted not to "come back, perhaps with a hundred or two of [his fellow savages], and make a feast upon me"? Such are the anxieties of empire.)

In his rewriting of the Crusoe myth, *Foe* (1986), South African author J. M. Coetzee draws attention to this arrogant presumption by having Cruso [*sic*] blithely commend God for not extending his beneficence too far:

> If Providence were to watch over all of us, who would be left to pick
> the cotton and cut the sugar-cane? For the business of the world to
> prosper, Providence must sometimes wake and sometimes sleep.

---

* In his memoir, William Dampier asserted that some "savages" of the New World didn't have a
  tradition of taking personal names themselves and would ask colonists to give them one.

Many of the older illustrations of Crusoe's meeting with Friday left no doubt about their master-servant relationship.

Of course, Defoe's readers would have seen the hierarchy presented in *Crusoe* as right and proper. But the book goes further by providing a parable that dramatizes and justifies that assumption. As he pulls himself from the breakers, dazed and exhausted, Crusoe has no more resources than any other man cast naked and alone on some remote shore. (True, he is soon able to procure some from the wreck, building a raft to carry out all the provisions, tools, and materials he can salvage. But those are already the products of his superior civilization.) And yet Crusoe will not descend to the primitive state of the savage and cannibal. From almost nothing he constructs a solitary empire, with a palace and country houses (as he refers to the shelters and fortifications he builds), pen and ink and a desk to write on, a well-stocked storehouse, and cultivated land and pasture for food and grazing of domesticated beasts. In short, Crusoe creates a simulacrum of European civilization, and does not neglect his Christian duties either. This, one infers, is the birthright of the Englishman. *This* is how an Englishman responds to adversity: with the mental, moral, and intellectual resources that his superior breeding have conferred on him. Why do you think a copy of Defoe's book was likely to be found in every Victorian school?

In the process, Crusoe recapitulates human history: sleeping first in a tree, then a cave, then making tools and domesticating land, and finally becoming the master of nature. All the while the local savages remain in their primitive state, unless they can be weaned off their crude and sometimes disgusting practices (such as cannibalism) and taught to speak English and to worship God. Little by little, these people can be brought to enjoy the conveniences and to adopt the good morals of Europeans, albeit obviously in a relationship of servitude. It's a utilitarian arrangement; once he has escaped the island, Crusoe takes little further interest in Friday, who has simply proved "a most faithful Servant upon all Occasions." In contrast, Crusoe takes great pains to discharge his obligations to the Europeans who help return him to civilization.

Unlike Selkirk (if the accounts are to be believed), Crusoe remains as thoroughly European when he leaves the island as when he first arrived. This is no surprise if we assume that civilized behavior is in his blood: he can no more revert to the "savage" than he can to an apelike ancestry. By the mid-nineteenth century, *Robinson Crusoe* supported a sense of imperial entitlement: it was nothing less than a portrait of the superior traits that made the British natural governors of a quarter of the world. Crusoe, according to Leslie Stephen in the *Cornhill Magazine* in 1868, represents "the shrewd vigorous character of the Englishman thrown upon his own resources." He is

the broad-shouldered, beef-eating John Bull, who has been shoulder-
ing his way through the world ever since. Drop him on a desert island,
and he is just as sturdy and self-composed as if he were in Cheapside.
Instead of shrieking or writing poetry [as an effete continental European might], becoming a wild hunter or a religious hermit, he calmly
sets about building a house and making pottery and laying out a farm.
He does not accommodate himself to his surroundings; they have got
to accommodate themselves to him.

It was not until the imperial adventure began to wane—perhaps informed too by a pseudo-Darwinian belief in the possible evolutionary
reversion of human nature—that Europeans were willing to contemplate the spectacle of the colonist who does accommodate himself to
his surroundings, who "goes native": a Crusoe who, rather than shoot-
ing cannibals, is crowned as their king and becomes the savage, atavistic
Kurtz of Joseph Conrad's *Heart of Darkness* (1899). Conrad's book has
left an equally complex legacy, damned as a racist colonial narrative and
praised as an exposé of the evils of empire. But however you slice it, there
is no longer any way back to civilization for its Crusoe-*manqué*.

The Crusoe myth supplies a framework for postcolonial reflection
in French writer Michel Tournier's *Vendredi ou les Limbes du Pacifique*
(1967), translated in English simply as *Friday*. Here Crusoe does convert
to Friday's way of life, and chooses not to return to European civilization
when the chance arises. Influenced by Claude Lévi-Strauss's interpretation of the role of myth, Tournier felt that these stories exist to *challenge* social norms, and argued that a modern mythical hero need not
be constrained by the book in which he first appears. He's right; in fact,
that characteristic positively defines such a hero (who need be no more
bound by the gender or race in which s/he was created). But that doesn't
mean such reinvention can be arbitrary. Myths have an irreducible core,
created in a collective social process, and mutation of content needs to
serve a purpose.

An inversion like Tournier's *Friday* draws its power from that very
core: it shocks or surprises us into a new view of what the myth contains.
By turning the tables, Tournier puts the Crusoe myth itself under the

microscope. Yes, it says, *Robinson Crusoe* is pervaded with the prejudices and, we might fairly say today, the racism of Enlightenment Europe. At face value it legitimizes Western rule and colonialism, and it would surely have been read as doing so in its day. But the Crusoe myth does not *embody* or require those prejudices, and for Tournier and Coetzee it has become a vehicle for examining them.

Coetzee's Friday is surely a symbol of Africa: not allowed to speak for itself, but expected to conform to Western norms and Western dominance. His ways are odd and incomprehensible, his flute-playing pronounced monotonous and pointless in much the same way as indigenous music was once considered crude and primitive by Europeans. The only measure of his intelligence deemed meaningful by the society in which he finds himself is whether he can learn English; when he cannot, he is assumed to be stupid and childlike. But he is also vaguely feared for this intransigence and opacity—might he be a cannibal, biding his time until the chance arises to take his feast?

And yet it is Cruso who is really the fool in *Foe*. On his island he builds terraces for cultivation even though there is nothing to plant. He does it seemingly because that is all he knows how to do: he follows his European habits even when the local circumstances render them futile.

Robinson Crusoe continued to resonate as a foundational text of colonialism throughout the twentieth century. It was, according to the St. Lucian poet and playwright Derek Walcott, "part of the mythology of every West Indian child." There is a version of Crusoe, he wrote, that "gives his name to our brochures and hotels." In an ironic reversal of ownership, Crusoe "has become the property of the Trinidad and Tobago Tourist board." At the same time, he bestows status like a patron saint. "It is the figure of Crusoe," Walcott writes, "that supplies the anguish of authority, of the conscience of empire, rule, benign power."

I don't intend that we should hold Crusoe, any more than Defoe, to account for all this; Walcott criticizes the tendency "of analysing figures from literature as if they were guilty." That is precisely what, to comprehend and assess modern myth, we must not do. To condemn the maker, or what he or she made, is to evade the simple truth: the writer told a story, and *we* made it a myth.

Defoe leaves no room for ambiguity about Robinson Crusoe's accounting. The sum accrued from his plantation in 1687, less "buying Slaves" and other expenditure, comes to "470 Moidores of Gold, besides 60 Chests of Sugar, and 15 double Rolls of Tobacco." That kind of precision appears throughout the book. Money, gold, and silver are of course useless to Crusoe on his island, as he bitterly attests—but he still informs the reader that he has "about thirty six Pounds Sterling" of the stuff. At several points one feels one might be reading an annual report: targets set and achieved and assets acquired as the Crusoe Company (employees: 1) expands its market share.

It sounds dull, but (to this reader, at least) there's something oddly satisfying about it. As Crusoe's stores expand and his investments (in labor) become more secure, you feel heartened: all that effort is paying off! Despite flurries of maritime action at the beginning and end of the story, it's not at all obvious that the reason the book became an alleged staple of children's literature was its derring-do; for that, you need Long John Silver's island, not Crusoe's. But the obsessive list-making and enumeration of possessions, and the single-minded way in which these stocks are accumulated, are familiar traits of childhood activity.

This exercise in responsible accountancy might have flattered a middle-class Enlightenment audience, but it reaches beyond the specifics of a newly affluent and aspirational age. *Robinson Crusoe* is a novel about capitalism itself, and this was one of the most significant ways in which it was read until the twentieth century.

For Defoe, shipwreck was a metaphor for insolvency, which he had known and greatly feared. And of course, for many a merchant the two really did go hand in hand. In his 1725 handbook *The Complete English Tradesman*, Defoe said that bankruptcy was like drowning, while keeping one's business afloat (the expression is of course still in use) was a matter of navigating around the rocks. The new mercantile economy emphasized the importance of trust between its actors, of the sort that is clearly central to the dealings Crusoe has with the sea captains who transport him: regardless of nationality, they are gentlemen of their word.

Crusoe is the economic agent stripped to essentials: a minimal model for the theory of labor. He starts with nothing except raw resources and an ax. Everything he wants, he must make himself; everything he makes is his alone. In his labors, he confines himself to his needs. Karl Marx recognized this lesson for economics in *Capital*, where he notes that already it has been widely cited by earlier economists:

> Moderate though he be, yet some few wants he has to satisfy, and must therefore do a little useful work of various sorts, such as making tools and furniture, taming goats, fishing and hunting. . . . In spite of the variety of his work, he knows that his labour, whatever its form, is but the activity of one and the same Robinson, and consequently, that it consists of nothing but different modes of human labour.
>
> Necessity itself compels him to apportion his time accurately between his different kinds of work. . . . This our friend Robinson soon learns by experience, and having rescued a watch, ledger, and pen and ink from the wreck, commences, like a true-born Briton, to keep a set of books. His stock-book contains a list of the objects of utility that belong to him, of the operations necessary for their production; and lastly, of the labour time that definite quantities of those objects have, on an average, cost him. All the relations between Robinson and the objects that form this wealth of his own creation, are here so simple and clear as to be intelligible without exertion. . . . And yet those relations contain all that is essential to the determination of value.

Of course, Marx adds, everything produced on Crusoe's island is the work of an individual, not a community*—for sociologist Max Weber,

---

* That's not as unrealistic, at least for economists, as you might think: modern economic theory is often criticized for being a theory of an economy with only one person in it.

Crusoe is "the isolated economic man." Yet still his situation expresses what we can see too in "a community of free individuals, carrying on their work with the means of production in common." The problem, said Marx, arises when those individuals no longer own the means of production or the rights to what is produced.

Crusoe was an equally useful figure for the founders of neoclassical economics. The English economist Alfred Marshall evoked him in his highly influential *Principles of Economics* (1890) as an agent who chooses to allocate his resources so as to "maximize his utility"—the central motive of *homo economicus*. And this is an example of how myths become adapted to the purpose at hand, for Crusoe himself didn't really work this way at all—look at the time and energy he wastes building a boat that he then can't use. This "Robinson Crusoe economics" is splendidly mocked by economist Homa Katouzian, who says that "if the Crusoe model is correct . . . we should not expect a man in these conditions to commit suicide, go crazy or become a mystic; rather we should expect him to allocate his material resources efficiently by considering (for example) the opportunity cost of talking with his parrot." For economist Stephen Hymer, the contrast between the Crusoe said to master nature through reason, diligence, and intellect and the real one, who relies on conquest, slavery, murder, and force, all too closely parallels the "mythical description of international trade found in economic textbooks" when set against the exploitative and coercive realities of the real world.

There's no reason to think that Defoe intended any allegory of this sort, although economic policy certainly interested him. It was the topic of his earlier book *An Essay upon Projects* (1697), but his views were complex, weighing human interests, social welfare, and education alongside the rule of the marketplace. Always a man with an eye on profit-making enterprises, he was fully at home in the emerging mercantile economy. For a time he ran a business exporting hosiery to Spain, and he had shares in the South Sea Company (which he was canny enough to sell before the infamous South Sea Bubble burst in the year of *Crusoe*'s publication).

Crusoe doesn't exactly exemplify the slogan Marx popularized: "From each according to his ability, to each according to his needs."

Rather, he accumulates and stockpiles: he is the sensible investor hedging against risk, who first strips the wreck of anything that might conceivably be valuable in the future (including money). All the same, when essayist Richard Steele, a friend of Defoe's, interviewed Alexander Selkirk* for his account of 1713, he discerned an economic moral in the Scotsman's exploits on the Juan Fernández Islands: "This plain Man's Story is a memorable Example, that he is happiest who confines his Wants to natural Necessities." Steele reported that Selkirk told him, "I am now worth 800 Pounds, but shall never be so happy, as when I was not worth a Farthing."

This back-to-basics scenario chimes with Jean-Jacques Rousseau's celebration of the "noble savage." Such isolation, Rousseau says, clarifies our judgment:

> The surest way to rise superior to all prejudice, and to form our
> judgment upon the true report of things, is to place ourselves in the
> situation of a man cut off from all society, and to judge of every thing
> as that man must naturally judge.

With his austere vision of a noble, pristine state of nature, Rousseau lamented all the "unnecessary rubbish" Defoe had included to bookend Crusoe's solitude. He wanted a pure desert-island story, and in some ways the myth has afforded him his wish. In that primal predicament, Rousseau says, we discover what *really* matters. There can be no frivolity in Crusoe; rather, he must school himself, become resourceful despite his initial practical ineptitude:

> The practice of simple manual arts, to the exercise of which the abil-
> ities of the individual are equal, leads to the invention of the arts of
> industry.

---

* There has been speculation as to whether Defoe might ever have met Selkirk, although he
denied it himself—he wouldn't want anyone thinking he had so directly lifted the story.

In this way, Rousseau says, we might learn to appreciate a philosophy of utility, in which the ironmonger is of more value than the "best furnished toy-shop in Europe."

It doesn't matter that Crusoe is unskilled at the mechanical arts, though, for what he makes has only to be good enough to do the job. "I had never handled a Tool in my Life," he admits, "and yet in time by Labour, Application, and Contrivance, I found at last that I wanted nothing but I could have made it." Samuel Coleridge, who adored the book as "a vision of a happy nightmare," wrote that in *Crusoe*

> nothing is done, thought, or suffered, or desired, but what every
> man can imagine himself doing, thinking, feeling, or wishing for. . . .
> The carpentering, tailoring, pottery, all are just what will answer his
> purpose.

After all, why make a thing finer or prettier than it need be? Defoe's biographer, the Scottish antiquarian George Chalmers, asserted that Crusoe's handiwork showed "how much utility ought to be preferred to ornament."

It all seemed highly meritorious. "Say 'Robinson Crusoe,'" wrote an anonymous author in the *National Review* in 1856, "and you see a desert island, with a man upon it ingeniously adapting his mode of life to his resources." In his moderation and graft, his humble application to the necessities of life, *Crusoe* was thus seen for the first hundred years or so after its publication as offering a moral lesson. On the lonely island we discover what is truly important to us—just as the "castaways" on BBC Radio's long-running *Desert Island Discs* have to select the eight pieces of music, and a single book, they could not do without.* The fact that so many of the guests on the show choose music that connects them to loved ones, however, suggests that in the end what

---

* Regular listeners will know that the castaways get to choose a luxury too. I like to think that
Rousseau would have firmly rejected that offer.

really matters, *pace* Rousseau, is not a hand ax or a treehouse but human company.

◆

The writer and children's educator Johann Heinrich Campe obliged Rousseau's wish for a stripped-down *Crusoe* by publishing *Robinson der Jüngere* (1779–1780), translated variously as *The New Crusoe* or *Robinson the Younger*. Here the moral lesson observed by Rousseau is emphasized for young readers by reducing Crusoe to something much closer to a state of nature, unassisted by any "European tools" rescued from the stranded ship. The purpose of Friday's appearance, meanwhile, is to show the young reader the value of fellow society in procuring happiness. Only when a second European vessel is wrecked off the shore does Crusoe acquire tools to improve his lot; thus the reader might see how we shouldn't take these things for granted. In Campe's retelling the whole tale is narrated to a group of children by their parents, who carefully explain the moral at each turn, implanting in the breasts of their offspring a resolution to do what Crusoe did: to persevere and trust to God.

It was an educational enterprise fraught with ambivalence: the Crusoe myth mobilized to convey to children that you get on in life through patience and diligent application—while what they surely loved about the story was the romance and adventure, the invitation to impulsive exploration. Whatever impression young readers took from Campe's book, it sold so well that some people believed it to be the original text. And it does have a certain charm, despite its rather hectoring tone. But the main value of *The New Crusoe* now is to show us how an attempt to rewrite a modern myth to fit a particular, preordained moral is the surest way of turning it into a mere yarn.

The Crusoe narrative was so often retold that it became a genre of its own: the so-called Robinsonade. By the end of the nineteenth century, according to critic and writer Robert McCrum, "no book in English literary history had enjoyed more editions, spin-offs and translations than *Robinson Crusoe*"; McCrum counts more than seven hundred of them.

The myth of the island beyond the farthest horizon has existed for as long as people have set out to sea: Atlantis, Ultima Thule, Elysium. But only in the Age of Empire did the island become a place ripe for profit and domination, as well as a locale of isolation and survival. Tales of shipwreck held a particular appeal for the nineteenth-century French fantastical writer Jules Verne. It's said that his schoolteacher in Nantes regaled her pupils with stories of how her husband, a naval captain lost at sea, was still surviving on some island as a Crusoe-like castaway. Verne's Robinsonades include *The Mysterious Island* (1874) (initially titled *Shipwrecked Family: Marooned with Uncle Robinson**) and *The School for Robinsons* (1882), which closely shadows Defoe's plot.

In marooning a "family" of sorts (an engineer, an ex-slave, a sailor and his adopted son, a journalist, and a dog), *The Mysterious Island* was also indebted to *The Swiss Family Robinson* (1812) by the Swiss pastor Johann David Wyss. For Wyss, an educator who wrote "improving" tales for children, the Crusoe template was an ideal vehicle for promoting Protestant Christian values of patience and hard work. The popularity of his little novel among young readers can, I suppose, be explained by its inclusion of child characters to whom the intended audience might relate. But its moral message of diligence is rather undermined by the magical ease with which the insufferably prim family overcomes all obstacles, mastering the art of spearing fish in an afternoon and erecting a bijou treehouse in a few days. Victorian children might not have known much more natural history than Wyss, but children today (in my experience) find it hilarious that on this remote desert island penguins rub shoulders with flamingoes, lions, walruses, giraffes, and bears—none of them remotely threatening to the resourceful family.

The happy, self-sufficient family has been an equally useful image for securing normative values in modern times. The 1940 RKO movie based on Wyss's book was a straightforward melodrama, but Disney's 1960 version starring John Mills is a cloying paean to the nuclear family

---

* The island in the South Pacific on which the American castaways are stranded turns out to be the secret port of the *Nautilus*, the undersea craft of Captain Nemo from Verne's *Twenty Thousand Leagues under the Sea* (1870).

in an age when that wholesome unit was starting to look threatened by decadent elements in American society. Animal races and pirate attacks keep the kids happy, but we're left in no doubt that the family is the bastion of decency and virtue against the lewdness and savagery of the world outside. A 1976 adaptation replays that moral, with flares.

Crusoe might be a characterless egotist, but he's still an entrepreneur tailored to flatter Enlightenment capitalists. By the nineteenth century, such ruthless individualism needed to be bound to a cohesive society by the glue of sound moral education and a respect for authority and tradition; the revolution in France had shown where you might otherwise end up. And while castaway children could enhance the appeal to a young audience, without guidance from parental authority *Robinson Crusoe* would not really become R. M. Ballantyne's *The Coral Island* (1858), in which three boys marooned in the South Pacific draw on their pluck, courage, and Christian values to survive cannibals and pirates. On the contrary, said William Golding in *Lord of the Flies* (1954) as the British imperial dream was withering, these children would soon enough become savages themselves.

Lord of the Flies reaches further than that, however. According to literary scholars John Fitzgerald and John Kayser, who call Golding's novel a myth in itself,* the dominant critical view is that the plight of the boys is an allegory for that of modern man, who denies and fears the irrational. Their regression into barbarism comes not from a mere breakdown of civilized order among the boys, but from their very attempt to impose it: to manufacture an imitation of European culture in the wild (which of course is entirely Crusoe's project). It won't work, Golding implies, for it means denying our deepest nature—which, outside the buttressing structures of civilized society, is sure to emerge. Assumptions of cultural superiority, taken for granted in *Crusoe*, will only make matters worse. "We're English, and the English are best at everything," says the leader of the rebel boys, Jack—and while that was a self-evident truth for *Crusoe*'s

---

* In response to critics' suggestions that *Lord of the Flies* is a fable, Golding too said he preferred to think of it as mythic.

first readers, it appears all too ironic amid the embers of empire. The irony is emphasized when the British navy does eventually show up to rescue the boys. "I should have thought," the officer loftily declares, "that a pack of British boys—you're all British, aren't you?—would have been able to put up a better show than that." For American writer Lois Lowry, this officer exposes all that is hateful and deceitful about his way of life: "He above all the others makes me question what is representative of civilization: a spotless uniform, a dignified posture, and a set of elaborate rules?"

The Island myth that *Robinson Crusoe* inaugurated holds all of the tension introduced by the Age of Reason, and which arguably all of our modern myths touch on: where, in a world that values science, logic, and intellect, do our darker, savage, inchoate impulses go? The narrative of a castaway holding it together until a rescue party arrives to carry him back to civilization often gives way now to one in which isolation from the normalizing forces of regular society presage a descent into madness and abomination, from the obscene experiments of H. G. Wells's Doctor Moreau (1896) to the murderously dysfunctional hippy commune of Alex Garland's *The Beach* (1996). Islands were once seen as refuges where well-regulated ideal societies could flourish: Thomas More's original Utopia (1551), Francis Bacon's scientific theocracy of Bensalem in *New Atlantis* (1626). But those were dreams of social harmony at a time when Western culture was just starting to glimpse an alternative to its bellicose and often tyrannical age. Now the island exposes the atavistic tendencies that the supposed Age of Reason has never vanquished.

Aldous Huxley's last novel, *Island* (1962), was a rare modern effort to resurrect the utopian rather than dystopian island. On the Polynesian island state of Pala, the English castaway Will discovers a society that embraces freedom of mind, body, and spirit, in contrast both to the repressed West and to the militaristic and materialist kingdom of Rendang-Lobo, which looks covetously at Pala's oil resources. As a counterpart to the conformist dictatorship of *Brave New World*, the novel describes an interesting experiment in running a genuinely more healthy society; as a work of fiction, however, it reads more like a manifesto for Huxley's enthusiasm for drugs and free love. All the same, it continues the tradition of making the island a place where familiar laws may be suspended. We are never sure if that offers escape or nightmare.

The real *Crusoe* for our times is surely J. G. Ballard's *Concrete Island* (1974), which explores the isolation possible amid, and indeed created by, the insane bustle of the urban jungle. "The Pacific atoll may not be available, but there are other islands far nearer to home, some of them only a few steps from the pavements we tread every day," writes Ballard in the novel's introduction. "They are surrounded, not by sea, but by concrete, ringed by chain-mail fences and walled off by bomb-proof glass."

In *Concrete Island*, Robert Maitland, a pampered and wealthy architect who echoes Defoe's bourgeois seafarer, is marooned on the traffic island of a motorway intersection in London, surviving on meager provisions salvaged from his crashed Jaguar. He never escapes, but resigns himself to his predicament almost with relief: "Already he felt no real need to leave the island, and this alone confirmed that he had established dominion over it." Maitland is an anti-Crusoe, reveling in his accidental freedom. "Perhaps, secretly, we hoped to be marooned, to escape our families, lovers and responsibilities," says Ballard. Set against the stifling pressures of modernity, the desert island doesn't look so bad after all.

By the second half of the twentieth century the remote desert island had become a cliché, rendered ever less plausible now that the well-to-do could fly to exotic beaches rather than seeing them as mythical realms. To be a real castaway in a remote, unpopulated place, you would have to go much farther afield.

The 1964 movie *Robinson Crusoe on Mars* got straight to the point: "A lone US astronaut pitted against all the odds beyond this earth . . . only his nerve and know-how to save him!" A rather literal recasting of Defoe's tale, it ducked the problematic Friday by making him an alien.* Updating Wyss's book for the space age, *Space Family Robinson* began as a comic-book series published by Gold Key in 1962, a year after Yuri

---

* The alien was played by Victor Lundin, who also portrayed the first Klingon seen on-screen in *Star Trek*. The second crew member, killed in the descent, was played by Adam West, who we will most certainly encounter again later in his most legendary role.

One of Gold Key's comics in the Space Family Robinson *series of the 1960s.*

Gagarin, in the first manned spaceflight, made a single orbit around the earth. It was the source material for the hipster space colonists of the TV series *Lost in Space* (1965–1968). Even in that predicament, this family of Robinsons remains obedient to prescribed gender roles, while the robot standing in for Friday could be ridiculed as a "bubble-headed booby" without causing anyone offence.

Island shipwreck tales had in fact become linked to space travel even in Defoe's own time. *A Trip to the Moon* (1728), by the pseudonymous Irish writer Murtagh McDermot, begins with the narrator taking to sea against his mother's wishes and being driven by a storm to land at Tenerife, where he is seized by a whirlwind at the top of the island's volcanic mountain and transported to the moon. Shenanigans in the courts and coffeehouses of the moon people ensue—for extraterrestrial civilizations in these early modern science-fiction tales always mirror those on Earth, the better to act as vehicles for Swiftian satire. *The Life and Astonishing Adventures of John Daniel* (1751) by Ralph Morris is a bizarrely entertaining fancy, obviously modeled on *Crusoe*, in which the eponymous hero is shipwrecked with a shipmate, Thomas, on a desolate island, only to discover that Thomas is in fact a woman called Ruth (I bet no one saw that coming). They marry and have a family, whereupon their son Jacob invents a flying machine he calls an "eagle," on which they have all manner of adventures, including a flight to the moon. Defoe even wrote a "moon tale" himself, *Consolidator* (1705), a social satire "translated from the lunar language" in which the device named in the title is a machine for making voyages through space.

All this isn't necessarily surprising. Tales of travel to the moon were almost a century old by the time *Crusoe* was written, having begun in earnest with *The Man in the Moone* by the bishop of Hereford Francis Godwin, published posthumously in 1638 but written the 1620s.* In the middle of that century, the French writer Cyrano de Bergerac gave a detailed account of the courtly and philosophical intrigues of other words in *The States and Empires of the Moon* and its sequel *The States and Empires of the Sun*, both published after his death, in 1657 and 1662 respectively. These books were inspired by the telescopic observations of astronomers like Johannes Kepler and especially Galileo, whose obser-

---

* The astronomer Johannes Kepler's account of travels on the moon, *Somnium*, was written earlier, around 1608, but wasn't published until 1634, also posthumously. The events he relates happen in a dream, however, so whether the book truly qualifies as a fantasy of spaceflight is open to question.

vation of mountains on the moon persuaded many natural philosophers that this celestial body, and perhaps the other planets too, were worlds like ours.

Public reception of those revelations was conditioned by the European experience of other cultures initiated by the era of maritime exploration. Voyages made by sailing ships, and the exploits of traders and explorers, supplied the natural point of reference for journeys into space: one has the sense that space was regarded as like the Indies or the Americas, if a little further. The themes of colonization and empire are apparent in a speculative account of possible space travel, *Discovery of a World in the Moone* (1638) by the philosopher and mathematician John Wilkins, a founding member of the Royal Society.

A rose-tinted view of the Age of Exploration—the voyages of Columbus in particular—still supplies today the predominant popular narrative for space exploration. A more honest and historically accurate telling—one that made war, trade, and profit the prime motives for voyages rather than a noble spirit of curiosity and adventure—would better fit the reality of the space race, whether in the 1960s or today. For however brave, noble, and technologically impressive the Apollo moon landings may have been, Neil Armstrong and Buzz Aldrin planting the American flag (with some difficulty) in the thin lunar "soil" sent out a message of territorial dominion, in an intellectual if not in a legal sense.

Still, the quest to push back the cosmic frontier holds tremendous public appeal. The image of a journey across the fathomless and inhospitable ocean of space to adventures in "strange new worlds" among the distant archipelagos of stars and planets supplies a conceptual connection to the old tales of seafaring: the Viking sagas, the *Odyssey*, the Argonauts. Naming space shuttles *Columbia* and *Enterprise* links them into that same narrative.

Thus space science fiction offers us mutinies (*Outland*), piracy and swashbuckling privateers (Han Solo of *Star Wars*, *Guardians of the Galaxy*), naval battles (most of the *Star Trek* franchise), sea monsters (*Alien*), storms and shipwrecks (via black holes, asteroid showers, supernovae), ghost ships (*Sunshine*, *Event Horizon*), and colonialism (*Avatar*). It's natural that the Island myth should become part of our narrative of interstellar travel too. *Forbidden Planet* (1956) was explicit about that, recapitulating on the distant planet Altair IV the shipwreck of Antonio, Alonso, and their entourage on Prospero's island in that other keystone text of the myth, *The Tempest*.

But in many of these borrowings from the tropes of exploration and conflict, space is a busy ocean thoroughfare, crisscrossed by trade routes and patrolled and contested by fleets. Only since the Apollo missions showed us the real desolation and harshness of the cosmos beyond our atmosphere has the image of the lone traveler eking out a precarious survival truly taken hold: witness the solitary Bruce Dern losing his mind in a leafy space-capsule paradise in Douglas Trumbull's *Silent Running* (1972). Duncan Jones's *Moon* (2009), with Sam Rockwell kept company only by his clone and his computer, makes it impossible not to think of the director's famous father supplying the backing track for the 1969 moon landings with the lonely refrain of "Space Oddity": "Here am I sitting in a tin can . . . planet Earth is blue, and there's nothing I can do."

These are tales in which survival depends not just on physical but on psychic endurance. As Walcott said of Crusoe, "[his] survival is not purely physical, not a question of the desolation of his environment, but a triumph of will." One must suspect, in watching the mental disintegration of the characters portrayed by Dern and Rockwell, that dealing with such isolation is a lot easier if you are as dull and unimaginative as Crusoe.

But the survivalist space story most obviously indebted to Defoe's tale of resourcefulness is Andy Weir's novel *The Martian* (2014), best known from the 2015 film adaptation by Ridley Scott. Mark Watney (played by Matt Damon in the movie) is a NASA astronaut left for dead on Mars. Struck by falling equipment in a heavy sandstorm, he fails to reach the ascent vehicle that will take the lander crew back to its spaceship, *Hermes*, and is left with just the habitation pod and the spent descent vehicle. In Scott's film, Watney wakes after the storm has passed as if washed ashore in a desiccated world of sand dunes by some invisible tide. He labors to stay alive under the pink sky of a pseudotropical sunset, improvising with his scant salvaged resources to communicate with

Hermes, keep himself in food and water, and ultimately get off the planet and back to the mothership. In this harsh place where the environment itself is barren and lethal, a Protestant work ethic is no longer enough; the castaway now needs technology too. To survive, he has to "science the shit out of this."

If plainness of expression and pedestrian, pedantic quantification of resources is intrinsic to the Crusoe myth, Weir's book qualifies as a true successor. "I need 1500 calories every day," Watney tells us. "I have 400 days of food to start off with. So how many calories do I need to generate per day along the entire time period to stay alive for around 1425 days? I'll spare you the math. The answer is about 1100." Watney shares with Crusoe a capacity—who knows, perhaps it is vital under these circumstances?—to respond to absolute isolation and existential peril with much the same "manly independence" and "slow yet efficient intelligence" as someone discovering their garden shed has collapsed. ("You know what?! Fuck this!") James Joyce's assessment of the Anglo-Saxon imperialist is probably no less apt to describe the archetypal astronaut, selected for the mission precisely because of the alleged stability conferred by a narrow, pragmatic mind.

With his obsessive inventory-making and stocktaking, and total lack of self-reflection, aesthetic response, or imagination directed toward anything other than fixing technical problems, Watney is indeed a space-age Crusoe. The slightest deviation in his monologue from practical issues is derided as useless philosophizing and curtailed with the same alacrity with which brigadiers in British India might have repressed poetic thoughts as a sign of latent homosexuality. He is not John Bull on a beach but John Doe on Mars, the British grit supplemented now with an American penchant for feeble wisecracks.

In these ways, Watney personifies the spiritual emptiness that so dismayed Norman Mailer when he visited NASA's Manned Spacecraft Center in Houston to research his 1970 book on the Apollo program, *Of a Fire on the Moon*. He found the whole enterprise "severe, ascetic," and soulless. To its denizens, Mailer felt, the moon landing was a mere technical achievement, devoid of sacred significance, much as Crusoe's voyages are all in the name of commerce and profit, the traveler seeming blind to the natural beauty he traverses.

You might conclude that I don't like *The Martian*, but I like it no more or less than *Robinson Crusoe*. It offers the same compulsive cataloguing of necessity, the same vicarious thrill of accumulation (as Watney grows his potatoes in feces-fertilized Martian soil) under the threat of storm and natural catastrophe. It is—give or take—the same story. Make no mistake: you *do* want to know what becomes of Watney, more so indeed than you care whether Crusoe returns to profit from his slave-tended plantations. His relentless calculations and problem-solving are as hypnotic and as satisfying as Crusoe's ledgers of assets.

To what end, though? There is in *The Martian* not even a hint of the dutiful spirituality that sends Crusoe to his Bible in his spare moments. Weir makes a stab at a secular inspirational message—"Every human being has a basic instinct to help each other out"*—but you sense that Watney has more terror of appearing "philosophical" than of perishing on the Red Planet.

The Martian gives the Crusoe myth a rather simple task: to reassure us that the Columbus narrative will suffice to take us to the stars. Of course there will be shipwrecks, but the indomitable human will—aided by the brilliance of our science and engineering—can find a way through them. In its unintentionally nihilistic way it is revealing about human spaceflight, dramatizing the vacuum at its heart.

Perhaps we *will* someday stand on Mars. But that will only bring home, even more clearly than the Apollo missions did, that space is not really another ocean, somewhat larger than the Atlantic, and its other worlds are considerably less clement than the Juan Fernández Islands.

We have known this since Neil Armstrong stepped out onto a lifeless, gray world, and that is one reason Crusoe has gone into space: because

---

* In the characteristically circular logic that sustains human spaceflight, this is elided into a justification for going into space in the first place.

we are anxious about doing so ourselves, and we are soothed by tales of resilience and survival. We can't really begin to imagine the loneliness of being marooned in so alien a place, apart from all of Earth's people and beaches and seas. We sense in our hearts how inhospitable the universe becomes once we are more than a few dozen miles above the ground. Like Crusoe, we dream of great voyages, but we are not sure we are meant to be taking them.

✿

Chapter 3

THE REANIMATOR

FRANKENSTEIN (1818)

An obsessive scientist creates a living being,
usurping the role of God or nature. But what
he creates will eventually destroy him.

═══════════════════════

In November 2018, Chinese biologist He Jiankui announced that he had used gene editing to alter the DNA of human embryos before implanting them in an in vitro fertilization procedure, a process that led to the birth of two girls. This use of a technique of unproven safety for human reproduction was widely condemned as irresponsible, reckless, and unnecessary. It earned He the sobriquet "China's Frankenstein."

Well it would, wouldn't it? Two hundred years after the publication of Mary Shelley's novel, the name of Victor Frankenstein is still invoked whenever science seems impetuously to have defied the bounds of nature, particularly when its actions have anything to do with the creation of new human beings. The same trope was routinely paraded in the early days of IVF in the 1960s and 1970s; it serves as a warning flag in debates on human cloning; and it was duly trotted out to accompany media reports in the 2010s of the new technique of mitochondrial replacement, luridly (and tendentiously) said to involve the creation of "three-parent babies."

The story of He Jiankui, however, fitted the template better than most. Here was that figure who many thought had vanished from contemporary science: the lone maverick, obsessively following a path of questionable morality for personal glory, heedless of the consequences. (And in the end, suffering them: he received a three-year jail sentence for violating ethical laws.) Manipulation of human genes is often condemned as tampering with the "stuff of life"—or, in the popular (albeit theologically barren) phraseology of our age, playing God.

There is powerful mythology at work here.

Frankenstein is the most obvious and most enduring exemplar of the modern myth—a tale deeply embedded in cultural discourse and readily mobilized, with no need to elaborate, for rhetorical and allegorical purposes. The story escaped its creator, and its source text, almost as soon as Mary Shelley gave birth to her "hideous progeny." It grew from a novel into a *worldview*, with an iconography that is instantly recognizable and endlessly malleable. To describe the story's creation using so gendered a metaphor as birth is intentional and necessary: it is Shelley's own metaphor, and gender is at the heart of *Frankenstein*.

It is, in the words of literary scholars George Levine and Ulrich Knoepflmacher, "the perfect myth of the secular, carrying within it all the ambivalences of the life we lead here, of civilization and its discontents, of the mind and the body, of the self and society." Here is the full mythic package: sex, violence, monstrosity, procreation, incest, technology, body horror, God, the devil, and the Fall. Only a misconceived critical focus on style and execution could prevent—and has prevented—the book from being accepted as one of the most phenomenal achievements in modern literature. I daresay the fact that it was achieved by a young woman barely out of her teens has had something to do with that neglect too. How could that be possible, male critics (even in relatively recent scholarship) have asked? The ludicrous chauvinism in the judgment of Italian critic Mario Praz was not uncommon: "All Mrs Shelley did was to provide a passive reflection of some of the wild fantasies which were living in the air about her." Why, anyone can write a myth!

Frankenstein will long outlive such silliness. There is not the slightest sign that we have exhausted the material in the tale of the obsessed scientist and his creature; the myth is, as we can see, as strong today as ever it was. If you still have any doubts about what I mean by the notion of modern myth, *Frankenstein* is the answer.

The Bride of Frankenstein, the 1935 sequel to the Boris Karloff movie—it's the one with Elsa Lanchester as the "bride" with her marvelous "lightning bolt" hairdo—begins in metanarrative mode with Lord Byron, Percy Shelley, and Mary "Shelley" (she and Percy were not yet married when Mary conceived of the novel) gathered around a fire on a stormy night. Mary reveals that there is more to the story, something beyond the apparent death of the creature in a blazing mill, which we, the cinema audience, had witnessed at the conclusion of James Whale's *Frankenstein* four years earlier. "That wasn't the end at all," Mary (cunningly played by Lanchester too) tells her comically effete companions. "Would you like to know what happened after that?"

The screenwriting history of *The Bride of Frankenstein* was an unholy mess, but in at least this aspect of what emerged there was genuine inspiration. For it folds into the myth of Frankenstein the tale of his creator, beginning what became a full-blown elision of the two. Mary Shelley, *née* Godwin, was not the author of a myth but a part of it.

That part of the story can be given its own condensed plot summary, which goes something like this:

> A girl half-orphaned at birth by the death of her famous and rebellious mother is brought up by her austere father, who sends her away to a childhood of isolation in a remote land, during which she has to educate herself. She learns about things young ladies were not meant to know. Then, as a teenager, she runs off with her lover—a married,

dashingly romantic poet—and in the company of his louche entourage during a summer of gloom and rain, she has a nightmare and writes a tale of horror that shocks and chills society. Meanwhile, she endures the death of three children and her husband.

All of the Frankenstein myth is true, but this part actually happened—more or less.

Mary Shelley's parents were political radicals living at the fringe of what might be considered respectable society, intellectually and socially. Her father was the philosopher and writer William Godwin. Her mother was Mary Wollstonecraft, whose book *A Vindication of the Rights of Women* (1792) argued that women are equal to men but held back by lack of education and opportunity. Wollstonecraft was on friendly terms with William Blake and Thomas Paine, and she and Godwin supported the French Revolution and advocated free love. Before her marriage to Godwin, she had an affair with the Swiss painter Henry Fuseli (whose most famous work, *The Nightmare*, is emblematic of the Gothic sensibility). She proposed that she live in a *ménage à trois* with the painter and his wife, who was not so keen. She had an illegitimate daughter, Fanny, by the American Gilbert Imlay; her second daughter, by Godwin, was christened Mary Wollstonecraft Godwin. Wollstonecraft died in 1797, eleven days after Mary was born, from septicemia caused by complications of the birth.

William Godwin was devastated by his wife's death, but he remarried in 1801 to Mary Jane Clairmont, a widow with two children. There followed the all-too-familiar struggle between daughter and stepmother for the affections of the father—who in any case seems to have preferred his first wife's daughter Fanny. Mary Godwin's childhood sounds rather desolate; she admitted that as a child she felt "lonely as Crusoe," brought up by a father who tried more to mold than to nurture her. She became listless, depressed, and argumentative; in 1812 William brought an end to the tensions and squabbles between Mary and his second wife by sending his fifteen-year-old daughter away to live with the Baxters, a family in Dundee, Scotland, who he knew only rather distantly. There she became almost an autodidact, leaving her forever lacking confidence about her learning. When she became Percy Shelley's lover and then his wife, he undertook to set that right, and even after his death in 1822 she could still write: "I am beginning to seriously educate myself."*

Returning to London for a visit in late 1812, Mary met a young man who had taken to frequenting the Godwin household: the aspiring but still little-known writer and poet Percy Bysshe Shelley. Within a year or so, William Godwin, who by habit lived beyond his means, had become dependent on the money that Shelley, the well-to-do son of a Whig MP, lent him. By the summer of 1814, the young man dined every day with the Godwins—and at the end of June he and Mary declared that they were in love.

Godwin was furious, insisting that the two should never again meet. Shelley was already married—his wife Harriet Westbrook had been only sixteen (and Shelley just three years her senior) when they eloped in 1811. Of course, Godwin was hardly respecting his liberal principles, given his (and Mary Wollstonecraft's) public views on marriage, but in any case, Mary was not to be deterred. In July the young couple eloped to France, taking with them Mary's half sister Jane Clairmont. Mary, like Harriet at the start of her union with Percy, was not yet seventeen years old.

The trio traveled in Europe for two months before returning to London, by which time Mary was pregnant. Her relationship with Shelley was complicated to say the least. Percy, another supporter of free love, may well have been conducting a sexual affair with Jane (who later took the name "Claire," apparently considering it more poetic), while encouraging Mary to do the same with Shelley's friend Thomas Hogg, the future author of the Gothic doppelgänger tale *Confessions of a Justified Sinner*. Mary's child, a girl she named Clara, was born prematurely in February 1815 and died within a month. Percy seemed little concerned; the day after Mary bluntly recorded in her journal "Find my baby dead," he was off on a trip with Claire.

---

* Compare the remark of Robert Walton, the metanarrator of *Frankenstein*: "My education was neglected, yet I was passionately fond of reading." Victor Frankenstein too was initially self-taught.

Mary Shelley, painted by Richard Rothwell, ca.1840.

Within months Mary was pregnant again. Her son William was born in January 1816 (and would die of malaria three years later in Italy, after *Frankenstein* was published). That summer, she traveled with Percy, William, and Claire to meet her stepsister's new lover, Lord Byron, in a residence on Lake Geneva in Switzerland, where *Frankenstein* began.

Mary married Shelley in London in December of that year, just twenty days after Percy's former marriage was dissolved by the suicide of his estranged wife Harriet: heavily pregnant, she drowned herself in the Serpentine. That October, William Godwin's favorite child Fanny also killed herself, and her stepfather would not claim or even identify her body. Mary had a third child, who again she called Clara, in September 1817; the child died a year later while Mary was traveling to Venice to join her husband.

Amid all this dysfunction and tragedy, the monster emerged.

***

It used to be a fairly common critical view that Mary Shelley created *Frankenstein* by accident, with little idea of what she was doing: that she was merely channeling something "in the air," as Praz asserted. Wasn't the book as much Percy's work as hers anyway? How, after all, could a young woman of eighteen possibly produce material this powerful, rich, and enduring?

This is nonsense. It's true that Shelley does not weave all the strands of her story into a beautifully formed shape. She sometimes lapses into melodrama, purple prose, and outright absurdity; occasionally she (literally) loses the plot. And she may have done herself few favors when, in 1831, she came to revise what was by then a famous book. Not only did she impose a stripped-down and rather simplistic moral reading of the text, but she concocted that overcooked, unreliable myth about the story's origins. And if she had been asked to expound on the extraordinary psychosexual dynamics of *Frankenstein*, it would not just have been the conventions of her time that would have inhibited the response; in all probability, she barely sensed, let alone articulated, them. All the same, we do not retrospectively read these elements into the text: she put them there herself. She lacked literary experience and technique, but she was not some kind of naïf unwittingly channeling the *Zeitgeist*. *Frankenstein* shows Mary Shelley to be a truly great writer, if we don't insist on too narrow a notion of what that may mean.

What *Frankenstein* achieves is not just remarkable but baffling and slightly scary. As literary critic Lawrence Lipking puts it, the question—to which we should admit right away that we don't have an answer—is this: "How did such a young and inexperienced writer, who in the remaining three decades of her life never again showed any phenomenal talent, manage to create a work that still haunts the dreams of much of the human race, a work whose title alone evokes a vision of history that has yet to run its course?"

That's how it is with modern myth. It does work that we cannot put into reasonable-sounding words, and which we can't quite identify in the texts themselves. It is not to diminish Shelley's authorship one whit to say that *Frankenstein* does things it was never intended to do. It does so precisely because of what Shelley threw into the stew, and how she mixed and seasoned it. Do taste the original, if you never have. It is still weirdly sublime.

So here it is then: *Frankenstein* is said to have been born one dark and stormy night near Lake Geneva, where Mary and her conspirators—Percy Shelley, Byron, his infatuated mistress Claire Clairmont, and his physician John Polidori—were encamped in that terrible summer of 1816. It was cold and wet, thanks to dust erupted from the volcanic Mount Tambora in Indonesia the previous year that shrouded the atmosphere and briefly cooled the climate. The group had settled in a holiday mansion called the Villa Diodati, and they clustered around the fire in the unseasonable chill and discussed many things: philosophy and science, poetry, the occult. Byron proposed that, to while away the evenings, each of them should compose a ghost story to chill the bones of the others.

He himself came up with a fragment of a tale,* and Percy Shelley began another based on some experience in his early life. As for Polidori, Mary seemed to hold both the man and his storytelling effort

---

* He also later began his Faustian Gothic fantasy *Manfred*, a tale of a Promethean rebel with vampiric traits, at the villa.

in rather low regard.* "Poor Polidori," she says, "had some terrible idea about a skull-headed lady who was so punished for peeping through a keyhole—what to see I forget—something very shocking and wrong of course."

Mary Godwin was keen to come up with a tale that "would speak to the mysterious fears of our nature and awaken thrilling horror—one to make the reader dread to look round, to curdle the blood, and quicken the beatings of the heart." Her companions were doggedly insistent that she contribute, asking her each morning if she'd had an idea, only for Mary to confess guiltily that no inspiration had come. Percy seemed to want to know if she had it in her, Mary later attested.

Then one night, after he and Byron had discoursed on life itself, with Mary a "devout" but mostly silent witness, she retired to bed and was beset by terrible imaginings. As she wrote in her 1831 account:

> When I placed my head on my pillow, I did not sleep, nor could I be
> said to think. My imagination, unbidden, possessed and guided me,
> gifting the successive images that arose in my mind with a vividness far
> beyond the usual bounds of reverie. I saw—with shut eyes but acute
> mental vision—I saw the pale student of unhallowed arts kneeling
> beside the thing he had put together. I saw the hideous phantasm of a
> man stretched out, and then, on the working of some powerful engine,
> show signs of life, and stir with an uneasy, half-vital motion. . . . The
> idea so possessed my mind that a thrill of fear ran through me, and
> I wished to exchange the ghastly image of my fancy for the realities
> around. . . . 'I have found it! What terrified me will terrify others;
> and I need only describe the spectre which had haunted my midnight
> pillow.'

---

* Byron was worse. "I never was much more disgusted with any human production" was his harsh assessment of his physician. Yet given the fate of Polidori's own story, so meanly traduced by Mary Shelley, you might argue that he had the last laugh, in a roundabout way of which we'll hear more in chapter 5.

A most uncanny genesis, wouldn't you say? That was presumably just what Mary wanted us to think. Yet one wonders how, fifteen years after the event, she could recall it so well. Besides, her account is demonstrably wrong in some details, which raises the suspicion that it takes liberties with others too; for example (and this might of course have been more than mere oversight), she writes as if her stepsister Claire wasn't even present.

It's in the 1831 account that Shelley speaks of *Frankenstein* as her "hideous progeny," and the allusions both to Victor Frankenstein's creation of a monster and to childbirth are surely intended. "I have an affection for [*Frankenstein*]," she wrote, "for it was the offspring of happy days, when death and grief were but words which found no true echo in my heart." (This was surely not true either, given the death of her first child Clara by this point—but here she presumably has in mind Percy's death by drowning off Italy in 1822 when his sailing boat was caught in a storm.)

What Mary Shelley sketched out in 1816 were just the bones of her book: the tale of Victor and his ghastly deed. The narrative at that point lacked the framing device supplied by the arctic explorer Robert Walton, who recounts Frankenstein's tale in a letter sent to his sister. Shelley finished writing the novel in May 1817, by which time she and Percy had married in England. Percy read through the manuscript and made extensive changes: around a thousand of the words are his, and while he eliminated some grammatical errors, not all of his changes are for the better. Her husband also had the necessary literary contacts: he sent the finished manuscript to his own publisher, John Murray, although in the end it was published by Lackington, Hughes, Harding, Mavor and Jones in three volumes, in March 1818.

Frankenstein is a glorious mess. Science-fiction expert Robert Philmus is right to suggest that Mary Shelley "glanc[es] about everywhere for scenes to inspire emotion, while leaving the plot in abeyance for pages, and even chapters, at a time." The proportions are even less balanced

than those of Frankenstein's creature: important sections of the story
are over almost as soon as they have begun, while digressions take forever. Some key scenes barely make sense. According to cultural critic
James Twitchell, the book is "awkwardly written, inconsistently plotted,
peopled with a host of seemingly superfluous cipher-characters" and
generally redolent of "artistic insecurity."

Given Mary Shelley's inexperience and youth, none of this is surprising. But neither is it necessarily a problem, provided that readers adjust
their expectations. It's no literary masterpiece they're reading, but another form of invention altogether. And invention "does not consist in
creating out of the void," Shelley asserted in her 1831 preface, "but out
of chaos." As far as mythopoeia goes, it's not a bad thing if a little chaos
remains.

The book begins (slowly!) with Walton's letters to his sister in England, in which he describes how the crew of his ship came across a man
with a sledge and dog team (mostly dead) trapped on a floating block of
ice. The stranger—a European, half-dead himself from cold, fatigue, and
starvation—is, of course, Victor Frankenstein, whose tale Walton now
commences to relate.

From a distinguished Genevese family, Frankenstein begins this
history as a young, idealistic medical student at the University of Ingolstadt. A slightly anxious and shy youth, he is engaged to marry his
adoptive sister Elizabeth. (Shelve that for now; we'll be coming back to
it.) As a boy he read books on alchemy, thinking them to be profound.
When his father's denounces these as "sad trash" it only strengthens his
belief in them, although to his dismay and annoyance his first university
tutor, Krempe, also regards them as nonsense. But then he meets another
professor, Waldman, who has a more enlightened view. Those old philosophers, Waldman says, never managed to transmute lead into gold, but
through their efforts to do so they discovered some useful things. Under this sympathetic guidance, Victor throws himself with passion into
his studies in chemistry and anatomy. One question in particular taunts
him: "Whence, I often asked myself, did the principle of life proceed?"

Once he has confessed to that obsession, things go downhill fast.
Victor is led "to spend days and nights in vaults and charnel houses" in

order to "observe the natural decay and corruption of the human body." From these observations, the untutored student finds he has discovered nothing less than "the cause of generation and life."

Yes, this is absurd—but you should no more expect a realistic, logical course of events in a myth than in a dream. Myths *are* dreamlike, because they are doing their work in deep strata of the mind. They have us wind up in odd places with no idea how we got there. A poorly constructed story simply leaves us dissatisfied and inclined to abandon the book. But myths are not illogical or implausible so much as—if I may be allowed some linguistic improvisation—alogical and aplausible. They step outside rules of logic and plausibility. Shelley sees no need to persuade her readers of the reasonableness of her tale, for it defies reason from beginning to end. It no more matters how the naive young Victor could possibly have discovered the secret of life and animation on his own by poking about in dead bodies than it matters how giants live at the top of beanstalks.

By the same token, it doesn't matter what that secret is. Some readers feel short-changed by the fact that Shelley makes not the slightest attempt to tell us. A science-minded fellow like H. G. Wells would have none of that, and he complained that Frankenstein "used some jiggery-pokery magic to animate his artificial monster." But in truth we don't know quite *what* Victor got up to in his "workshop of filthy creation." Perhaps electricity held the key—Shelley throws hints in that direction, although nothing to quite justify the spark-strewn pyrotechnics Hollywood employed. But here the devil is anywhere but in the details.

Suffice it to say that, as Victor explains, "on a dreary night of November . . . I beheld the accomplishment of my toils." There lies his creature; a "dull yellow eye" opens, and "a convulsive motion agitated its limbs":

> His yellow skin scarcely covered the work of muscles and arteries beneath; his hair was of lustrous black, and flowing; his teeth of a pearly whiteness; but these luxuriances only formed a more horrid contrast with his watery eyes, that seemed almost of the same colour as the dun white sockets in which they were set, his shriveled complexion, and straight black lips.

The description, again, is barely comprehensible. We are led back and forth between wonder and horror, beauty and repulsion: the skin, like translucent oiled paper over the veins, gives a horrid impression, but what then of that "lustrous" and "pearly"? The creature, we're told, is nigh eight feet tall; how has such a gigantic frame been assembled from ordinary human parts? The palette is all confusion: *white* sockets? *black* lips? And why does the thing look so repulsive if Victor has taken "infinite pains and care" to select only the most beautiful body parts? Why hasn't he noticed that until the thing stirs to life?

What comes next is equally incoherent. As the creature awakens to life, Victor flees to his bedroom. He can't compose himself enough to sleep. He throws himself onto his bed, then is suddenly inside such a dream as to warn you, the reader, that there is something very dark going on in Mary Shelley's mind:

> I thought I saw Elizabeth, in the bloom of health, walking in the streets of Ingolstadt. Delighted and surprised, I embraced her; but as I imprinted the first kiss on her lips, they became livid with the hue of death; her features appeared to change, and I thought that I held the corpse of my dead mother in my arms; a shroud enveloped her form, and I saw the grave-worms crawling in the folds of the flannel.

Victor wakes from one nightmare to another: his creature stands over him, staring down with a twisted grin on his face. Frankenstein rushes out of the chamber and spends the night in the courtyard. Then by degrees, he resumes his daily business, as if it were all indeed a bad dream.

The whole episode has the quality of a nightmare. It's hard to make out quite what Victor witnessed; he seems to lurch in and out of sleep as if swooning. And is he now going to carry on as if the creature never existed? After a "nervous fever" lasting several months, through which he is nursed by his fiancée Elizabeth and his friend Henry Clerval, he does precisely that. Even for a man frightened and appalled at what he has done, this childlike refusal to engage with the deed seems bizarre. Some critics complain that Victor's actions just don't make sense. He seems constantly in the throes of what Freud later diagnosed as male hysteria;

he doesn't so much abdicate responsibility as simply scream, faint, fall asleep, or run away.

Yet a dreamlike style of narration* is characteristic of the mythic, full of improbable coincidences and unmotivated actions. It is one of the factors that has vexed and foxed critics who are used to narrative logic and realistic psychological grounding. But as Roland Barthes has said, "The absence of motivation does not embarrass myth." It is beside the point to demand (as we might be tempted to) why Frankenstein made a creature at all, or what made him so egotistical. Was he spoiled or neglected as a child? Only a bad writer would have used that device. The ontology that matters here is not that of the world but that of the myth.

Victor's father writes to tell him that his much younger brother William has been murdered by strangulation close to the family home in Geneva. Suspicion falls on the boy's governess, Justine—but Victor is visited by his creature, who admits that he committed the crime to repay Victor for his cruel abandonment. Justine is sentenced to hang, and Victor, telling himself he won't be believed, says nothing. Indeed, he knows that the blame for William's death is ultimately on him.

Tormented by his creature, Victor agrees to his demand for a mate to give him companionship and love. He makes a female creature in a remote workshop on the Orkney Islands, only to fear that the two artificial beings might breed, and their progeny wipe out the human race.† As the creature watches his labors with an expression of "the utmost extent of malice and treachery," Victor tears apart the bride-to-be. The incensed creature promises revenge: "I will be with you on your wedding-night."

---

* Remember too that we're hearing Frankenstein's own account as narrated to Walton, not the perspective of an omniscient narrator. "The whole series of my life appeared to me as a dream," he tells the captain.

† The obvious solution would be to omit the womb from the female creature. But such logic would subvert the mythic function of the tale as an exploration of the anxieties of procreation. The female must appear as a potential Eve to the creature's hellish Adam, and practicalities of surgery cannot be allowed to interfere with that. Once again, it's a mistake to try to read the plot rationally.

With more dream-logic, Victor seizes a fishing boat and is carried by the sea to the coast of Ireland, where he is immediately arrested on suspicion of having murdered his friend Clerval, whose body has been found on the shore. Now there is no denying it: Frankenstein confesses to himself that "my friend and dearest companion had fallen a victim to me and the monster of my creation.... I, not in deed, but in effect, was the true murderer." For Victor and his creature are really one and the same—or rather, doppelgängers like Jekyll and Hyde, an ego and id combination harrowing one another to their mutual doom. "My form is a filthy type of yours," the creature tells its maker, while Frankenstein admits, "I considered the being whom I had cast among mankind... nearly in the light of my own vampire, my own spirit let loose from the grave, and forced to destroy all that was dear to me." This duality of identity, always latent in the Frankenstein myth, was made explicit by director Danny Boyle in his 2011 production of *Frankenstein* for London's National Theatre, in which the two lead actors (Benedict Cumberbatch and Jonny Lee Miller) alternated the roles of Victor and the creature on different nights.

Victor is eventually released, the jury being satisfied that he was on the Orkneys when Clerval's corpse was found, and he returns to Geneva, where preparations are made for his marriage to Elizabeth. The creature is, of course, as good as his word—on the wedding night he attacks not his creator but the bride, whom he strangles. Victor, determined now to bring an end to his creation, sets out in pursuit. He follows the "daemon" to the arctic, where eventually, close to death, he is picked up by Walton's expedition. He confesses all to the captain, whose narrative resumes at the end of the book, before expiring on the ship.

Walton, now forced by his crew to abandon his foolhardy quest to find a northwest passage to the Pacific, hears in the night a hoarse human voice coming from the cabin in which Victor's body lies. Entering, he is confronted with the "final and wonderful catastrophe." The creature looms over Frankenstein's body "utter[ing] exclamations of grief and horror"; to Walton he professes his bitterness at his maker's rejection and cruelty, but also his remorse at the crimes he has committed. "I was the slave, not the master of an impulse, which I detested, yet could not disobey," he says. But soon, the creature tells Walton,

I shall die, and what I feel now will be no longer felt. Soon these miseries will be extinct. I shall ascend my funeral pile triumphantly, and exult in the agony of the torturing flames.

With that, he leaps from the window onto a floating raft of ice, and is "soon borne away by the waves, and lost in the darkness and distance."

Such is the fate of Frankenstein and his creation. But the emotional heart of the story—and the segment that lifts it onto the mythical plane—comes in the middle of the novel, when the narration is taken over by the creature himself. Having fled after Frankenstein's rejection and been shunned and attacked by frightened villagers in the region, he finds refuge with the De Lacey family, who live in an isolated cottage in the countryside. Taking care to stay hidden from them for fear of how they might respond to his appearance, the creature brings the family firewood. In turn they leave for this unseen helper a pile of books—Plutarch's *Lives*, Goethe's *The Sorrows of Young Werther*, and Milton's *Paradise Lost*. The books deepen his appreciation of human culture and teach him to reflect on his own existence. "Like Adam," he later tells Victor,

> I was created apparently united by no link to any other being in existence; but . . . many times I considered Satan as the fitter emblem of my condition; for often, like him, when I viewed the bliss of my protectors, the bitter gall of envy rose within me.

Eventually the creature builds up the courage to reveal himself to the family and hope that they will, despite his fearsome appearance, trust he means no harm. When he enters the cottage, only the blind old father is there, and they converse. But then the son Felix, daughter Agatha, and son's wife Safie return and are immediately thrown into panic. Felix attacks their terrifying visitor with a stick, and he flees. The De Laceys decide that their father is no longer safe left alone in the house while they are out at work, and they abandon the place—whereupon the enraged creature burns it to the ground.

This perspective is unexpected—and revolutionary. When, in classical mythology, do we hear the story from the point of view of Polyphemus, the Gorgon, or the Minotaur? These ancient monsters were not just beyond nature but beyond reach, their minds as inscrutable and alien as those of an octopus. They did not register even as living things, exactly—they were portents, warnings, signs of moral aberration: they were *de-monstrations*. Saint Augustine asserted in *City of God* that a monster signifies God's displeasure at vice and folly: it does not need to *do*, but only to *be*.

By putting us inside the demon's head, Mary Shelley delivers us into the territory of fallen angels and corrupted innocence. Indeed, she invokes the resonant words of Milton's Satan in her epigraph:

> *Did I request thee, Maker, from my clay*
> *To mould me man? Did I solicit thee*
> *From darkness to promote me?*

But while he can know jealousy and anger, Satan never evinced any moral awareness. We never imagine for a moment that there is a spark of humanity in him. What is truly innovative—and transgressive—in *Frankenstein* is not that the monster is human-made, but that it is humanized.

This is so much more daring than anything the Romantic poets attempted in their glorification of Milton's Satan for what William Godwin called his "principled opposition to tyranny." While to Byron and Percy Shelley this Satan was a heroic rebel, Mary makes the creature all too recognizably flawed, his rebellion against his maker understandable and yet hardly something a decent person could admire. She transforms what a monster can be—*and makes it modern.*

Mary's ambivalence and skepticism toward the Romantic rebellion so lionized by Byron and Percy Shelley struck a blow at their narcissistic project, whether they realized it or not. It doubtless would have meant more to her than to them that the prime architect of this Romantic idealism, Jean-Jacques Rousseau, left his five children in a foundling hospital.

Frankenstein was first published anonymously, and its preface, signed by Percy Shelley, created a general assumption that he was its author.

Given his already rather dissolute reputation, some critics saw the novel as an opportunity to take the poet to task. "Our taste and our judgement alike revolt at this kind of writing," said John Croker in *Quarterly Review*. "It inculcates no lesson of conduct, manners, or morality; it cannot mend, and will not even amuse its readers, unless their taste have been deplorably vitiated." The reader is left guessing, Croker concluded, "whether the head or the heart of the author [assumed to be Percy] be the most diseased." *Edinburgh Magazine*, taking aim not just at Shelley but that other notorious rascal, "his great model Mr Godwin"

Frontispiece of the 1831 edition of Frankenstein, *by Theodore von Holst.*

(to whom the book was dedicated), boomed that "they would make a great improvement in their writings, if they would rather study the established order of nature as it appears, both in the world of matter and of mind, than continue to revolt our feelings by hazardous innovations in either of these departments." The reviewer found the book "bordering too closely on impiety." In other words, he was not just repelled but unsettled by *Frankenstein*—it had evidently done its mythical work on him. Walter Scott, reviewing for *Blackwood's Edinburgh Magazine*, was more favorable, writing, "Upon the whole, the work impresses us with a high idea of the author's original genius and happy power of expression."

Perhaps because of its mixed reviews, the book didn't exactly leap off the shelves. Although the first printing of five hundred eventually sold out, the publisher didn't see fit to reprint it. Some critics didn't even bother to track down a copy but based their assessments on the reviews. "*Frankenstein*, by Godwin's son-in-law," wrote Thomas Carlyle, "seems to be another unnatural disgusting fiction." It may have been only because of the successful stage adaptations in the early 1820s that a second edition appeared in 1823, negotiated (and edited) by William Godwin himself. All the same, after the revised edition in 1831, *Frankenstein* fell out print until the 1870s, creating a void that the myth could start to fill.

It is because *Frankenstein*'s themes are so rich and diverse that the conventional interpretation of it today as a cautionary tale about scientific hubris is desperately inadequate. The novel is deeply engaged with the sociopolitics of its day: it comments on the French Revolution and slavery, social injustice and imperialism. It is about birth and death, family and kinship and their complications and contradictions. It spoke as well to the scientific debates of the early nineteenth century: on electricity, the principle of life, arctic exploration, even evolution. A secular tale, it nonetheless confronts profound religious questions. It offers, as we've seen, a piercing critique of the Romantic idealism championed by Percy Shelley and Byron. It is saturated with sexual anxiety. All of these things are not projections after the fact, but are to some degree infused into the text by Mary Shelley. If some of the modern interpretations seem anachronistic—it has been analyzed from the perspective of queer theory, feminism, transhumanism, and ecocriticism, among others— nonetheless Shelley seeded the story with the substance needed to carry it into these (post)modern times.

The problem is, then, where to start. Shelley, says Chris Baldick, "overload[s] the novel with approximately parallel 'codes' of signification— psychological, pedagogic, sexual, Miltonic, political—which overlap and interfere with one another at so many points that no single line of interpretation can convincingly fend off all the others." It is one of the most overdetermined novels in all of fiction.

Let's take that scientific hubris. Certainly this is a major theme, but in far less simplistic a manner than is often asserted. The assessment of Brian Aldiss in his recasting of the tale, *Frankenstein Unbound* (1973), is typical:

> Frankenstein was the archetype of the scientist whose research, pursued in the sacred name of increasing knowledge, takes on a life of its own and causes untold misery before being brought under control. How many of the ills of the modern world were not due precisely to Frankenstein's folly.

This interpretation is understandable, for at one level *Frankenstein* recapitulates the Faust legend without the divine elements and the moralizing. In the old myth, the ambitious wizard-scholar Faust makes a pact with the devil, trading knowledge and power in the worldly sphere for his soul in the afterlife. With diabolical assistance, Faust acquires magical abilities and pursues a life of trickery and deceit—but the devil arrives at the end of his life to collect his due. The myth goes back at least to the New Testament, when the Faustian sorcerer Simon Magus competes with St. Peter in a magic bout. The story became hugely popular in the Middle Ages and the Renaissance, when several vernacular versions were printed in Germany; Christopher Marlowe used these as the basis of his 1592 play *Doctor Faustus*. Marlowe's relatively simple morality tale remains popular in modern times—witness the legends of occultist Aleister Crowley and of Delta bluesman Robert Johnson, who allegedly paid the devil for his musical wizardry down at the crossroads. But the Faust legend was already being woven, in Mary Shelley's day, into a far more complex story by Wolfgang von Goethe, who was concerned with Romantic notions of creativity rather than with the pros and cons of demonic pacts. (The first volume of Goethe's play was published in 1808, but Shelley claimed to have been unaware of it until after her book was written.)

In the Middle Ages a craving for knowledge was not considered a virtue, and curiosity was still apt to arouse suspicion throughout the Renaissance. This idea of dangerous knowledge harks back to the myth of Prometheus, the Greek god who gave fire (and technology) to humankind and paid the price by being chained to a rock and having his liver pecked out each day by carrion birds. By acknowledging the debt in her book's subtitle, *The Modern Prometheus*, Mary Shelley invited the idea that Victor is another hubristic overreacher, seeking things mankind is not supposed to know. His retribution now comes not from God or the devil, but from his own creation: a resolution suited to the age of science and reason.

We often call such narratives "Faustian" today: Faust is commonly seen as the prototype of the mad scientist, unleashing ungovernable forces without heeding the consequences. There is, however, little justification for that view in the old Faust legends themselves, where he is simply seduced by the devil's offer of worldly power, pleasure, and wealth. And whether Mary Shelley meant Frankenstein to fit this traditional Faustian mold is not clear either. I certainly don't think she would recognize the popular interpretation that makes her novel a feminist critique of a "male science" obsessed with disturbing and distorting nature. There's no good reason to suppose, as Mary Shelley's biographer Anne Mellor argues, that she divvied up the science of her day into good—"the detailed and reverent description of the workings of nature," in Mellor's words—and bad—"the hubristic manipulation of the elemental forces of nature to serve man's private ends." Far less is there reason to imagine she saw the former as "feminine" and the latter as "masculine." There is (to label it anachronistically) surely a feminist aspect to her work—the apple did not fall far from the tree—but it's not to be found here.

Shelley seems in fact to have held science in high regard. She certainly knew plenty about it, affording the most attention to the primary science of the Romantic era: chemistry. It is, Waldman tells the young Frankenstein, "that branch of natural philosophy in which the greatest improvements have been and may be made. . . . [Its] modern masters . . . penetrate into the recesses of nature and show how she works in her hiding-places." They even ascend into the heavens, the professor explains—a reference to the balloon flights demonstrated in France in the 1780s, some of them using the lighter-than-air gas hydrogen. Modern chemists, Waldman explains,

> have acquired new and almost unlimited powers. They can command
> the thunders of heaven, mimic the earthquake, and even mock the
> invisible world with its own shadows.

This is certainly a Promethean portrait. There seemed almost nothing scientists could not achieve by means of chemistry, no degree of mastery they could not attain over nature. But there is no Faustian hubris to Waldman: "His gentleness was never tinged with dogmatism," Frankenstein attests, "and his instructions were given with an air of frankness and good nature that banished every idea of pedantry." Even the crabby Krempe at Ingolstadt, who mocked Victor for his early interest in alchemy, is not without his virtues, offering "a great deal of sound sense and real information."

The humble, thoughtful Waldman seems to be based on Humphry Davy, the dashing chemist who directed experiments at the Royal Institution in London and who probably visited the Godwin household. There's a possibility that Mary was the young girl mentioned in the Institution's records as having been in the audience at a public lecture given by Davy, although sadly it's impossible to be sure. She read his works, probably including his *Elements of Chemical Philosophy* (1812) and *A Discourse, Introductory to a Course of Lectures on Chemistry* (1802),

Sir Humphry Davy, possible model for Frankenstein's *Dr. Waldman. Portrait by Thomas Phillips, 1821.*

while working on her book in late 1816. In his *Discourse* Davy delivers a Waldman-like paean about how the chemist has acquired the knowledge

> to modify and change the beings surrounding him, and by his experiments to interrogate nature with power, not simply as a scholar, passive and seeking only to understand her operations, but rather as a master, active with his own instruments.

The gendered language, in which a female nature is probed by men with their instruments of enquiry—a habit harking back at least to Francis Bacon at the dawn of the so-called Scientific Revolution—has a lot to answer for and is rightly taken to task today by feminist critics. But there's no evidence that Shelley took issue with it.

Davy's vision of mastery does not make him the archetypal mad scientist seeking to control that which he does not understand. Nor does it entail some megalomaniac quest to control or create life—Davy was in fact skeptical about how much of physiology could be given a chemical explanation. Rather, the kinds of experiment he envisaged for the scientist "active with his own instruments" were those he conducted himself at the Royal Institution, in which he used electricity to pry apart the intimate unions of elements in chemical compounds. In 1807–1808 he used electrolysis—passing an electrical current through a solution or a molten substance, generated by a primitive battery called a voltaic pile—to discover new chemical elements, including sodium, potassium, magnesium, and calcium, present in common salts and minerals such as potash and lime. Davy's colorful language merely emphasizes how these artful manipulations in the laboratory, especially using the newfound power of electricity, could disclose hitherto unknown aspects of nature.

Electricity was still a mysterious force at the start of the nineteenth century, and it crackles through the Frankenstein myth. Shelley's 1831 introduction to the book explains how, on the night of her monstrous vision, Byron and Percy Shelley had been discussing "the nature of the principle of life, and whether there was any probability of its ever being discovered or communicated." She recounts their discussion of what sounds like a bizarre and rather unlikely experiment conducted by Erasmus Darwin, the grandfather of Charles, "who preserved a piece of vermicelli in a glass case till by some extraordinary means it began to move with voluntary motion."

Perhaps then, she wonders, "a corpse would be reanimated; galvanism had given token of such things." This is a reference to the experiments in the early 1790s of the Italian physician Luigi Galvani, who found that static electricity created by a rotating wheel, when discharged through the recently dissected limb of a frog, would cause it to twitch as if animated. These observations led Galvani to speculate that electricity is a kind of vital fluid that awakens life itself: the theory became known as galvanism.

Galvani's nephew Giovanni Aldini took these experiments into more grotesque territory. In 1802 he demonstrated before various British nobles in London the use of electricity to reanimate the severed head of an ox. The following year, he passed currents through the corpse of a recently hanged criminal brought to the College of Surgeons, with alarming results:

> The jaw began to quiver, the adjoining muscles were horribly contorted, and the left eye actually opened. . . . Vitality might, perhaps, have been restored, if many other circumstances had not rendered it impossible.

This brief reference to galvanism is all the explicit indication Shelley ever provided that electricity had anything to do with the animation of Frankenstein's creature. In the text itself she is cryptic about the secret knowledge that comes to Victor in his laboratory:

> A sudden light broke in upon me—a light so brilliant and wondrous, yet so simple, that while I became dizzy with the immensity of the prospect which it illustrated, I was surprised that among so many men of genius who had directed their enquiries towards the same science, that I alone should be reserved to discover so astonishing a secret.

Victor himself explains (as he relates his tale to Walton) that he can't reveal this secret for fear that others will make the same mistake as he. Very convenient, you might say; but the ploy saves Shelley from having to concoct some pseudoscientific explanation for what is really rather irrelevant to the tale anyway.

And yet . . . surely there's electricity involved in this dark art. Victor goes on to record how, on that dreary November night, "I collected the instruments of life around me, that I might infuse a spark of being into the lifeless thing that lay at my feet." "Infuse" speaks of a fluid like that which Galvani held electricity to be; "spark" evokes the discharges that made his frogs' legs twitch into a semblance of life.

What's more, imagery of lightning pervades the text. As a boy, Victor witnesses a great thunderstorm in the Jura mountains and sees an old oak tree struck by lightning and reduced to a "blasted stump." He says that already he knew about "the more obvious laws of electricity," and that a visitor with his family explained after the storm his own theory of "electricity and galvanism." The creature first reappears before Frankenstein during a storm in Geneva, illuminated by a bolt from the sky. However much impelled by Hollywood's demand for grand spectacle, the fizzing sparks and crashing thunderbolts in the famous animation scene of the 1931 movie have ample justification.

The deepest engagement of Shelley's book with the science of her time, however, comes elsewhere. Frankenstein's obsession with life consists not so much in the mechanics of how to awaken it, but in its fundamental nature: "Whence, I often asked myself, did the principle of life proceed?"

"To examine the causes of life," he concludes, "we must first have recourse to death." It seems an odd assertion—what could you learn from dead things?—until you appreciate that Frankenstein's early, precocious interest in alchemy had led him to venerate the sixteenth-century Swiss alchemist and physician Paracelsus, who believed that matter had first to undergo death and putrefaction before it could be reborn. Frankenstein resorts to studying "the natural decay and corruption of the human body," lurking in churchyards and vaults to witness how "the corruption of death succeed[s] to the blooming cheek of life."

Paracelsus is hardly a household name, less still to a teenage girl. But this is the daughter of William Godwin, who wrote about Paracelsus, as well as Victor's other youthful idol, Cornelius Agrippa, and further pre-scientific "natural magicians" (Faust too), in his *Lives of the Necromancers* (1834). As Victor learns from Professor Waldman, the ideas of those men were no longer taken seriously now that chemistry had become a sober, quantitative science. But the mystery of life was not lessened in the nineteenth century, and the real debate now centered on whether it could be understood as purely a material affair—that it somehow arose from the nature or organization of matter—or whether something more, call it a spark or a soul, needed to be injected into substance, perhaps by God.

That question was hotly debated when Shelley began writing her story. The distinguished surgeon John Abernathy argued that life could not be understood solely on a material basis but depended also on some "subtile, mobile, invisible substance," which might be akin to electricity but could also be equated with a soul. Abernathy's former student and now colleague at the Royal College of Surgeons, William Lawrence, demurred, asserting that there was no need to impose anything so mysterious. Lawrence was, during the period of this dispute (1814–1819), Percy Shelley's personal physician.

Lawrence laid out his objections to Abernathy's view in two lectures of 1816, which were published as a pamphlet. In subsequent lectures the following year, he argued that Abernathy's "vital force" was in any event still a material agent, however subtle its constitution. Despite being in many respects a more compelling debater, Lawrence faced censure when he set out the materialist position in his 1819 book *Lectures on Physiology, Zoology and the Natural History of Man*, which was widely regarded as atheistic and therefore blasphemous. He was forced to recant in order to keep his place at the Royal College of Surgeons.

Frankenstein doesn't obviously take sides in this debate but instead rather brilliantly explores the question. The creature is assembled from dead, even disgusting matter—but then that unexplained "spark of life" is needed to give it animation. Some of the changes Shelley made to the 1831 edition of *Frankenstein* might, however, have been motivated by a desire to distance herself from Lawrence's now disgraced materialism and its taint of impiety.

◆━◆

The idea that Mary Shelley voiced, as Anne Mellor puts it, an "implicit warning against the possible dangers inherent in the technological developments of modern science" is often taken for granted now in discussions of her book. Literary scholar Charles Robinson echoes that view in his introduction to a 2017 version of the text "annotated for scientists, engineers, and creators of all kinds." He reminds us that the evils released from the mythical "box" by Pandora, wife of Prometheus's brother Epimetheus, are reflected in "the pesticide DDT, the atom bomb, Three Mile Island, Chernobyl, and the British government's permission . . . that a stem cell scientist could perform genome editing despite objections that ethical issues were being ignored." It is almost as though, whenever science and technology have unanticipated repercussions, we must point to *Frankenstein* and say that Mary Shelley warned us this would happen.

Leaving aside the question of whether that's what Shelley would have wanted (although almost certainly it is not), this isn't a very useful way to frame the dangers of modern science. Each of the situations listed by Robinson involved a complex and case-specific chain of events, and incurred a delicate balance of pros and cons. Some, such as the Chernobyl nuclear accident, had little to do with the intrinsic ethics of the underlying technology but were, rather, a consequence of political and bureaucratic failures. To imply that they all unambiguously show a lack of forethought, or indeed of responsibility, on the part of the scientists whose work made them possible would be to cheapen the discourse and to evade the real issues.

So why do we do it? The mythical Frankenstein represents a fear of science itself: his creature sits like a gargoyle guarding knowledge too hazardous to contemplate. In the extreme case this view becomes a reactive Luddite fear of any technological innovation, but more commonly the implication is that there are certain directions science should never take: steps that are beyond the pale and guaranteed to end badly. When *Frankenstein* is deployed in this disapproving manner, those steps very often involve what might be seen as mimicking Victor's desire to control and manipulate life itself. In this reading, the terrible outcome is preordained by the transgression that led to it. As Mellor says, "Nature prevents Victor from constructing a normal human being: his unnatural method of reproduction spawns an unnatural being, a freak." The perilous, Faustian step is the violation of nature itself.

It is with this reading in mind that Frankenstein's name is invoked whenever any significant new reproductive technology arrives. The advent of in vitro fertilization—so-called test-tube babies, a phrase that imputes a spurious chemical and thus artificial aspect to the embryos so made—led to headlines about "Frankenstein scientists." Likewise when the possibility of human cloning hit the news after the creation of Dolly the sheep in 1997, *Time* magazine called the leader of that cloning project "Dr. Frankenstein . . . in a wool sweater,* baggy parka, soft British accent and the face of a bank clerk." The researchers, based at the Roslin Institute in Scotland, had moved the genetic material of cells taken from an adult ewe into an egg that had had its own genes removed, and stimulated the egg to grow into an embryo that was then implanted in a surrogate mother ewe. Were these cloned animals "monsters or miracles," asked the *Daily Mail*? For the widespread perception was still that intervening in the "natural" process of procreation must necessarily produce a monster, as Frankenstein did. The British biologist J. B. S. Haldane anticipated such a response in 1924:

> There is no great invention, from fire to flying, which has not been hailed as an insult to some god. But if every physical and chemical invention is a blasphemy, every biological invention is a perversion.

---

* Evidently, Victor never in fact held the title "Doctor"—but it seems we must qualify him as a scientist before damning him as one.

By implying that Victor's work is inherently perverted and bound to end hideously, Mellor's accusation leaves us wondering what exactly is meant by "unnatural." Which interventions are guaranteed to produce a freak? Might that be so with IVF, as its early detractors insisted? Is it the case for so-called three-parent babies made by mitochondrial replacement, a misleading term apparently invented for the very purpose of insisting on its unnaturalness? Would the first human clone be the next "unnatural freak"? "Unnatural" is not a neutral description but a morally laden term, and dangerous for that reason: its use threatens to prejudice or shut down discussion before it begins.

"Franken-" has now become a prefix that one can attach to any technological innovation, particularly in the life sciences, as an invitation to disapproval. We have Franken-crops (genetically modified), Franken-bugs (made with altered or chemically synthesized DNA), Franken-foods: a US lawsuit brought against the alleged harmfulness of Chicken McNuggets saw them denounced as "a McFrankenstein creation." It's no wonder scientists working in these areas of the life and biomedical sciences often feel the same way as IVF pioneer Robert Edwards, who complained in 1989, "Whatever today's embryologists may do, Frankenstein or Faust or Jekyll will have foreshadowed, looming over every biological debate." But nothing will change until we grasp the reasons why the Frankenstein myth seems (rightly or wrongly) so relevant here.

The superficial notion, however, that Shelley herself was warning us away from "tampering with nature" undermines the subtlety of her story. We might, for example, simply ask what would have happened if Victor had lived up to his responsibilities by choosing to nurture his creature rather than rejecting it. You might answer that the result would have been a comparatively dull and short novel. But I'm not so sure. Imagine the story of Victor struggling to get his creature accepted by a society that shunned it as hideous and unnatural. That too would have become a narrative about social prejudice and an interrogation of our preconceptions about nature and life—indeed, it could be even more relevant to the likely predicament of a human born by cloning, should such as thing ever become scientifically possible and ethically permissible (neither is yet the case). The moral and philosophical landscape it might have explored would be no less rich.

That Victor did not do this—that he spurned his creation the moment he had made it, merely because he judged it ugly—means that the conclusion we should reach is more plausibly that articulated by speculative-fiction writer Elizabeth Bear. It is for Victor's "failure of empathy and his moral cowardice," she says, for his overweening egotism and narcissism, that we should think ill of him—and not because of what he discovered or created.

All the same, the question of how we should judge Frankenstein is complicated. Look, we might want to say: it is not the creature's fault! Victor is the real monster! After all, consider what Percy Shelley said in a review of his wife's book:*

> In this the direct moral of the book consists. . . . Treat a person ill, and
> he will become wicked. Requite affection with scorn;—let one being
> be selected, for whatever cause, as the refuse of his kind—divide him,
> a social being, from society, and you impose upon him the irresistible
> obligations—malevolence and selfishness.

Percy insists that there is nothing behind the evil deeds of the creature than "causes fully adequate to their production." The creature does what he does because of how he is treated by Victor, not because of some inherent fault of his nature.

There's ample reason to read the text this way, but it's not so clear that Mary Shelley wanted us in consequence to consider Victor a terrible person. I believe we might aptly say that she herself reacted much as he did: she could not help but look at what she had made and feel repelled by it. Look, for instance, at how wildly inconsistent the book is in

---

* What was Percy doing reviewing Mary's book, you might well ask? But the short essay, written in 1817, was unpublished in his lifetime and only released posthumously in 1832.

its portrayal of the creature. At one moment he is eloquent, victimized, and outcast, longing for the companionship and love that humans refuse him. But the next instant we see a hideous creature, a malevolent devil, tormenting the haunted Frankenstein with cruel glee. We might want to insist, as Mellor does, that Victor deserves nothing but opprobrium for his behavior—but Shelley doesn't demand that judgment from us. All of her characters think highly of "poor, dear Victor." Even Robert Walton sees in him a noble and tragic figure, "amiable and attractive" despite his wrecked and emaciated state. Frankenstein's only real critic is his creature.*

As for that creature, he is said to be so hideous that we're almost persuaded he deserves his fate. Walton confesses when he sets eyes on the being that "I dared not again raise my eyes to his face, there was something so scaring and unearthly in his ugliness." In other words, Shelley denies us the neat morality tale that Percy offers us—and thereby opens up an ambiguous space in which myth can grow. Maybe the creature really *is* a monstrosity, a perversion, by its very existence? Maybe (I think it likely) Shelley was too repelled herself by her creature to award it the sympathy she intended?

Nonetheless, the "modern" reading of the book insists that we regard Victor as the villain and his creature as the victim. This interpretation has hardened into such an orthodoxy that it can seem insensitive, perhaps even politically incorrect, to challenge it. Lipking skewers this view delightfully in his 1996 essay "*Frankenstein*, the True Story," where he lists the general consensus among his peers:

> Item: Frankenstein is the degenerate offspring of a dysfunctional
> family; Not Worth Mentioning: every character in the book loves and
> admires him. . . . Item: Frankenstein's abandonment of his Creature is

---

* Mellor points out that some of the changes to the text made by Percy Shelley present Victor more sympathetically and the creature more monstrously, possibly because Percy identified with the Promethean Frankenstein himself. This is true, but it doesn't by itself rewrite Mary's intentions (although it complicates how Mellor wants us to understand them).

an act of unforgivable irresponsibility; Not Worth Mentioning: the Creature murders a small child and frames an innocent woman for the crime. . . . Item: Frankenstein's incredible stupidity, narcissism, and insensitivity to women lead to the death of his bride on their wedding night; Not Worth Mentioning: the Creature kills that bride for no other reason than to get back at Frankenstein. Etc., etc.

The problem, as Lipking ironically puts it, is that "despite the consensus of sophisticated critics, ordinary readers keep looking at the wrong evidence and coming to the wrong conclusions." He recounts the testimony of professor of English Mary Lowe Evans, who says, "My attempts to arouse the students' indignation at Frankenstein's failure to take responsibility for Clerval's death and his reluctance to warn Elizabeth of imminent danger, for example, proved futile." Anne Mellor goes even further with such coercion (Lipking says), warning students that "if they do not draw the moral that she does from the story, they may be responsible for the end of life as we know it." Only "by reading the sublime, the unknowable, as lovable," she says, "can we prevent the creation of monsters, monsters both psychological and technological, monsters capable of destroying all human civilization." Lipking wonders "how many students disagree at this point; or if they do, how many dare to speak up."

Lipking's perspective on *Frankenstein* is worth bearing in mind with all the myths in this book: *They have neither a moral nor a single "meaning."* That, as I suggested earlier, is one of their qualifying features. To explain what a myth "means" necessarily neutralizes its function as a vehicle for exploring irresolvable questions and anxieties. "Despite the modern consensus," Lipking says,

> *Frankenstein* does not admit of a resolution. The basic, defining questions it raises in the mind of a reader—is Victor an idealistic hero or a destructive egoist? is the Creature a natural man or an unnatural monster? what moral are we to draw from this strange story?—never receive a satisfactory answer, or, rather, receive strong answers that directly contradict one another. This impasse . . . is the very heart of the novel. . . . *Frankenstein* teaches its readers to live with uncertainty,

in a world where moral absolutes—even the ones we cling to—may cancel each other out.

We should, he says, grant Shelley "the courage of her lack of convictions."

In other words, the problems in *Frankenstein* that critics often perceive—its inconsistency, clumsiness, digressive nature—are positive qualities from the perspective of mythopoesis. The reason it is so depressing to see *Frankenstein* trotted out as a cautionary fable about the alleged unnaturalness and dangers of new reproductive technologies is not just that this distorts the message of the book but that *Frankenstein* could actually be an excellent vehicle for all the ambivalence, the open-ended possibilities and perils, the challenges to preconception and prejudice, that such developments usually generate.

* * *

During the nineteenth century, the perceived allegorical content of *Frankenstein* was in any case not about science at all, but about politics. The anxieties of that time were concerned less with technological than with social change. In particular, the French Revolution had disturbed political thinkers all over Europe. That radical uprising had gone sour before spawning the pseudomonarchy of Napoleon Bonaparte, which in turn ended with his defeat at Waterloo in 1815. Political radicals and idealists were disillusioned by this failure to challenge the status quo; conservatives regarded the events as a dire warning. Much of their talk was of a political monstrosity that had arisen and fallen apart.

The notion of an embodied politics—the body politic—goes back to ancient Greece, but it was popularized after the English Civil War by Thomas Hobbes, who argued in *Leviathan* (1651) that elective but absolute monarchy created a ruler who was the embodiment of the collective will of the people. This image of a ruler as some gargantuan composite of the public will was given graphic expression in the famous frontispiece of the book by Wenceslas Hollar: a crowned giant composed of innumerable human figures looms over the countryside. The state, said Hobbes, "is but an Artificiall Man."

In the ideal Hobbesian state, public opinion cohered into a stable mode of governance, personified by its monarch. But in France it had mutated into something raging and uncontrolled. That was the view of Edmund Burke, who wrote in 1790 that we

> should approach to the faults of the state as to the wounds of a father, with pious awe and trembling solicitude. By this wise prejudice we are taught to look with horror on those children of their country who are prompt rashly to hack that aged parent in pieces, and put him in the kettle of magicians, in hopes that by their poisonous weeds, and wild incantations, they may regenerate the paternal constitution, and renovate their father's life.

You can't make a healthy, vibrant state out of hacked-up parts (however carefully chosen), but only something deformed, terrible, and unnatural. "Out of the tomb of the murdered monarchy in France," Burke wrote, "has arisen a vast, tremendous unformed spectre." In the Romantic view, beauty, order, and perfection are not produced piecemeal but arise from an organic internal harmony. Mary Shelley will have known of Burke's conservative opinions, for they were the very ones her radical father and mother had publicly opposed. It's reasonable to suppose that her choice of Ingolstadt as the location of Frankenstein's education was influenced by *Memoirs Illustrating the History of Jacobinism* (1797) by Abbé Barruel, in which that city was said to have "engendered that disastrous monster called Jacobin, raging uncontrolled, and almost unopposed, in these days of horror." Barruel's tract was in truth little more than a popular conspiracy theory, although apparently it made a deep impression on Percy Shelley.

The image of the French Revolution as a monstrous body appealed too to Thomas Carlyle, another influential conservative thinker, for whom post-Revolutionary France was "as a Galvanic Mass, where all sorts of far stranger than chemical galvanic or electric forces and substances are at work." One of the earliest and most explicit uses of *Frankenstein* as sociopolitical allegory appears in an article of 1830 in *Fraser's Magazine*, the author of which says:

> A state without religion is like a human body without a soul, or rather like an unnatural body of the species of the Frankenstein monster, without a pure and vivifying principle; for the limbs are of different natures, and form a horrible heterogeneous compound, full of corruption and exciting our disgust.

Many in the well-to-do classes feared the uprising of the monstrous mob, and never more so than after Marx and Engels published their Communist Manifesto during the social unrest of 1848. In that same year, Elizabeth Gaskell wrote in her most famous novel, *Mary Barton*, "The actions of the uneducated seem to me typified in those of Frankenstein, that monster of many human qualities, ungifted with a soul, a knowledge of the difference between good and evil." The people, she went on,

> rise up to life; they irritate us, they terrify us, and we become their enemies. Then, in the sorrowful moment of our triumphant power, their eyes gaze on us with a mute reproach. Why have we made them what they are; a powerful monster, yet without the inner means for peace and happiness?

See here what has already become of Shelley's novel, even in the hands of the educated: the creature has been confused with his creator, and, in line with the popular stage adaptations of that time, has been rendered amoral and mute. It symbolizes the masses of the lower classes, ungovernable and full of inarticulate rage.

In keeping with the creature's otherness, this vision of monstrous social rebellion touched also on issues of race. Several nineteenth-century cartoons in England depicted the creature in the form of a gigantic, uncouth Irishman, while in the United States it represented fears of uprising by Black slaves. For conservative commentators, *Frankenstein* metaphorically exposed the dangers of allowing Black people too much liberty—in which context it was the debased, childlike conception of the creature from stage adaptations that was needed to mobilize the racist view that these people were not sophisticated enough to handle freedom. "In dealing with a negro," wrote the British foreign secretary George Canning in the wake of slave rebellions in the American south in the 1830s,

> we must remember that we are dealing with a being possessing the form and strength of a man, but the intellect only of a child. To turn him loose in the manhood of his physical passions . . . would be to raise up a creature resembling the splendid fiction of a recent romance; the hero of which constructs a human form with all the physical capabilities of man, and with the thews and sinews of a giant, but being able to impart to the work of his hands a perception of right and wrong, he finds too late that he has only created a more than mortal power of doing mischief, and himself recoils from the monster which he has made.

This image of the Black man as an immense, brutish threat was grotesquely illustrated in a cartoon of 1852 in the periodical *The Lantern*, showing abolitionist Horace Greeley in the role of Frankenstein creating a monster he can't control. And here we can see how *necessary* it was for these analogies that the creature be stripped of nobility, intelligence, and indeed humanity. This is the myth already being reconfigured to do the ugly work required of it. A creature capable of learning from Milton and rendered murderous by the mistreatment and neglect of his master would not do at all. It must be attributed the ungovernable "manhood of physical passions" with which Black men were stereotyped. Didn't the creature, after all, assault the innocent and virginal Elizabeth in her bedchamber?

Eventually, however, *Frankenstein* supplied a vehicle for a more positive metaphor about race: the creature became a symbol of unjust repression, and "black Frankenstein" stories took on a progressive slant. In his 1898 novella *The Monster*, American writer Stephen Crane described an African American man, disfigured by a fire from which he

THE MODERN FRANKENSTEIN.

Frankenstein's creature shown as a monstrous, rebellious Irishman (left) *and an ungovernable Black slave* (right) *in the metaphorical political imagery of the nineteenth century.*

rescues the son of the town physician, who is subsequently shunned as a monster by the local residents. Such racial readings of *Frankenstein* have continued to modern times: civil-rights activist Dick Gregory saw in James Whale's iconic 1931 movie "a monster created by a white man, turning upon his creator." Whether Mary Shelley intended any social comment on slavery and social injustice isn't clear; she and Percy were abolitionists, albeit probably of the sort who saw slaves condescendingly as pitiable victims. But the inequities of human society were not lost on the creature. "I heard of the division of property," he says, "of immense wealth and squalid poverty; of rank, descent, and noble blood."

The identification of Frankenstein's creature with the uncontrollable foreigner persists today. Writing of the political "monsters" the United States has created and then destroyed—the Shah of Iran, General Pinochet, Osama bin Laden—filmmaker and political activist Michael Moore indulges the habitual confusion of man and monster when he says that "just like the mythical Frankenstein, Saddam eventually spun out of control. He would no longer do what he was told by his master. Saddam had to be caught."

This analogy also alludes to the common perception that Frankenstein's creature was made to *serve* its master but, like Milton's Satan, rebelled. That was evidently not so, but it is now a feature of the myth because it is so useful an image for characterizing a lack of political foresight in the construction of disobedient puppet leaders. "The United States is extraordinarily gifted in creating monsters like Frankenstein," says Mexican novelist Carlos Fuentes. "Then one day they discover that these Frankensteins are dreadful."

Whatever social and political resonances it might have had, *Frankenstein* was also a deeply personal project for Mary Shelley. Here, after all, is a young woman who has been disowned by her father, writing of a fatherless creature spurned by its creator. It's simplistic to imagine that Victor is merely a representation of William Godwin (there's much that is Byronic about him too), although the imprint of Godwin's ideas — alchemy, social rebellion, Gothic imagery*—is plain enough in the text, and, after all, the novel was dedicated to him. But in the autodidactic creature who has to read Frankenstein's laboratory notebooks to find out about his own origins, it's tempting to suppose that Mary saw aspects of herself: she is said (although it's not established beyond doubt) to have read the love letters between her father and her dead mother.

There are scarcely any major human relationships in the book that are not warped in some manner. James Twitchell has rather tendentiously claimed that all classic horror is fundamentally about incest; whether you buy that or not, there's no avoiding the discomfiting fact that Vic-

---

* Godwin's 1799 tale *St Leon* is particularly germane: a fantasy about a French aristocrat (Count Reginald) who acquires an elixir of immortality and is destined to wander the world, torn between a desire for greatness and the responsibilities of social and family bonds.

tor is betrothed to his "more than sister" Elizabeth, an orphan adopted
by the Frankensteins. They call one another "cousin," but even without
any blood relation there is clearly more familial closeness between them
than that. Victor's mother presents the young Elizabeth to her son as
a "pretty present" and all but commands him to marry her, saying on
her deathbed that "my firmest hopes of future happiness were placed
on the prospect of your union." At the same time, she adds as she wastes
away, Elizabeth must now take *her* place as mother to her younger children. Is it any surprise that Victor later has that disturbing vision of his
bride-to-be mutating into his dead and moldering mother? This is not a
healthy family.

Even by nineteenth-century standards, when cousin marriage was
not uncommon,* the union of brother and adoptive sister made for a
problematic match. The tension on the eve of Victor's marriage to Elizabeth creates misgivings for which the threat made by the creature—"I
will be with you on your wedding-night"—seems all too plausibly a
mere pretext. "What is it that agitates you, my dear Victor?" his bride
asks. "What is it you fear?" To which Victor replies: "this night is dreadful, very dreadful."

Victor thought that the creature meant to come and murder *him*.
Only too late does he realize that the intended victim is Elizabeth—
for if he denied the creature sexual gratification with a mate, then the
creature has determined to visit the same fate on him. But play these
episodes back through the filter in which the creature represents Victor's
id, and a different story emerges—one in which his anxiety about the
incestuous character of his relationship with Elizabeth has summoned
up a demon to put her out of the way, and to prevent her from usurping
his own self-love. When Victor admits that "suddenly I reflected how
dreadful the combat which I momentarily expected would be to my

---

* Only in the latter part of the century did cousin marriage begin to seem amiss, perhaps due to
the influence of Charles Darwin's ideas on the fitness consequences of interbreeding. (Darwin
himself married his cousin Emma Wedgwood, and fretted if this had damaged his children's
health.) But this doesn't mean anxieties about incest were absent from such unions before then.

wife," you have to read it more than once to figure out what this "combat" is: fighting the monster, or sexual grappling with his wife? Literary
critic Marilyn Butler makes the disturbing but not unjustified assertion
that "Frankenstein seems more aroused when making and dismembering
the female Monster than at the prospect of joining Elizabeth on his own
wedding night."

In short, Victor spends the novel in desperate flight from sexuality.
Anne Mellor says that "*Frankenstein* is a book about what happens when
a man tries to have a baby without a woman," and professor of English
Judith Wilt puts it nicely when she calls Victor an "anti-husband." Wilt
points out that in his destruction of the mate he makes for his creature,
he is again denying a possibility of sexual union, frightened that *they* will
procreate.* Mellor is surely right to claim that what Victor "truly fears
is female sexuality as such" (although her suggestion that, by making a
creature without sex, he is intent on eliminating the need for women at
all seems to be overstating the case).

What Mary Shelley did, says literary scholar Ellen Moers, "was to
transform the standard Romantic matter of incest, infanticide, and
patricide into a phantasmagoric of the nursery." Nothing quite like it
was done again in English literature, she says, until *Wuthering Heights*
(we'll get to that later). It is to risk crude gender essentialism to say that
Frankenstein could only have been written by a woman, but I'll take
that risk. Moers does too, calling *Frankenstein* "a horror story of teenage
motherhood" and a "birth myth." Although a few male authors in the
nineteenth century, such as William Thackeray, could write on childbirth with sensitivity and tenderness, for men of that time it was always
a (painless, bloodless) joy to be celebrated. If the child died, meanwhile,
as was all too tragically common (and if you were not a self-absorbed

---

* By having Frankenstein make the creature's "mate" from the corpse of Elizabeth after she is
murdered, Kenneth Branagh's 1994 film *Mary Shelley's Frankenstein* makes a mythically valid
change to the plot: he is now passing his sexual partner to his doppelgänger. Only Elizabeth
can rescue herself from that nightmarish, abusive male fantasy: as the female creature, she
pours oil over her head and sets it alight.

chauvinist like Percy Shelley), the sadness of the father shared little of
the wrenching anguish or self-recrimination of the mother. The process
through which Victor Frankenstein gives birth to his creature, in contrast, is visceral, prolonged, fraught, and ambivalent: "often did my human nature turn with loathing from my occupation," he admits. Moers
says *Frankenstein* is at its "most feminine" in its "motif of revulsion
against newborn life, and the drama of guilt, dread and flight surrounding birth and its consequences." The book brought birth into fiction as
a kind of Gothic fantasy in itself. When Frankenstein sees his creature
initially as "the hideous corpse which he had looked upon as the cradle
of life," Moers claims, the description is really of a stillbirth.

For the mother, a child was not an achievement to be occasionally
paraded in front of relatives but a being with a mind of its own, a responsibility and, yes, at times a burden. What if the child were ugly or
deformed? What if it did not love you, or you it? "Remember," the child
could always say, "that I am thy creature . . . to whom thy justice, and
even thy clemency and affection, is most due." Then again, it might kill
you even as it is born. Mary knew that could happen.*

A mother who voiced such fears would then have been deemed monstrous herself. And really, things are not so different now.

Meanwhile, a man might have grand notions of making life, but for
Mary Shelley the vision of reanimating a corpse was real. "Dream that
my little baby came to life again," she wrote in her journal, weeks after
her first daughter Clara died—"that it had only been cold, and that we
rubbed it before the fire, and it lived. Awake and find no baby."

Behind all the trappings of science and horror, *Frankenstein* is then
at its core a personal story of human relationships. It is about the fears

---

* The story of Safie, the young Turkish woman married to Felix De Lacey, is often seen (with
some reason) as a pointless digression—but buried within it we find this: "Safie related that
her mother was a Christian Arab. . . . She instructed her daughter in the tenets of her religion
and taught her to aspire to higher powers of intellect and an independence of spirit forbidden
to the female followers of Muhammad. This lady died, but her lessons were indelibly impressed
on the mind of Safie." As if in a dream, the narrative drifts repeatedly toward its author's life.

and failures of procreation and parenting: Victor is the original "bad parent." If it did not have these dimensions, *Frankenstein* would be a more slight affair, perhaps just one of the many Gothic novels of the period that we have long ceased to read. Percy understood at least that much about his lover; as he wrote in his introduction, the book "affords a point of view to the imagination for the delineating of human passions more comprehensive and commanding than any which the ordinary relations of existing events can yield." This is in fact an eloquent defense, *avant la lettre*, of science fiction and fantasy as a whole, which at its best creates a fantastically extreme scenario over which human dilemmas can be stretched to magnify them in ways that mundane existence cannot. *In extremis*, Percy argued, "the elementary feelings of the human mind are exposed to view."

<hr/>

Adaptations of *Frankenstein* for the theater began to appear only a few years after the book was published. The earliest, called *Presumption* and written by Richard Brinsley Peake, was staged at the English Opera House in London in 1823. Already the production changed the plot in crucial ways that anticipate the reworkings of Hollywood a century later. As his new title indicates, Peake made the story a rather strident morality tale about the hubris of usurping God—a story closer to its ancestral form in the Faust legend. "It's the Devil—for I'm sure he's at the bottom of it," says Frankenstein's burlesque assistant Fritz.

Fritz is a traditional stage clown, one of the stock character types that these adaptations imposed on the story. Presaging the lumbering Igor of the movies, he was the light relief, probably close to the comic character played by Marty Feldman in Mel Brooks's 1977 spoof *Young Frankenstein*. *Presumption* also followed theatrical convention in having songs and musical interludes. Other burlesques appeared in the Victorian music hall, notably *Frankenstein; or, the Vampire's Victim* (1887), in which the creature has a vampire for a wife. An outright spoof in that same year, *The Model Man*, had the monster wearing a monocle and singing (like *Young Frankenstein*'s Peter Boyle) in a story that was now close to pantomime, full of in-jokes and absurdly named characters. The assistant Igor is of course now indispensible: the Sancho Panza to Frankenstein's Quixote, the Watson to his Holmes. He has even become the central protagonist, Victor's friend and voice of conscience, in the fairly disastrous movie *Victor Frankenstein* (2015), where he is played by Daniel Radcliffe.

Theatrical adaptations never felt under much obligation to the source material. Henry Milner's *Frankenstein; or, The Man and the Monster* (1826) was set in Sicily; John Kerr's *The Monster and Magician; or, The Fate of Frankenstein* made Victor an alchemist. Most importantly, *Presumption* and its ilk turned the creature into a dumb brute with the mind of an infant, probably drawing from the stock theatrical character of the Wild Man. Thomas P. Cooke, the popular actor who took this role in *Presumption*, was allegedly made up with yellow and green greasepaint, a wild wig, blue legs, and black lips. Similar changes feature in Milner's play, where the creature, to judge from the engraving on the frontispiece of the published script, was a dark-skinned, shaggy "savage." He looked much the same—hirsute, gurning, and caked in greasepaint—in the fourteen-minute silent movie shot by J. Searle Dawley for Edison Studios in 1910. The message is clear: human-made monsters are not supposed to be eloquent, for then we're not just playing God but rivaling him. Propriety demands that our creations be visibly inferior to those of nature or the divine. "We can *create* nothing," asserted D. H. Lawrence apropos *Frankenstein*, "and the thing we can make of our own natures, by our own will, is at the most a pure mechanism, an automaton."

Humanoid monsters made by artificial and "unnatural" means in the Victorian era, then, could never have much to say for themselves. Their status is symbolized rather bluntly in "The Monster Maker" (1897) by William G. Morrow, a writer of horror and suspense fiction who was as celebrated in his day as Edgar Allan Poe and Bram Stoker. Here a depressive young man turns up on the doorstep of an old doctor and asks to be killed with "extreme anaesthesia." At first the doctor, outraged, refuses—but then he realizes this is his opportunity to conduct an experiment about which he has fantasized for forty years: "After a struggle, mind had vanquished conscience." Three years later the doctor's frantic wife tells a policeman and a detective that her husband has something

An advertisement for the 1910 silent film of Frankenstein *made by Edison Studios, showing how differently the creature was initially presented, compared to the traditional persona today. This portrayal was probably much closer to those that appeared on the Victorian stage.*

locked in the house behind a heavily barred door. When they come to investigate, the creature escapes and crushes the woman to death. It is "huge, awkward, shapeless; a squirming, lurching, stumbling mass, completely naked"—and has no head at all, but instead "a small metallic ball surmounting its massive neck." As the doctor attacks his creature with a knife, an oil lamp is knocked over and the house goes up in flames.

This creature is literally brainless (and more horrid than many a modern movie ghoul)—the kind of travesty that will always result when humankind tries to play God.

Surprisingly, perhaps, Mary Shelley didn't seem to object to what her creature had become. She went to see Peake's play in 1823 with her father, and approved of the way Cooke's role as the creature was listed in the program as merely a blank: "This nameless mode of naming the unnameable is rather good." Then again, Shelley was not averse to adapting her story herself to fit public perception—or perhaps also, to temper the impulsiveness of her youth to the conventions of a society against which she was no longer rebelling. She made many changes in the 1831 edition, the most significant of which was to alter the nature of Frankenstein's transgression. While in the original version we are left in no doubt that his woes arise from his refusal to love and take responsibility for his creature, the revision implies that it is in the hubristic act of trying to make a human artificially that his sin lies. As Shelley wrote in her famous preface, "Supremely frightful would be the effect of any human endeavor to mock the stupendous mechanism of the Creator of the world." She even nodded to Peake's adaptation by having Victor state that his creature is "the living monument of presumption and rash ignorance which I had let loose upon the world."*

These theater productions played a large role in embedding *Franken-*

---

* It wasn't until James Whale's 1931 film, however, that charges of impiety became a serious problem for *Frankenstein*. The censors in charge of compliance with Hollywood's self-imposed Production Code, created the previous year to respect the sensitivities of the Catholic Church and other Christian organizations, requested that Colin Clive's Frankenstein proclaim not "Now I know what it feels like to be God!" but the less specific "Now I know what it feels like to be a god." The studio ignored the request. When the film was rejected by some censor boards in the United States and Canada because of the alleged blasphemy in—as the Kansas board put it—"the act of claiming the attributes or prerogatives of the deity," the studio responded by adding a prologue in which Edward Van Sloan, who played Waldman, explained that the story was "pure fiction" and "delves into the physically impossible and the fantastic." This ploy of adding a disclaimer, then getting on with business as before, would be repeated for other films threatened with censorship on the grounds of religious morality. Even in the twentieth century, then, modern myths had not become totally secular affairs.

stein within popular culture, and for many people they *were* the story. The tale became so well known, in one form or another, that when Charles Dickens alludes to it in *Great Expectations* (1861) he doesn't need to give the title. As Pip comes to realize that it's been Magwitch all along who has been shaping and "manufacturing" him, he exclaims in dismay, "The imaginary student pursued by the misshapen creature he had impiously made, was not more wretched than I, pursued by the creature who made me." Now the circle closes: the human is made by the creature, and indeed Pip is forced to wonder if it was not *he* all along who—despising Magwitch and rejecting his adoptive father Joe Gargery—is the real monster.

By and by, *Frankenstein* became a kind of code word that didn't even need to be spoken. It hardly takes much deduction to figure out the context into which Dr. Frankstone, a female physician in D. H. Lawrence's *The Rainbow*, is speaking when she says to the heroine Ursula Brangwen:

> I don't see why we should attribute some special mystery to life—do you? We don't understand it as we understand electricity, even, but that doesn't warrant our saying it is something special, something different in kind and distinct from everything else in the universe—do you think it does? May it not be that life consists in a complexity of physical and chemical activities we already know in science? I don't see, really, why we should imagine there is a special order of life, and life alone.

This of course is nothing else but the materialism of that other Lawrence—William, Shelley's physician, which got him into such trouble after *Frankenstein* was published. Brangwen rejects it: making the human cell her object of study, she concludes that life must be more than matter, for it has purpose.

Since Mary Shelley's book was out of print and hard to procure for two decades from around the middle of the nineteenth century, it was largely through the stage adaptations that people knew of it. "By the end of the century," writes historian William St. Clair, "the simplicities of the stage versions had decisively defeated the complexities of the literary text." This meant that "the impact has been, for the most part, the opposite of what the author . . . had hoped for and intended." Rather, the story was what society demanded it be.

The theater also shaped the most famous of retellings in the age of the motion picture. James Whale's 1931 movie was based on *Frankenstein: An Adventure in the Macabre*, a 1927 play by the English writer Penny Webling for actor-director Hamilton Deane. The actor had recently starred in a successful adaptation of *Dracula*, and this was the natural choice of a successor in the same Gothic-horror mode. After a successful run in London, the play was adapted for Broadway by John Balderston (who had done the same job on Deane's *Dracula*, creating the version that led to the Bela Lugosi film.) When Universal acquired the rights to Balderston's script, Lugosi was set to play Frankenstein's creature too until he discovered that he would have no lines. He went off instead to make the Poe adaptation *Murders in the Rue Morgue*, along with the movie's intended director, Robert Florey. The English director Whale was Florey's replacement; the unknown Boris Karloff (the stage name of British actor William Henry Pratt) stepped into Lugosi's oversize shoes.

Perhaps more than any other movie except Fritz Lang's *Metropolis* (itself a product of the Frankenstein myth), Whale's *Frankenstein* showed how modern myths are told as much through iconic imagery as through words. When we think of Frankenstein's monster, it is flat-headed, bolt-necked Karloff who looms into view. When we consider how the monster was made, we see the cinematic sparks of the dungeon laboratory devised by Colin Clive's "Henry" Frankenstein, the levers and switches manipulated by the lumbering Fritz (played by Dwight Frye, riding the wave of his camp turn as Renfield in Universal's *Dracula* two years earlier).

Much of the subtlety and complexity of Shelley's novel was sacrificed. The creature is again a dumb, pathetic brute, while Frankenstein in his white lab coat is the personification of the amoral, lunatic scientist, famously cackling, "It's alive!" as the creature stirs. And the creature's

The classic "Frankenstein": Boris Karloff as the Creature, in a publicity photo for Universal's Bride of Frankenstein *(1935).*

homicidal tendencies can no longer be attributed to acts of revenge after his creator has spurned him. With almost comic crudeness, the blundering Fritz steals from "Goldstadt Medical College" a bottled brain bearing a big warning label: "Abnormal"*—belonging originally to a man whose life, Dr. Waldman has informed us, "was one of violence and brutality and murder."

These may sound like the familiar complaints of the purist about the

---

* Or as Marty Feldman puts it in *Young Frankenstein*, "Abbie Normal."

coarsening and dumbing down of a movie adaptation. But they are not really that, for Whale's movie works brilliantly on its own terms, and not only sustains but modulates the myth in interesting and revealing ways.

Whale was one of the few openly gay directors working in Hollywood, and some have claimed that both *Frankenstein* and its sequel, *Bride of Frankenstein* (1935), explore themes of alienation and empathy for the Other. Certainly, Whale seeks to allow the audience some sympathy toward both the creature—seen to be ill-treated in *Bride*—and Victor himself, who in the sequel is replaced as the head mad scientist by the scheming Dr. Pretorius (Ernest Thesiger).* Whale "was clearly attempting to suggest further ways, technical and psychological, in which the Frankenstein myth might be explained and recast for our times," says Gothic expert David Punter. Ignore the hammy acting and look again at some of the fresh elements he introduced: the end sequence, for example, in which the mill looms like some gigantic mechanical being, with creator and creature pursuing one another amid the cogs and gears. *Bride of Frankenstein* further enriches the myth with its unforgettable visuals, its dreamlike images of homunculi in jars, and the partial rehabilitation of Frankenstein himself. (The title contributes further to the confusion between creator and creation, for there's no doubt that the "bride" in question is Elsa Lanchester's screeching creature, not Frankenstein's demure wife Elizabeth.†)

The habit in Hollywood's early days of signaling genre with something like a repertory ensemble adds to the porosity of the horror-tinged myths. Karloff reprised the creature in *Son of Frankenstein* (1939), in which Bela Lugosi (in bizarre wig and beard) played a new assistant,

---

* Frankenstein's escape from the destruction of his laboratory at the end of *Frankenstein* was even narrower than it looked. Only at a late stage was a sequel planned that necessitated his survival. A scene had already been shot in which he dies in the explosion. It was too late to reshoot it, and his demise can still be glimpsed.

† Why, you might wonder, was Elizabeth not killed by the creature in the original movie, as she was in the book? James Twitchell probably has the right answer: "No monster could touch a Hollywood heroine in 1931 and get away with it."

Ygor, to Henry/Victor's son Wolf von Frankenstein (Basil Rathbone of Sherlock Holmes fame). *House of Frankenstein* (1944) was one of the first *Van Helsing*–style mash-ups, including Lon Chaney as the Wolf Man and John Carradine as Dracula—but here Karloff is out of grease-paint as the Frankenstein figure, the archetypically mad and amoral scientist Dr. Gustav Niemann. Karloff took the dual title roles in *Abbott and Costello Meet Dr. Jekyll and Mr. Hyde* (1953), and when the comedy duo met Frankenstein in 1948, Lugosi's Dracula and Chaney's Wolf Man were again there for the ride. In all honesty there is little going on in these spin-offs (better to call them cash-ins) beyond a cultural cementing of the iconography: all they really needed was a doctor mad enough to send Frankenstein's creature on the rampage again. By 1966, when *Jesse James Meets Frankenstein's Daughter* was released on a double bill with *Billy the Kid Versus Dracula*, it seems clear that Hollywood was simply seeking news ways to eke out a brand name.

Elsa Lanchester is the Bride of Frankenstein *(1935).*

Hammer Films came to the rescue, introducing its own repertory of stars. Christopher Lee was the tall one: elegant as Dracula, ominous as Frankenstein's monster. Peter Cushing was the older, more cerebral presence: paternal and reassuring as Van Helsing, cool, ruthless, and Byronic as Frankenstein. It's too little remarked how Frankenstein aged over time: the mad scientist tends by common consensus now to be a graying eminence, his years seemingly helping to transform him from the impassioned, obsessive figure in Shelley's novel to the dispassionate, unprincipled genius of the myth.

It was only in the 1970s that a taste began to develop for "authenticity"—which is perhaps to say, there was deemed to be potential for kudos in selling movies touting a return to the source. It happened for *Bram Stoker's Dracula* (1974) featuring Jack Palance, and *Frankenstein* was awarded the attentions of literary writer Christopher Isherwood, assisted by his partner Don Bachardy, for *Frankenstein: The True Story* (1973). The film did not really tell "the true story" at all, but instead mixed elements of the "original" with an artful seasoning of the myth surrounding the genesis of the book. The real villain was now none other than "Dr. Polidori," plucked from the Villa Diodati and inserted into Shelley's tale as a replacement for *Bride of Frankenstein's* wicked Dr. Pretorius. Played by James Mason, he is the model of suave wickedness, with ominous black gloves concealing his deformed, clawlike hands.

Here it is Henry Clerval, Polidori's student, who assembles and animates a creature from the body parts of miners killed in an accident; his friend Victor merely helps him. When Clerval dies of a heart attack just before the business is completed, the enterprising Victor elects to use his friend's brain to finish the job. The result is no frightful Karloff-like creature that, like Shelley's original, provokes horror in all who see it. Rather, it is—to begin with—a smolderingly good-looking, if inarticulate, young man, played by Michael Sarrazin, whom Victor begins to introduce to the circles of London society.

But alas!—Victor discovers the hard way what Clerval had realized just before he died: the process will not last. The creature begins to deteriorate, and with an explicitly homoerotic subtext, Victor rejects it as it loses its looks. The creature flees, and . . . frankly, you don't need to know much more beyond the predictable fact that it ends badly for all concerned. Polidori is killed by lightning, strung up on the mast of a ship that is supposed to be bearing Victor and Elizabeth to America; the creature kills Elizabeth and steers toward the north pole, where both it and Victor are last seen being buried in an ice avalanche unleashed by Victor's final cry: "Forgive me!" Well, there's a lot to forgive, for the authors of this "true story" seem to have given rather little consideration to the themes of the source material or (beyond the ambiguous love/hate response of Victor to his creature) how it might be extended in fruitful new directions.*

The selling power of "authenticity" was revisited in the 1990s with Kenneth Branagh's *Mary Shelley's Frankenstein* (1994), hard on the heels of Francis Ford Coppola's (equally mistitled) *Bram Stoker's Dracula* (1992). In many respects, Branagh's film is all noise and confusion.† But what it really shows us is Mary Shelley's novel refracted through the prism of its own myth. The creature, played against type by Robert de Niro, is made with the brain of Victor's mentor Professor Waldman (an even more unlikely John Cleese), and he is finally allowed once again the eloquence he was so long denied.‡ Branagh himself is Victor, reclaimed from Cushing's austere ascetic as a youthful, tousle-haired, and rather sympathetically tragic hero. As in the 1973 *True Story*, Victor is

---

* Being the first Frankenstein I ever saw, it left me deeply baffled for years about where Polidori fitted into the picture.

† Dissociating himself from it after writing a rather different original screenplay, Frank Darabont (now best known as the creator of the zombie series *Walking Dead*) complained that "it has no patience for subtlety. It has no patience for the quiet moments. It has no patience period. It's big and loud and blunt and rephrased by the director at every possible turn." Shelley's book, Darabont said, "was way out there in a lot of ways, but it's also very subtle." For him, watching the film seems to have been eerily like Victor's experience when his creation came to life: a terrible distortion of what he had intended, indeed what Darabont called a "weird doppelgänger."

‡ Still there remains, however, the implication that the creature inherits his evil nature from his constitution: his body is that of the criminal who murdered Waldman. It seems we're still unsure about the relative roles of nature and nurture in the creature's disposition.

also granted partial exoneration for his transgressions by way of a motivational backstory: the death of his mother after giving birth to his brother William (you see what they did there?) makes him vow on her grave to conquer death.

From the Hammer era onward, sex could be made an explicit part of the tale. For all their schlocky, exploitative surface, a subtle intent sometimes moved beneath in the Hammer films of Terence Fisher. His 1967 outing *Frankenstein Created Woman* doesn't lead you to expect much ("A beautiful woman with the soul of the devil!"), but there are surely sexual anxieties at play when Cushing's Baron Frankenstein takes it upon himself to transfer the "life force" ("Call it a soul if you want to") from his young assistant Hans (who is executed for murder) into the body of his buxom lover Christina. We are left to decide for ourselves what fantasy the aging Cushing is vicariously fulfilling via his virile assistant.

The lewd innuendos of Mel Brooks's spoof *Young Frankenstein* don't beat around the bush: the monster gets the girl, who expresses delight in finding that he has been supersized in all respects. And there is of course very little hidden from view in Tim Curry's Frank N. Furter from *The Rocky Horror Picture Show* (1975): he is an architect of anarchy who has created the dumb-blond, gay-centerfold Rocky for his delectation. For Twitchell, *Rocky Horror* is not so much a parody (or a return to the old burlesque treatments) as an alternative *Frankenstein*, in which all the repressed libido finds orgiastic release in what is—and has certainly become for many college students—"almost the exit ceremony from adolescence." It doesn't just let it all out, but exposes the hypocrisy of repression: for all his madness and violence, Frank N. Furter feels a good deal more healthy than the upright, buttoned-up Brad, wide-eyed in terror at his awakening homosexual urges.

The celebration of monstrosity and otherness in cultural theorist Donna Haraway's *Cyborg Manifesto* presages the adoption of *Frankenstein* within the canon of queer theory. An extended essay on "science, technology, and socialist-feminism in the late twentieth century," Haraway's influential text argues that traditional boundaries between humans, other animals, and machines have been eroded: "we are all chimeras, theorized and fabricated hybrids of machine and organism."

Similarly, Haraway argues, sexual dualities have broken down, and we are free to reconfigure them as we wish. There's a lot going for the argument that Frankenstein's creature challenges boundaries of sexuality and gender. "This creature given life by the mad scientist that then disowns him," writes queer theorist George Haggerty, "this calls to mind the struggles of a young gay man, monstrous to himself in so many ways, confronting the man who has perhaps first seduced him but now refused to support or even acknowledge him." To call a reading like this anachronistic would be to miss the point: modern myths hold within them the potential for interpretations that could never have occurred to their originators. There's nothing magical or even prescient in that; it's simply a feature that some stories possess. Jeanette Winterson has taken full advantage of this potential in her modern retelling *FranKissStein: A Love Story* (2019), in which transgender Ry (once short for "Mary") Shelley narrates their exploits alongside their sometime lover, Silicon Valley transhumanist Victor Stein, and sleazy misogynist and sex-robot salesman Ron Lord. Pretty much anything is technically possible here—except, perhaps, real love.

The arrival of Frankenstein in Silicon Valley brings us to those bolts. The metallic protrusions from Boris Karloff's neck are as much a part of the iconic look of the creature as the flat head with its ring of crude stitches. But they're not there (as often assumed) to hold the head in place: they were intended to be the electrical contacts via which Colin Clive's Frankenstein infused his creation with a spark of being.

Yet they do a lot more semiotic work. They signal to us that there is something not entirely organic about this being, hinting that if we were to strip away his green skin we might find not flesh and blood but the mechanical frame of a robot.

The bolts are indeed connectors. They connect *Frankenstein* to the iconography, more or less contemporaneous with Whale's film, of a strictly mechanical being: the gleaming metallic body made by the mad scientist Rotwang in Fritz Lang's *Metropolis* (1927). The mechanical being has a long history reaching back to classical times, arriving in modernity via the ingenious automata of the Enlightenment. Historically, it has run on a track parallel to that of the Faust and Frankenstein stories. But the two streams begin to merge toward the end of the nineteenth century, most notably in the strange Symbolist novel *L'Eve future* (*Tomorrow's Eve*) by Auguste Villiers de l'Isle-Adam, published in 1886. Here a thoroughly fictionalized version of the American inventor and electrical impresario Thomas Alva Edison creates an artificial android woman called Hadaly from marvelous new materials—a "magnetoelectric entity" animated by quasi-scientific forces. This being becomes a kind of idealized avatar of a human woman named Alicia—a robotic Alicia created by Edison for an English aristocrat named Lord Ewald, who is helplessly in love with her. Yes, to put not too fine a point on it: this is the first sex robot. Rotwang makes his robotic woman for much the same purpose in *Metropolis*, and eventually it is made over in the likeness of the movie's heroine, Maria.

The advent of the modern robot heralded fears that industrialization was turning humans into soulless machines enslaved by the regimented demands of routine work. While Halady and Rotwang's android are bespoke beings, Karel Čapek's 1921 play *R.U.R.*—which gave us the Czech word *robot*, meaning laborer—transformed Frankenstein's solitary experiments into a system of mass production of artificial creatures. Čapek's robots are still made of flesh, however, not metal: in the factories of the company Rossum's Universal Robots their body parts are manufactured individually from a kind of fleshy dough and then put together on an assembly line. This vision of organic yet synthetic beings seems to owe its inspiration to the *in vitro* culturing of biological tissue samples that became possible in the laboratories of the 1910s and 1920s (Čapek was a keen observer of scientific developments). But it was the image of the metallic robot, a Frankenstein's creature for the age of chrome-and-steel mechanization, that lodged in the public imagination.

There is little in the recurring narratives of robot fiction in the mid-twentieth century that is not foreshadowed in *Frankenstein*. The robots of *R.U.R.* stage a rebellion and wipe out humanity, just as Victor feared might happen if he made his creature a mate with which it could breed.

The android Maria and her Frankensteinian creator Rotwang in Fritz Lang's Metropolis
(1927).

Robots could be cute (Robby from *The Forbidden Planet*, R2D2 from *Star Wars*) or comic (C3PO looking like a desexualized, freshly buffed version of Rotwang's android)—but we had best keep a watchful eye on them, as not even Isaac Asimov's Three Laws of Robotics, from his short-story collection *I, Robot* (1950), will guarantee our safety in the face of their inexorable and sometimes idiotic logic. At the same time we can't help but wonder what is going on in the robot's powerful, synthetic frame—and whether it houses a spark of consciousness that would leave us with moral responsibilities toward our creations.

By the time the fleshy robot returns in Philip K. Dick's *Do Androids Dream of Electric Sheep?* (1968)—made much more famous, of course, by Ridley Scott's movie adaptation *Blade Runner* (1982)—the debt to *Frankenstein* couldn't be more clear. From the novel's slightly drawn antagonist, replicant Roy Batty, *Blade Runner* constructs the most compelling of all modern reinventions of Frankenstein's creature, sharing not only that being's tremendous strength and murderous rage but also its existential angst and pain at the uncaring attitude of its creator. When Rutger Hauer's Batty kills his maker, Eldon Tyrell, we can't help but feel his act of Miltonian rebellion is somehow just. The Frankenstein figure, however, is now not merely Tyrell but the entire corporation he heads and represents: a company that has set itself above society and moral accountability.

Batty and his fellow replicants are no longer marked out and shunned as hideous; they walk indistinguishably, indeed beautifully, among us. And what Batty demands from his maker is not love but life: an extension of the short span that replicants have been allotted. "More life" here really stands for "eternal life," which is to say, a human soul. That is something the robots of *R.U.R.* do finally acquire "by means of pain and horror," and it guarantees them a capacity to love, to feel, and thus to propagate. They won't, as Frankenstein feared, create a "race of devils"; they are simply the new humans.

Yet the robot company's director-general, Harry Domin, comments bleakly that the true mark of humanity is not love but hate: "Nobody could hate man as much as man!" We (like Victor, after all?) are the real monsters. When, knowing his own death is nigh, *Blade Runner*'s Roy Batty offers mercy to his would-be assassin, Rick Deckard, he illustrates the recurring trope of the modern Frankenstein-robot myth: our creations may become *more* human than we are.

Conversely, the subtext of *Blade Runner* (and the main theme in Dick's novel) is the fear that *we* are already the robots. Deckard is at one stage explicitly suspected of being an android with implanted memories, and when he takes the Voight-Kampf Test, a series of questions designed to probe for the empathy only humans are thought to possess, the result is ambiguous.* Dick asks us to consider: how do we know that we are

---

* The test is an adaptation of one proposed in 1950 by the British mathematician Alan Turing to try to distinguish human from machine. He called it the Imitation Game; it is now known simply as the Turing Test.

not such beings? *And would it matter if we were?* If we create our sense of self out of (probably misremembered, half-constructed) experience and memories, what difference does it make if those are implanted?

Such questions, implicit already in Shelley's interrogation of materialism, have come to a head today in debates about the origins of consciousness and identity. Some philosophers of mind are quite content to assert that the self and consciousness are illusions spun from networks of electrical signals exchanged between neurons. Others see no reason to doubt that making self-aware machines is just a matter of adding more bytes and flops to our computer circuitry. The question is whether what we humans experience as consciousness is merely a by-product of the kinds of computation our brains perform—or whether consciousness can be expressed in computational terms at all. It's not necessary to impute some mysterious agency, some nonphysical or pseudobiologized soul, to acknowledge that no one knows the answer. In the face of this ignorance, myth can show us pathways through the void.

By making the creature a narrator for part of her book, Mary Shelley answered the question already—at least insofar as it relates to the capacity of the bland, blancmange-like material of a salvaged human brain to become again a mind and a person. Yes, she seems to say, we are physical substance—awakened to itself. But how swiftly we then moved to deny that conclusion, reducing Frankenstein's creature to a blundering hulk and then, in time, to a mechanical robot. Yet Shelley's voice remained insistent: from this *thing* you have created, sentience may yet emerge to challenge your certainties and your presumption. Not just that: this created mind might supersede you, physically and perhaps morally. How much nobler, suddenly, is the once-murderous Roy Batty, bloodied and naked in the smoke and the rain, than the pitiful, crumpled Deckard. How quickly our AI minds might tire of us as they soar beyond our puny abilities, much as the hypersmart virtual mind Samantha leaves "her" pathetic human operator Theodore Twombly—and, he discovers, hundreds of other would-be lovers—to their pitifully circumscribed meat-space in Spike Jonze's *Her* (2013). In the age of the computer, precociously glimpsed by Byron's daughter Ada Lovelace, Frankenstein's creature is apt to become the disembodied Samantha or *2001*'s

HAL2000: either benignly indifferent to us or killing us softly, for our own good.

Yes, this amoral, robotic AI can morph into an actively malevolent Terminator—and there is surely no one better suited than Arnold Schwarznegger to convey the creature's combination of brute physical might and clumsy, monosyllabic bluntness. Yet it is Mary Shelley's exploration of the *making of a mind*—and not apocalyptic visions of Skynet and robot overlords—that is the real legacy of *Frankenstein* for contemporary debates about artificial intelligence.

Just as with our "machine learning" systems today, the creature acquires its way of knowing not by design but by imitation. In consequence, it reflects back to us our own faults, but also our aspirations and dreams—the poetics of Milton and Goethe blended with the rage of a child. If the creature can't find its way to morality, that is because we have been unable to show it the path.

And if we can make a mind, what then becomes of us? Will we simply be turning ourselves into inferior machines, obsolete forms of cognition? As Karl Marx said in 1856, "All our inventions and progress seem to result in endowing material forces with intellectual life, and in stultifying human life into a material force." In this way, says Chris Baldick, "the myth of Frankenstein has become the great fact of nineteenth [and surely twenty-first] century life": the *things* are taking over.

The nightmare that humans will become mere servants of the machines has been with us since the heyday of industrialization, and was explored in Samuel Butler's 1863 essay "Darwin among the Machines." "Day by day," he wrote,

> the machines are gaining ground upon us; day by day we are becoming more subservient to them; more men are daily bound down as slaves to tend them, more men are daily devoting the energies of their whole lives to the development of mechanical life. The upshot is simply a question of time, but that the time will come when the machines will hold the real supremacy over the world and its inhabitants is what no person of a truly philosophic mind can for a moment question.

Butler expanded on this fear in "The Book of the Machines," a section of his 1872 utopian fantasy *Erewhon*. Here he imagined a society that had forsworn machines altogether, mindful that they might otherwise evolve not only to reproduce themselves but to acquire consciousness, dominate us, and "rule us with a rod of iron." E. M. Forster described a society of that kind in his short story "The Machine Stops," first published in 1909. Here humans live underground (in a manner inspired by H. G. Wells's Morlocks in *The Time Machine*) and are totally dependent on a vast machine. When it breaks, civilization stops—for no one any longer knows how to fix it. "In all the worlds there was not one who understood the monster as a whole." It sounds today all too plausibly like the predicament the internet is placing us in.

It might seem a stretch to suppose that *Frankenstein*, which after all makes no reference to mechanisms and came out of an age before mass industrialization, foreshadowed such ideas. But what matters in a story that becomes a myth is the *affordance* it permits. All that is needed for the myth to fit is that the monster, our nemesis, is our own creation, the fruit of knowledge injudiciously applied to generate an entity too large and unruly to control. By trying to remake ourselves, we seed our own destruction.

Chapter 4

UNCHAINING THE BEAST

THE STRANGE CASE OF DR. JEKYLL AND MR. HYDE (1886)

Beneath a man's urbane and cultivated exterior lurks
a savage and depraved brute who bursts forth and
wreaks havoc. He must either be vanquished or destroyed
by the suicide of his better half.

The subconscious rarely speaks in genteel language. It gnashes its teeth, raises hairy paws, roars, and commits terrible and obscene acts. That can't be helped.

We don't mind watching it, though. Horror movies are one of the most popular genres in cinema, and this is harder to explain, as well as more unsettling, than the public taste for pornography. It is one thing to suggest that fairy tales help children process and deal with the anxieties of childhood: feelings of jealousy and fury toward siblings and parents, the difficulty of socializing egotistical and selfish impulses, emerging sexuality, and so forth. It is quite another to understand why adults like to watch people turn into werewolves and tear others limb from limb.

Oh, we've tried. The British psychiatrist Ernest Jones considered vampires, werewolves, witches, and demons in his 1910 book *On the Nightmare*, where he imagined—as any good Freudian would—that all such visions are rooted in sexual drives. By cloaking something illicitly desired in a repulsive form, he said, we create camouflage for our attraction. Literary theorist Rosemary Jackson, meanwhile, argues that fantasy—including tales of horror and the supernatural—expresses an unconscious urge to subvert natural categories. (It's not too clear why we should feel compelled to do that, however.) For philosopher of art Cynthia Freeland, the genre simply piques our curiosity: "Horror in general attracts us because we want to *understand* the monster."

Well, I'm not sure how much we really want to understand, say, Godzilla or *Hellraiser*'s Pinhead. But there are certainly monsters in the realm of horror fiction that are more likely to repay a bit of attention. Mr. Hyde is one of them.

Mr. Hyde is a brute and a bully—pretty much a total psychopath. If that were all there were to him, we might be content for him to be locked away and never heard from again. But of course the only place he was locked away was in the psyche of Dr. Henry Jekyll, and that makes him far more interesting.

Robert Louis Stevenson's tale seems at face value to have a rather transparent reading: Hyde is the violent, ugly side of ourselves that we would prefer to keep hidden but who insists every now and then on seizing control of the psyche and causing mayhem. This is in truth not so bad a summary of what the Jekyll and Hyde myth "means," and its psychological validity suffices to explain why the story has currency still today. But that meaning alone would seem to reduce Stevenson's novella to a rather simple allegory, and hardly explains how it could become mythic. To truly understand this monster, we have to look more carefully at where the story came from—in a time when the idea of a buried and disreputable unconscious was barely formed—and at the many guises it has taken since. Is he no more than the inchoate, childlike, raging Hulk, suffering from severe anger-management issues? Or is there something more malevolent and perverse that fuels his rampages? Can he ever be tamed? What is it, really, that we fear from Hyde?

———◆———

Robert Louis Stevenson was never a well man. Born in Edinburgh to a lighthouse keeper in 1850, he suffered from tuberculosis and related pulmonary problems from an early age and died when he was just forty-four. He studied engineering in his home city, but quit and went to London to pursue law, passing the bar in 1875 after moving back to Scotland. Yet he was never quite respectable. In his youth he had a reputation for rowdy conduct, and in 1877 he began an affair with a married (but separated) American woman a decade his senior named Fanny Osbourne. When she returned to the United States the next year, Stevenson took off for a trip through France (allegedly by donkey)—but, his passion undiminished, he sought her out in California in 1879, where they married. Once the couple (with Fanny's son Lloyd) returned to Scotland, Fanny found herself as much the caretaker as the wife of the constantly ailing writer.

Treasure Island, published in 1883, made Stevenson's name. His reputation as an author who could spin a fine yarn was sealed by *Kidnapped*, serialized three years later, which drew praise from his friend Henry James. *Jekyll and Hyde* followed soon after, also in 1886, and the following year Louis (as his friends called him), Fanny, and Lloyd returned to America. Given the state of Stevenson's health, it seemed an ambitious plan, bordering on reckless, when the family bought a yacht and set sail into the South Pacific, finally landing on the Samoan island of Upolu. Stevenson bought a large plot of land, and there the family remained until his death in 1894.

The authors of *Frankenstein*, *Jekyll and Hyde*, and *Dracula* all claimed that their tales came to them in dream visions.* What are we to make of that? Possibly not too much. "It came to me in a dream" was the roman-

---

* "What myths are to the race," claimed psychologist James Hadfield, "dreams are to the individual."

tic and clichéd refrain for many inspirations and insights in the nineteenth century, used by scientists (such as the chemists August Friedrich von Kekulé and Dmitri Mendeleev) as much as by artists. The implication is that the product of a dream is somehow greater than the individual who acted as the vessel—it gives a discovery or work of art the stamp of timeless authority. Yet these claims can't be dismissed as mere convention. It's banal but undeniable that dreams and reverie loosen the bonds of thought and let association and the subconscious do their work; scientists even today attest to the value of that. And most often, in both art and science, the creations of dreams are not de novo flights of the imagination but spring from well-prepared ground. Stories that give

Robert Louis Stevenson (1850–1894), photographed in 1893 by Henry Walter Barnett.

fresh shape and meaning to older ideas are precisely what you'd expect to emerge from oneiric reverie.

Stevenson, moreover, was ill and feverish, his body plagued by hemorrhages and filled with narcotic painkillers, when a nightmare woke him shrieking and presented him with "a fine bogey tale" about "a voluntary change becoming involuntary." This was not, he attested, the full story of *Jekyll and Hyde*, but three key scenes, including the one in which Hyde "pursued for some crime, took the powder and underwent the change in the presence of his pursuers." Here Stevenson was happy to attribute the inspiration, perhaps only half in jest, to "some Brownie, some Familiar, some unseen collaborator whom I keep locked in a back garrett" and to whom he acts only as a dutiful scribe. These elflike muses, he says, "do one-half my work for me while I am fast asleep, and in all human likelihood, do the rest for me as well, when I am wide awake and fondly suppose I do it for myself."

The story (unreliably related by his stepson Lloyd, who himself became an undistinguished writer) is that he worked at a frantic pace to capture the essence of his dark vision, completing the first draft in three days. He then promptly threw it on the fire after Fanny volunteered some strong criticisms, and returned to his room to write a second draft at a similarly rapid clip. The account is probably so much myth in itself; two full drafts of the novella still exist, and they were clearly prepared over several weeks. Still, Stevenson averred that the book "was conceived, written, re-written, re-re-written, and printed inside ten weeks"—a speed that would make even the fastest of today's authors envious, and which was apparently motivated by a desire to capitalize on the Christmas market.

It could be the rate of its composition, as much as the visionary nature of its concept, that makes a work of literature potentially mythical material. Such works should not be too well-formed, and should have plenty of gaps, even errors of coherence and continuity. Stevenson himself recognized that, with *Jekyll and Hyde*, his "Brownies" were not simply trotting out a morality tale. Sometimes, he said,

> I cannot but suppose my Brownies have been aping [John] Bunyan,
> and yet in no case with what would possibly be called a moral in the

tract; never with the ethical narrowness; conveying hints instead of life's larger limitations and that sort of sense which we seem to perceive in the arabesque of time and space . . . my Brownies are somewhat fantastic, like their stories hot and hot, full of passion and the picturesque alive with animating incident; and they have no prejudice against the supernatural.

With the help of such a disposition (whether from Brownies or not), Stevenson "created a living text," writes horror critic Leonard Wolf, "in which too much that happens is inexplicable." Stories are at their most mythically fecund when they don't entirely make sense. For which dreams ever do?

<hr />

For a short novella, *Jekyll and Hyde* has a complicated, even convoluted, structure. The main body of the narrative is related in third person from the perspective of Jekyll's friend Mr. Utterson. But the full story only unfolds later, when the reader is given two personal accounts of the events, the last by Dr. Jekyll himself.* As is commonly the case with modern myths, this framework obstructs any smooth unraveling of the tale and requires the reader to do some of the work, undermining any possibility of making a definitive interpretation.

The remote "yet somehow lovable" lawyer Mr. Utterson is out walking in London† with his friend, "man about town" Mr. Richard Enfield, when they pass a neglected doorway on a quiet backstreet. The sight prompts Enfield to recall to Utterson how, returning home through the

<hr />

* Vladimir Nabokov claimed that "Jekyll" comes from the Danish word *Jökull*, meaning ice cap or glacier. It sounds like a suspiciously Nabokovian derivation.

† It's been suggested that the story is really set in Edinburgh, where Calvinism produced an even stronger repression of desires and instincts. But it is hard to see what motive Stevenson might have had for disguising the location.

lamplit streets one winter night, he watched in horror as a man collided with a small girl and "trampled calmly over the child's body and left her screaming on the ground." Enfield collared the marauder and, with the girl's family and an angry crowd of onlookers gathering, demanded he pay the family a hundred pounds as compensation. The man acquiesced, leading the crowd to this street in which Utterson and Enfield now stand before unlocking that shabby door and emerging with the payment—part of which was a check made out in the name of a well-known and respectable gentleman. Why this person should be connected with so disreputable a fellow is a mystery, but Enfield suspects blackmail.

The man who assaulted the child so gratuitously is called Mr. Hyde, Enfield tells his companion, and he describes him:

> There is something wrong with his appearance; something displeasing, something downright detestable. I never saw a man I so disliked, and yet I scarce know why. He must be deformed somewhere; he gives a strong feeling of deformity, although I couldn't specify the point.

Utterson confesses that he knows the name Hyde already, and also the gentleman with whom he is apparently connected: Dr. Henry Jekyll. For Utterson is the guardian of Jekyll's will, in which the doctor says that if he were to die, all his possessions should pass to his "friend and benefactor" Edward Hyde. Learning now of Hyde's detestable nature, Utterson is alarmed and goes to seek advice from another medical man, Dr. Lanyon, who is a friend of Jekyll's. But Lanyon knows nothing about Hyde, and so Utterson decides to confront the miscreant himself, loitering by that unkempt door. "If he be Mr Hyde," he tells himself, "I shall be Mr Seek."

His patience is rewarded, for one night Hyde appears. But their brief and fractious exchange reveals nothing except to confirm Enfield's claim that there is something loathsome about Hyde. "God bless me, the man seems hardly human!" Utterson tells himself. "Something troglodytic... or is it the mere radiance of a foul soul that thus transpires through.... O my poor old Harry Jekyll, if ever I read Satan's signature upon a face, it is on that of your new friend."

Mr. Hyde locks the "blistered and distained" back entrance to Dr. Jekyll's residence in this illustration by Charles Raymond Macauley for a 1904 edition of Stevenson's novella.

At a dinner at Jekyll's two weeks later, Utterson broaches the subject of Hyde. "This is a private matter," says the discomfited doctor, "and I beg of you to let it sleep." Yet he secures from the lawyer a promise still to do justice by Hyde in the event of his own death. "I do sincerely take a great, a very great interest in that young man," he admits.

A month later, London society is shocked by the report of a maid-servant who has witnessed Hyde battering an "aged and beautiful gentle-man," Sir Danvers Carew, to death in the street after the old man stopped to say some word to him. The murder is shockingly brutal:

> With ape-like fury, he was trampling his victim under foot, and hailing down a storm of blows, under which the bones were audibly shattered and the body jumped upon the roadway.

Utterson is Carew's lawyer too, and he takes a policeman to the house where he understands Hyde to lodge: a dingy part of Soho shrouded in London fog. The murderer is not at home, and his room shows signs of having been left in a hurry. So they try instead at Jekyll's house, where the doctor is aghast to learn of this turn of events and swears he will never set eyes on Hyde again. But he gives Utterson a letter written by Hyde in an "odd, upright hand," in which the criminal assures his benefactor that he has made for himself a safe means of escape. "O God, Utterson, what a lesson I have had!" Jekyll moans.

As the hunt for Hyde goes on, Utterson visits Dr. Lanyon and finds the man transformed: pale, aged, haunted. When Jekyll's name comes up, Lanyon declares, "I am quite done with that person," and will say no more about him. Utterson writes to Jekyll to inquire about the rift, and the doctor replies that he himself has determined to lead a life of seclusion: "I have brought on myself a punishment and a danger that I cannot name." Less than two weeks later, Lanyon dies of some unspec-ified affliction. But he has left Utterson a sealed letter—inside which is another, saying that it should not be opened until the death or disap-pearance of Jekyll.

As Utterson is out for another walk with Enfield, they pass that ill-omened doorway again. Enfield tells his companion that he has dis-covered what the other already knew: it is a back way to Jekyll's house. They step through and see Jekyll at the window, who greets the two men below but is suddenly struck with "an expression of such abject terror and despair" that the pair, turning back after the window is closed, are horrified.

One evening after dinner, Utterson receives an unexpected visit from Jekyll's servant Poole, who tells him that he fears some foul play has happened at the doctor's household. The lawyer accompanies him back to Jekyll's house, where the master has locked himself away in the laboratory and will see no one. At least, he *claims* to be Jekyll, but barely sounds like him. Has the doctor been murdered by an intruder? Poole explains that Jekyll has for the past week "been crying night and day for some sort of medicine." And when Poole glimpsed him one day, he could not recognize his master, who scampered away from sight. Indeed, the figure that "jumped from among the chemicals," says the servant, looked for all the world like Hyde.

Utterson concludes that it is indeed Hyde who lurks in the locked room, and that he has murdered Jekyll. He breaks open the door to apprehend the killer—but finds Hyde lying on the floor twitching, a vial of poison by his side. Of Jekyll's body there is no sign, but the searchers discover a letter from him addressed to Utterson. It tells him to read Lanyon's account; but there is another sealed packet too, which tells Jekyll's version of the tale.

The story ends with the denouements of Lanyon's and Jekyll's narratives: a polyphony of voices that mirrors the title character's psychological fragmentation. First we are shown another letter from Jekyll, begging Lanyon to come to his house that night and take a drawer of chemicals from his laboratory cabinet, then to give them to a man who will present himself in Jekyll's name. This is of course Hyde, who duly arrives, seizes the chemicals, and mixes up the potion of transformation, which fumes and bubbles and changes color, first to dark purple and then to green. He warns Lanyon to be gone before he witnesses "a prodigy to stagger the unbelief of Satan." But the doctor stays, and to his utter horror watches Hyde—the creature he now knows (in his letter) to be Carew's murderer—drink the mixture and turn back to Jekyll.*

---

* For Oscar Wilde this transformation read "dangerously like an experiment out of the *Lancet*," as one character remarks in "The Decay of Lying" (1889)—an example of an author (in Wilde's view) "robbing a story of its reality by trying to make it too true."

Jekyll's account, meanwhile, brings to the tale a dimension beyond the quasi-scientific supernatural. He admits that ever since his youth he has hidden "a certain impatient gaiety of disposition" beneath a grave and respectable mask. By the time he rises to his current social status, he has become trapped in "a profound duplicity of life," and forced to conclude "that man is not truly one, but truly two"—one honorable and good, the other evil:

> If each, I told myself, could but be housed in separate identities, life would be relieved of all that was unbearable. . . . It was the curse of mankind that these incongruous faggots were thus bound together— that in the agonized womb of consciousness, these polar twins should be continuously struggling. How, then, were they dissociated?

His laboratory investigations eventually showed him how:

> Certain agents I found to have the power to shake and to pluck back that fleshly vestment, even as a wind might toss the curtains of a pavilion. . . . [I] managed to compound a drug by which these powers should be dethroned from their supremacy, and a second form and countenance substituted, none the less natural to me because they were the expression, and bore the stamp, of lower elements in my soul.

Jekyll then describes the anguish of this transformation in a manner that invited the melodramatic grimaces of actors who would later bring the doomed doctor to the screen: "The most racking pangs succeeded: a grinding in the bones, deadly nausea, and a horror of the spirit that cannot be exceeded at the hour of birth or death." Yet in time he becomes addicted to his sprees as Hyde: "my new power tempted me until I fell in slavery. . . . [I could], like a schoolboy, strip off these lendings and spring headlong into a sea of liberty." He could commit crimes with impunity, swallow the restorative draught, and become again the unimpeachable Henry Jekyll.

Hyde, however, begins to overwhelm him. One evening he falls asleep as Jekyll and wakes up as Hyde without having taken the potion at all. And whereas at first the difficulty had been to throw off the body of Jekyll—sometimes the drug worked only in double or triple doses—now the problem is to return to it. "Between these two [selves]," the doctor writes, "I now felt I had to choose." It is hard to imagine he could live forever as Jekyll, in chosen abstinence from his base impulses, rather than as Hyde—despised and friendless, but heedless of that social cost.

He tries anyhow, eschewing the potion for weeks. But like an addict, he finally succumbs—and Hyde, long denied expression, "came out roaring." Hyde bludgeons Carew to death, and Jekyll, faced with the enormity of his crime, determines never again to put himself or others in danger by taking the draught.

But to no avail. One day, sitting on a bench in Regent's Park and feeling pleased with his renewed respectability, he finds he has transformed once more into the brute. Grotesquely comical now in his oversize garments, he takes refuge in a room at an inn and writes the letter begging Lanyon to find the antidote in his cabinet and have it ready for him at his house; this is when Lanyon sees him transform. The gambit works, but the next morning, after breakfast at home as Jekyll, he once again spontaneously becomes Hyde. From that day on, it is only with great effort of will that he can resist the change; after sleeping as Jekyll, he invariably wakes as Hyde.

Jekyll is filled with great hatred for Hyde, who, it seems to him, is usurping not just the life of Jekyll but the force of life itself. In retaliation for this loathing, Hyde torments Jekyll, scrawling blasphemies on his books and destroying the portrait of his father. The conflict might have gone on this way for years, had Jekyll not run out of the chemical needed to change back from his Hyde-self. He sends Poole to get a fresh supply but finds that the draught no longer works; it must, he figures, have depended on some unknown impurity in the original batch. Using up the very last of his old powder, he finishes his account as Jekyll, knowing that when the potion wears off he will be doomed to remain as Hyde for the rest of his life. But that, he decides, will not last long: either Hyde will be captured and hung for his crime, or he might find the resolve to "release himself" with the poisons in Jekyll's laboratory. And that is precisely the fate that the savage, cornered Hyde chooses.

After the dream that precipitated *Jekyll and Hyde*, Stevenson allegedly told his doctor, "I've got my shilling shocker." Some critics saw the little book as nothing more, and indeed at one point in the writing Stevenson thought as much. "I am pouring forth a penny dreadful," he wrote. "It is dam dreadful." That it appeared during a time when penny-dreadful pamphlets were creating a moral panic was, however, not necessarily a bad thing for sales. The publisher, Longmans, seemed to recognize as much and cannily published two versions: a respectable, costly cloth-bound edition (for the Jekylls) and a shilling paperback (for the Hydes).

Fantastical stories were apt to incur critical snobbery and distaste in the late Victorian age, and sure enough *The Times* accused *Jekyll and Hyde* of displaying a vulgar materialism and excessive love of the grotesque. For the reliably snooty *Athanaeum* the story was "not merely strange, but impossible, and even absurd"—a sign, perhaps, that it cut too close to the bone for the buttoned-up, gentlemanly readers of that periodical. The *Saturday Review*, on the other hand, found the villain thrillingly nasty: "We would welcome a spectre, a ghoul, or even a vampire gladly, rather than meet Mr Edward Hyde."

The novella is, it must be said, no masterpiece of structure or plotting. It's a ramshackle affair, and the device of having two letters reveal everything at the end is a classic example of telling rather than showing. There are no real characters in the book at all—Lanyon and Enfield, for example, are more or less interchangeable bachelors, and the latter serves no narrative purpose beyond making Utterson first aware of Hyde.

But these failings ensure that we don't get distracted by elegant particulars from the substance of the story. In the Edwardian era Stevenson was regarded as little more than a spinner of rather childish yarns, and by the 1940s *Jekyll and Hyde* was seen as a somewhat crude morality tale: give in to your bad impulses and no good will come of it. But Stevenson's achievement lies in how, in today's critical parlance, he was able to complicate and problematize what could have been a fairly routine Poe-like shocker. How much of that complication was intentional might matter for Stevenson's reputation among literary critics, but not for the power of his myth.

It's in the ambiguities, the lacunae, the oddness, that this potency lies. Sure, this is in some measure the age-old tale of a struggle between good and evil—but Stevenson doesn't allow it to become a simple Manichean conflict. For is Jekyll really so virtuous, and Hyde so vile? Should we really regard Jekyll's repression and social hypocrisy as preferable to Hyde's unfettered liberty? Must instinct always be enslaved by civility? "Everything is true," Stevenson wrote to a friend, "only the opposite is true too; you must believe both equally or be damned." Whether by design or sheer lack of it, he left windows open in the narrative through which other things could enter.

***

Jekyll and Hyde is arguably the best known and most influential example of a common trope in nineteenth-century Gothic literature: the doppelgänger story. E. T. A. Hoffmann's "The Devil's Elixirs" (1816) and Thomas Jefferson Hogg's *Confessions of a Justified Sinner* (1824) established the template in the early part of the century. Hoffmann's tale, an adaptation of Matthew Gregory Lewis's 1796 Gothic classic *The Monk*, tells of a Capuchin monk called Medardus who is tormented by his lusts after drinking an elixir given to him by Satan. He encounters his doppelgänger, his half brother Count Viktorin, who is semi-lunatic and haunts Medardus's steps across the world, committing crimes as he goes. The roles become confused: at one point the Count is blamed for a murder committed by Medardus, at another Medardus believes he *is* his double. The tortuous plot is filled with such ambiguities of identity, with madness, sins, and vile acts blamed by one character on the other—all triggered by that diabolical potion.

Hogg's story is scarcely less sprawling. The author, a Scot, seems to have intended it as a critique of Calvinist ideas about predestination: if we believe we are already saved, what is to stop us from sinning without fear or guilt? Hogg's antihero Robert Wringham has a demonic familiar called Gil-Martin, who persuades him to murder sinners. Wringham is driven mad but Gil-Martin takes on his appearance and continues the homicidal spree. Wringham finally hangs himself in despair, although it is never quite clear—you could regard this either as psychologically modern or as an echo of *Macbeth*—whether his tormentor is just a figment of his imagination.

Some of the horror in these tales relies on the uncanniness of seeing another person with your own features. While for identical twins a "double" is a part of life, to glimpse a *stranger* with the same face as you is another matter, and generally regarded with horror.* In "An Unwritten Drama of Lord Byron," Washington Irving describes a story that Byron allegedly intended to write, about a Spaniard called Alfonso whose steps are dogged by a masked stranger. Finally driven to mad jealousy by the belief that this fiend is seducing the woman he loves, Alfonso tracks the man down and stabs him—whereupon "The mask and mantle of the unknown drop off, and Alfonso discovers his own image—the spectre of himself—he dies with horror!" Edgar Allan Poe stole the plot for his story "William Wilson," in which the title character, as a schoolboy, meets another lad with the same name who at first looks rather like him but comes to resemble him exactly. William is repelled by his double, but the boy continues to haunt him in later life. William falls into debauched behavior, before he too stabs his doppelgänger to death over a dispute about a lover. At that moment a mirror appears in which William sees his bloodstained self and seems to hear his victim say "see . . . how utterly thou hast murdered thyself." Poe made no bones about the source, sending a copy to Irving for comments (which were generously favorable).

It's very likely Stevenson had read Poe's work; he certainly knew of another contemporary doppelgänger tale, *Le Chevalier Double* (1840) by Théophile Gautier, in which a knight born under a "double star" duels with his own double. And we saw that the theme of the alter ego is present in *Frankenstein*: Victor himself recognizes his creature as a projection of his dark and violent thoughts, and the creature's crimes as his own. (Mary Shelley was, as we saw, at one time close to Thomas Hogg.)

All of these tales make the double with wicked intentions a distinct

---

* And it is not, in fact, uncommon. Years ago I was told that there was a young man in Salisbury who was my double except for his being taller. When I saw him one day across the street, I understood at once the basis of the suggestion. I didn't dare say hello.

being. Stevenson's innovation was to make the duo coexist in a single individual, albeit one who is altered in appearance. Hyde's evil physiognomy was consistent with the crudely Darwinist belief that criminality is innate and that bad character is written on the face. This theory owed much to the Italian criminologist Cesare Lombroso, and Hyde bears Lombroso's indicators of a debased personality: there is something brutish in his visage that makes others shun him. Evil radiates from his features, and everything about him is coarsened. He inflicts pain— trampling a small girl, bludgeoning to death the elderly Sir Danvers Carew—out of sheer wanton malevolence.

Lombroso's insistence on a "criminal type" was an invitation to see crime in class-based terms. "Crime belongs exclusively to the lower orders," says Lord Henry Wotton superciliously in Oscar Wilde's *The Picture of Dorian Gray*, published four years after *Jekyll and Hyde* (which Wilde admired). But Wilde's vision of a man's double nature astutely refutes this, for it is the dissipated aesthete Gray who falls into wickedness while his misdemeanors are absorbed in the degenerating features of his hidden portrait. The degenerate and murderer, Wilde implies, exists in all of us, lord and laborer alike.

It seems obvious and inevitable to us now that Jekyll and Hyde should be played by the same actor on stage and screen. Yet the book doesn't obviously prepare us for that. Hyde is said to be physically quite different—smaller, for one thing—and those who encounter him seem to spot no resemblance to Jekyll. The dual role began with a stage adaptation by Thomas Sullivan in 1887, on which several subsequent movies were based. Such casting, of course, offered the lead actor a marvelous opportunity to show his versatility*—an appeal later amplified by the cinema star system—but it also helped to emphasize the psychoanalytical reading. The 1920 silent movie version starring John Barrymore, for example, offered this sub-Freudian "moral":

---

* The star of the Sullivan version, Richard Mansfield, doubtless had this in mind when he acquired the rights the year after the novel was published, and commissioned Sullivan to adapt it (see page 155).

In each of us, two natures are at war—the good and the evil. All our
lives the fight goes on between them, and one of them must conquer.
But in our own hands lies the power to choose—what we want most
to be, we are.

The double role was established by the almost twenty-year run of the
Sullivan play, and cemented by Fredric March in Rouben Mamoulian's
1931 movie. (Like *Frankenstein* and *Dracula*, *Jekyll and Hyde* reached
the screen via a theater adaptation, which meant not only that it was
"audience-tested" but that it would fit the rather stagey constraints most
movies had to accommodate.) March was heavily made up as Hyde, but
in 1941 Spencer Tracy repeated Barrymore's feat (at least in the early
part of the story) of achieving the transformation merely by contorting
his face, with scant use of makeup. Only his inner depravity (and some
early stop-motion special effects) turns the matinee idol into a demon.*

The double continues to unsettle us—perhaps more so than ever in
an age when human cloning has started to look like a real possibility.
Indeed, it is really the old aversion to the double that leads us mistakenly
to imagine that cloning would produce a second self, rather than just
another human who might look a lot like us. But why we are so unsettled
by that resemblance?

Any discussion of that question today is likely to be a mere footnote
to Sigmund Freud's seminal 1919 essay "The Uncanny" (*Das Unheim-
liche*), in which he argues that what makes the sense of uncanniness distinct from a feeling of horror is that it comes when we confront not a
monstrous anomaly but "a species of the familiar . . . something that
was long familiar to the psyche and was estranged from it only through
being repressed." In ancient times, he asserts, the double was regarded as
a good thing: "an insurance against the extinction of the self" (which
is as good a summary as any of the narcissistic wish to be cloned). But

---

* Mamoulian's cameraman Karl Struss regretted March's transformation to a hairy beast, feeling
the changes should be purely psychological.

once civilization left behind the infantile phase of "boundless self-love" from which this desire springs, the double becomes instead "the uncanny harbinger of death." And that, for sure, is what the doppelgänger so often represents in nineteenth-century literature. You can't help wondering if Jekyll knew he was doomed from the moment he first stepped into Hyde's skin.

The fecundity of Stevenson's tale hinges on the merging of these twin selves into a "split personality," consumed by mutual loathing and contempt. Jekyll is disgusted not only by Hyde's violence but by his vulgarity, his "apelike tricks"—in which we can all too easily recognize the base, ugly, and destructive acts we sometimes find ourselves perpetrating. When he despairs at how Hyde would scrawl "in my own hand blasphemies on the pages of my books," we can imagine a hundred other degrading actions we would like to blame on some inner demon that we can disown as no part of our true, pure self. And yet Jekyll can't wholly disown or denounce Hyde. His "love of life is wonderful," the doctor admits, and with his "greater boldness" and "solution of the bonds of obligation," Hyde has the freedom that the repressed Jekyll craves. A lifelong struggler against binge-drinking, Stevenson doubtless knew how that felt.

He had explored ideas of conflicted personalities before, most notably in the play he wrote in 1880 with William Ernest Henley called *Deacon Brodie; or, The Double Life*, the story of a real-life Scottish town councilor who committed crimes by night. Stevenson's 1885 short story "Markheim" also depicts a murderer forced to come to terms with his wasted life.

Although Stevenson flatly denied that he had been inspired by medical theories, *Jekyll and Hyde* was released into a milieu made receptive by the burgeoning scientific interest in "double" or "split" personalities—terms that still adhere (erroneously) to perceptions of schizophrenia and other mental disorders today. And despite Stevenson's own claim, his wife Fanny wrote in 1905 that "my husband was deeply impressed by a paper he read in a French scientific journal on sub-consciousness." If her account is to be believed (opinions vary on that), there are several candidates for this source of inspiration. In the 1870s the French physician Eugène Azam published a series of reports describing the "double personality" of a woman identified as Félida X, who could be transformed into a more confident, outgoing individual by hypnosis. Another French doctor, Ernest Mesnet, described how a soldier had developed several distinct personalities after being wounded in combat. Stevenson could certainly have come across these cases also via two articles on "dual consciousness" written by the science popularizer Richard Proctor and published in the 1870s in the *Cornhill Magazine*, to which Stevenson occasionally contributed. Proctor presented the soldier as an example of people "who live two lives, in the one of which they may be guilty of the most criminal acts, while in the other they are eminently virtuous and respectable."

One of the best-known cases of "multiple personality" at the time was that of the Frenchman Louis Vivet, a patient who was confined several times to asylums. His condition was described by a doctor named Camuset in a paper of 1882, and in another account three years later. Vivet was said to have been a quiet and obedient teenage street urchin, then to have undergone "morbid disintegration" that rendered him "violent, greedy, and quarrelsome," and to have exhibited as many as ten separate personalities. Several other cases were described by the eminent psychologist Frederick W. H. Myers, a friend of Stevenson's, in an 1886 paper published shortly after *Jekyll and Hyde*.

These supposed cases of dual, often opposed, personalities were typically ascribed to the division of the brain into two hemispheres. The notion of "hemispherical dominance" was proposed in the 1860s by the French anatomist Paul Broca; in line with long-standing prejudice, dominance of the left-hand functions of the body was deemed to be linked to moral weakness and insanity. This, then, was an anatomical rather than psychological interpretation of the behaviors observed in patients, reflecting the view of most medical professionals at the time that we do the things we do because we are built the way we are. The wicked alter ego was considered a pathological and potentially dangerous state, but one dictated by physiology. By situating Hyde in Jekyll's psyche rather than his brain, Stevenson took his tale in a different direction—with connotations that were far more unsettling, and perfectly attuned to Freud's emerging theory of the unconscious.

We can't attribute to Stevenson any real prescience of the Freudian view of human behavior—by his own account, the tale was simply an exploration of humans' internal struggle between good and evil impulses. Yet Jekyll's confession about hiding his baser desires to conform to society's expectation of virtue sounds in retrospect like a textbook case of Freudian repression. Stevenson made explicit the nominative coding in naming the bestial alter ego Hyde.

In the year that Stevenson wrote *Jekyll and Hyde*, Freud was only just setting up his clinical practice in Vienna and had not yet formulated his theory of psychic conflict between the impulsive id and repressive, civilizing superego. This is what makes the book so remarkable—for what better illustration could there be of the dawning recognition of unconscious, primitive drives within us all, held in check only by the restraints of conscience and convention? We're not led to suspect that there is anything *medically* "wrong" with Jekyll—it's not as if he has some anatomical disorder of the brain. He could be any of us.

Hyde is then no longer a malicious, spectral double in the tradition of Hogg and Gautier. He bursts out of honest, decent Jekyll himself, sabotaging his schemes, ruining his reputation, and leaving him feeling ashamed, ugly, and soiled. This is the price we pay for the repression and sublimation needed to maintain the fabric of society, as Freud would suggest in *Civilization and Its Discontents*.

In other words, the strange tale of Jekyll and Hyde was *the myth that the age of Freud needed*. It's well known that Freud's ideas were initially perceived as scandalous, seeming as they did to deny that deviant, dirty deeds were the sole preserve of Lombroso's criminal degenerates but instead to attribute them (potentially) to us all. Freud denied us denial; beneath our civilized veneer, he seemed to be saying, we seethe with forbidden desires and inclinations. None of us is truly safe from them, or the ruination they threaten. In *Jekyll and Hyde* the point is no longer that good may triumph (as it does in "Markheim"), nor even that we will be destroyed by our inner monster. Jekyll's final death hardly matters—in fact, we never see his dead body at all, for only Hyde lies poisoned on the floor of the lab. What matters is the awful exposure and shame that went before.

The agent of transformation—Jekyll's fateful potion—is really just a diversion; in the end, it is not needed anyway for Hyde to seize control. In this regard, *Jekyll and Hyde* evinces the attribute that the literary critic Tzvetan Todorov applies to all of fantastical literature: it creates ambiguity about whether what we are witnessing stems from supernatural (or pseudoscientific) causes or from psychological ones. As in *Frankenstein*, the mythos arises because the story denies us easy answers. We can't simply refuse to take the potion.

Fredric March's Hyde, in Mamoulian's 1931 movie, is indeed a fantastical creature, a fairy-tale demon with peltlike hair and protruding teeth that place him somewhere between simian and lupine. The story's exploration of post-Darwinian anxiety about regression to the alleged bestial origins of humankind could hardly be more clear.

It was in this context that Stevenson understood dual-personality behavior. Like many biologists of the time, he believed that Darwinian theory allowed the possibility that evolution could move backward—toward degeneration—as well as forward, by adaptation, to more "perfect" states of being. This idea was endorsed by the British zoologist Edwin Ray Lankester, by the Darwinist Herbert Spencer, and by the psychiatrist Henry Maudsley, who regarded degeneration as a physiological state of the nerves that could be inherited. Some felt that degeneration was responsible for traits such as homosexuality and "artistic decadence," and also for the emergence of the feminist New Woman. The degeneration theory supported the idea of a hierarchy of races—an almost inviolable scientific dogma of the time. With his engineer's technical training and his interest in science, Stevenson not only followed these debates but contributed to them. The evolutionary theme is mined in one of the more elliptical modern retellings of *Jekyll and Hyde*, Ken Russell's

Fredric March's Mr. Hyde in Dr. Jekyll and Mr. Hyde *(1931)*.

characteristically garish *Altered States* (1980), where drugs and sensory deprivation cause William Hurt to regress to a feral, apelike form.

The specter of degeneration—physical, moral, and societal— haunted European thought at the fin de siècle. Its influence is clear to see in mythopoeic tales such as *Dracula* and H. G. Wells's *The Time Machine* and *The Island of Doctor Moreau*. Wells crystallizes the two poles of the evolutionary spectrum—the civilized and the savage—into separate species in *The Time Machine*, in the form of the elegant, effete Eloi and the troglodytic Morlocks. Class anxiety played a role too in this fear of the evolutionary throwback. Edward Hyde is almost a caricature of a working-class thug, dwelling among grimy Soho tenements. Wells even compared his Morlocks explicitly with the "East-end worker" who lives "in such artificial conditions as practically to be cut off from the natural surface of the earth."

The surgically realized transformation of beasts into people in Well's *Doctor Moreau* makes its narrator Prendick recognize that an ancestral bestiality lurks within us: "Even it seemed that I, too, was not a reasonable creature, but only an animal tormented with some strange disorder in its brain." That there could be two-way travel between human and beast was hinted at in Welsh writer and mystic Arthur Machen's horror fantasy *The Great God Pan* (1894), in which the sinister femme fatale Helen changes between the two before she dies.

It's easy now to forget what demons Darwin's "dangerous idea" unleashed even among scientifically inclined souls. A debased Darwinian view of human variation, in which the progression from the bestial was considered to be a gradual affair across the human species, curdled in the early twentieth century into an enthusiasm for eugenic social engineering among the progressive intelligentsia—and after that, among circles that were anything but progressive. In fascist Germany, portraying some groups of people as scarcely more than beasts was the prelude to their extermination.

***

What does this beast inside us want? Well, what do you think?

In the Mamoulian movie, March's leering Hyde pursues the louche Ivy Pearson (Miriam Hopkins), a bar singer whose loose morals make her a more suitable target than Jekyll's virginal fiancée Muriel Carew. Ivy has little time for these attentions, calling Hyde "disgusting" and scolding him that "your conduct was positively indecent." So his pursuit ends how we know it must: with Ivy's murder. The deed is done in her bedroom, and as Hyde sinks out of shot with his hands around her throat, the camera lingers on a statue of winged Eros looming over a naked woman: there is no doubt what is really going on here. Some modern adaptations (such as the eminently forgettable 2003 version, starring John Hannah) have made Hyde's predilection for rape clumsily explicit.

The device of contrasting Jekyll's "good woman" with the "fallen woman" Hyde torments began with Thomas Sullivan's early stage adaptation, and features too in the 1920 John Barrymore film based on it. While Sullivan's adaptation was simply observing the conventions of the time in giving Jekyll a romantic interest, the change highlighted the sexual undercurrents of the original story. Now Hyde is continually subverting Jekyll's plans for a happy match with the daughter of Danvers Carew—actions we can read as ambivalent self-sabotage. Carew's disapproval of the liaison earns him the punishment that Jekyll—generally portrayed in these versions as an effete wimp—could only imagine but never dare to deliver: Hyde now has a clear motive for strangling the obstructive old man.

Stevenson impatiently dismissed suggestions that Hyde's depravity was sexual. "The harm was in Jekyll, because he was a hypocrite," he insisted, "not because he was fond of women." But, Stevenson lamented,

> people are so filled full of folly and inverted lust, that they can think of nothing but sexuality. The hypocrite let out the beast Hyde—who is no more sensual than another, but who is the essence of cruelty and malice, and selfishness and cowardice: and these are the diabolical in man—not this wish to have a woman, that they make such a cry about.

To which one can only reply that Stevenson protests too much. In the classic manner of mythmakers, he had given his unconscious too much liberty to have any hope of curating what it bodied forth. For what is his Hyde if not the adolescent Jekyll, hairy and horny and ready to go?

Sex is made all the more conspicuous an element in Stevenson's tale by the rigor with which it is excluded. There are no female roles of any note, merely a few stock characters like the witchy landlady and the terrified maid. None of the male characters, meanwhile, seems to have any romantic attachment at all—it is a world of clubbable men.

Yet it doesn't take much reading between the lines to find sex encoded in the text. Was this suppression, or repression?

The poet Gerard Manley Hopkins suggested that Hyde's "trampling" of a young girl ("leaving her screaming on the ground") was an allusion to a sexual violation, and the substantial compensation demanded for the girl's family does seem rather odd were it just in recompense for a few cuts and bruises. And what can Stevenson be driving at when he writes of Hyde "drinking pleasure with bestial avidity from any degree of torture to another"? To what is Jekyll referring when he declares that "into the details of the infamy at which I thus connived . . . I have no design of entering"? After all, there are plenty of "evil" deeds he might have been confident in describing without raising the eyebrows of Victorian moral guardians: alcohol abuse, thievery, even physical assault, would hardly be so taboo.

What makes the tale most suggestive of sexual repression is Jekyll's account of the side of his nature from which Hyde sprang. To the modern reader it seems decidedly odd that what begins Jekyll's double life is nothing but a "certain gaiety of disposition," as though he is mortified at exhibiting a bit of happiness and high spirits. But "gay" at the time (wait, I'm coming to that too) could refer to a sexually licentious nature: the "Gaiety Girl" of the 1890s was the sort of young woman who enjoyed the risqué shows for which London's Gaiety Theatre was notorious.

When he transforms, Jekyll might be considered to take on characteristics of his dissolute youth: he is smaller, and attests that "I felt younger, lighter, happier in body." Even the distaste he provokes in others might be seen (unkindly, it's true) as the typical response to a smelly, clumsy, hormone-soaked teenager. He is selfish and childishly destructive, taunting Jekyll and committing the classic oedipal act of obliterating a portrait of his father. Hyde is not just Jekyll's id but his impetuous, embarrassing youthful self.

Jekyll admits that in the "gay disposition" of his youth he indulged pleasures that he "hid with an almost morbid sense of shame." In an early draft of the manuscript these youthful vices are said to be "abhorrent in themselves," and to have made Jekyll "the slave of disgraceful pleasures." As Hyde he experiences "disordered sensual images running like a mill race in my fancy." And what denial these "incredibly sweet" sensations then elicit! "It was Hyde, after all, and Hyde alone, that was guilty," Jekyll insists to himself. "Jekyll was no worse; he woke again to his good qualities seemingly unimpaired."

There seems little doubt, then, that sexual deviance lurks in Hyde's shadow. So why did Stevenson not admit it? There was the simple matter of public taste, of course, which mattered to a writer keen to reach a wide audience. But it is not as though, in the age of Madame Bovary, sex per se was off limits. Maybe the problem is the very fact that Hyde is *not* simply a Lombrosian degenerate but inhabits the same mortal coil as a gentleman very much like those in Stevenson's target audience. Had he been too explicit, they might have felt obliged to find the tale odious. Leave the matter open and readers can reassure themselves that Hyde's vices are alien to *them*.

That, of course, was precisely the kind of hypocrisy Stevenson was talking about. Hollywood's fixation on drugs and potions in all their steaming glory as the agent of transformation becomes literally a smoke-screen to obscure the true source of Hyde's rampages. The mixture of prurience and prudishness in the legacy of Jekyll and Hyde shows that we know very well what the myth is about, but have perhaps not learned very much about the cost of duplicity from the example of the unfortunate doctor.

Even in the unlikely event that Stevenson had banished sexual depravities from his mind, the image of Hyde slouching around London's dingy backstreets committing horrible and violent acts would, soon after the book's publication, have conjured up for his readers a tale in which sexual perversion seemed central: the killings committed by Jack the Ripper in Whitechapel in 1888.

That Hyde concealed his crimes behind the mask of a well-bred and dignified member of society anticipated public suspicions about the Ripper: at one point even Queen Victoria's grandson Prince Albert Victor was rumored to be a suspect.* The parallels with the Ripper murders were uncanny—so much so that Richard Mansfield, the well-known actor who took the dual roles of Jekyll and Hyde in Sullivan's stage adapta-

---

* The prince's name was also rumored to have come up in the Cleveland Street scandal of 1889, when a police roundup of male prostitutes and pimps in London elicited the identity of some of their aristocratic clients. There is no evidence that Arthur Victor was homosexual, but an aura of sexual deviancy clung to him.

tion in the late 1880s, was suspected (purely on those grounds) of being the Whitechapel murderer.*

The Ripper was horribly real, but in the public consciousness he merged with an older Hyde-like figure of urban folklore: Spring-Heeled Jack, said to stalk London with fiery red eyes and a devilish visage searching for both men and women to assault. In the 1870s he was said to have reappeared in Peckham in south London, and his exploits were serialized in a penny dreadful of the 1880s.

By and by, Hyde has morphed into a lunatic stalker motivated wholly by warped sexual urges. *Edge of Sanity*, a 1989 adaptation of *Jekyll and Hyde*, starred Anthony Perkins, who had played the most famous "homicidal sexual pervert" in cinema history: *Psycho*'s Norman Bates. Perkins's Jekyll acquires sadomasochistic longings after seeing his father have extramarital sex in a barn and being whipped for it. When in later life he becomes a doctor and swallows a cocktail of cocaine and ether, he turns into a monstrous Ripper-style rapist/murderer. As with many *Jekyll and Hyde* vehicles, the way women are portrayed in this movie as mere victims of rape, misogyny, and ridicule makes one wonder whether they were not done a favor by simply being left out of Stevenson's tale. Or, to put it another way: *Jekyll and Hyde* offers a disturbingly popular vehicle for brutal male fantasies of sex and power.

There is nothing unusual about a tale of British late-Victorian or Edwardian chaps consorting in manly camaraderie with not a woman in sight except to bring in the tea. But the point about modern myths is that they not just permit but positively invite new readings beyond their author's horizon. It's natural, then, that *Jekyll and Hyde* has become ripe for examination by queer theorists.

If Stevenson rejected interpretations that invoke heterosexuality, we might imagine what his response would have been to gay readings. And yet . . . what could seem more relevant, in the era of Oscar Wilde, than

* The suspicion was apparently based on a single letter to the police from an audience member at one of Mansfield's performances. If he could display such murderous fury on stage, the letter's author wondered, what might he be capable of behind the scenes?

a story of a man concealing shameful, unspeakable urges while clinging, with ever loosening grip, to a life of bourgeois respectability?

But gay interpretations of *Jekyll and Hyde* are far from fanciful acts of revisionism. They revolve in particular around the central, rather perplexing incident in the novella: the murder of Danvers Carew. Here is the old man, strikingly described as "beautiful," who "accosts" Hyde "with a very pretty manner of politeness"—and says something that makes the villain explode into savagery. It is never explained what that remark was. Perhaps he was only asking the way, said the maid who witnessed the crime—but would an old man so well known in the city really have occasion to do that? Might he just possibly have been soliciting? As Leonard Wolf says, "For us, more than a hundred years after the book was written, the ambiguous diction describing the scene stirs our suspicion that we are witnessing a scene that begins as gay-bashing and ends in death." And it seems such suspicions were evident well before the modern day: in the 1920 film, "Sir George Carew" becomes a corrupter of morals reminiscent of Oscar Wilde's Henry Wotton, who introduces Jekyll to showgirls at the music hall.

True, we might be inclined now to read too much into Utterson's remark that "it turns me cold to think of this creature stealing like a thief to Harry's bedside; poor Harry, what an awakening!" That Hyde enters Jekyll's house via the "sordid negligence" of the back entrance might also be at risk of anachronistic overinterpretation. But these things add up. You have to wonder, for example, what motivated Utterson's immediate suspicion that Jekyll, apparently so decorously unattached, was being blackmailed by Hyde to pay off the family of the trampled girl. Their suspicion of blackmail would have immediately suggested to Stevenson's readers one of the most common causes of it: homosexual liaison. The trial of Oscar Wilde had yet to demonstrate what might befall a man found guilty of gay sexual relations, but the so-called Labouchere Amendment (the Criminal Law Amendment Act), passed in 1885, made sodomy an act of "gross indecency" punishable as a crime.

And what does Enfield make of this business? Yes, it might look like blackmail, he says, but "the more it looks like Queer Street, the less I ask." We shouldn't jump to conclusions, for "queer street" was commonly used in the nineteenth century to refer to financial difficulties; it could be a corruption of Carey Street, which was where London's bankruptcy courts were located. On the other hand, the use of "queer" to mean homosexual was already current by that time.

Then there is Hyde's left-handedness. Of course this could just connote that Hyde is Jekyll's reflection in a mirror darkly: a projection of his "sinister" self. We have seen already that left-side dominance was associated with degeneration—with the primitive, emotional, and feminine, in contrast to rational, masculine right-side dominance. But the inversion of the norm that produced left-handedness was considered also to engender a disruption of normative sexuality. "Effeminate men and masculine women," wrote the German psychoanalyst Wilhelm Fliess, a friend of Freud's, "are entirely or partly left-handed." For reasons such as these, literary critic Elaine Showalter believes that *Jekyll and Hyde* "can most persuasively be read as a fable of fin de siècle homosexual panic, the discovery and resistance of the homosexual self."

Stevenson's own sexuality has been a topic of some debate. His friend, the Scottish critic and folklorist Andrew Lang, wrote that he "possessed more than any man I ever met, the power of making other men fall in love with him." His circle of friends included Edmund Gosse and John Addington Symonds, both of them poets and critics—and living the married family life of many Victorian men who, like them, were gay. The Labouchere Amendment made it more necessary than ever for gay men to lead this double life. Symonds seemed to see in *Jekyll and Hyde* a terrible depiction of his own situation, and he complained to Stevenson about its "shutting out of hope." The author should, he said, have shown "more sympathy with our frailty"—to have given less brutal a shape to poor Jekyll's repressed impulses. "There is something not quite human in your genius," he told Stevenson.

Perhaps your judgment depends on how you regard that bubbling potion: is it an elixir of liberation or a draught of doom? The wild joy that even Jekyll recognizes in Hyde is that of the person who no longer has to dissemble and pretend. But the frustration and self-loathing is only doubled, then, when the mask must fall back into place. Any story about a personal transformation that releases one's hidden, shameful urges is bound to be fertile ground for debates about sexuality and queerness.

I hesitate to make too much hay from a movie described by one critic as "so crass, witless and misogynistic that it makes *Confessions of a Window Cleaner* look like Dostoevsky"—but such is the business of examining modern myths: often, the cruder the outpouring, the more revealing it is of the anxieties behind it. And so, while I don't wish to make recommended viewing of *Dr. Jekyll and Ms. Hyde* (1995), it has something to say about the fears associated with dissolving gender boundaries. Here a modern-day descendent of Jekyll, called Jacks, inherits his secret recipe but adds to it the hormone estrogen and becomes—you guessed it—Helen Hyde, who needless to say is not (on the outside) monstrous in the slightest but, rather, voluptuous.* Like Edward Hyde, she decides that she rather likes her new form and wants to stay that way.† There's a moment of trans liberation (that is, a confused 1990s version of it) at the film's climax when Jacks, wearing a dress at a business party, declares that he wanted to understand women by becoming one. Before this epiphany, we get the juvenile "comedy" of seeing Jacks/Helen randomly sprouting body parts of one sex or another (you can guess which they are) after imbibing a partial dose of the potion: a heavy-handed hint that some studio exec's bright idea of how to turn *Jekyll and Hyde* into a modern screwball gender-war comedy stirred up all kinds of fears about emasculation and feminization.

---

* This gender-crossing transformation was anticipated by Hammer's 1971 *Dr. Jekyll and Sister Hyde*, one of the earliest postmodern mash-ups, which throws into the mix Jack the Ripper, Frankenstein, and William Burke and William Hare (convicted of murdering sixteen people and selling their corpses to an anatomist for dissection in Edinburgh in the late 1820s). As a woman, Sister Hyde is prey to what the 1970s considered to be her instinctive impulses: the first thing she does is to head off on a shopping spree.

† We might ask, of course, why Hyde himself ever voluntarily takes the antidote to restore him to Jekyll. Perhaps some vestige of the doctor remains active within his bestial mind after all?

In his snarling savagery, Edward Hyde represents yet another familiar figure of supernatural horror: the werewolf. You can see from the hirsute Fredric March that the connection was swiftly made. In the B-movie *The Daughter of Dr. Jekyll* (1957), his daughter Janet is told that her father was a werewolf. ("Is there a ravaging beast hidden in her body?" asks the publicity poster breathlessly. Well no, actually—Janet is being framed by her ex-guardian, who is a werewolf himself.) And when *Abbott and Costello Meet Dr. Jekyll and Mr. Hyde* in 1953, they find that the doctor's worse half (played by Boris Karloff) is decidedly lupine. The Spanish movie *Dr. Jekyll and the Werewolf* (1971) just gets straight to the point.

This fabulous beast of folklore was already prowling the dark alleys of Victorian fiction. Most notably, George W. M. Reynolds, who began as a hack churning out penny dreadfuls and became as popular in his time as Dickens, produced *Wagner, the Wehr-Wolf* (1857)—hardly a distinguished work, but one of the first to explore the creature in English literature.* The potential for the lycanthrope to represent alter egos was explored in a startlingly feminist manner by Clemence Housman, sister of the poet Alfred Edward. Her novel *The Were-Wolf* (1896) features a female werewolf, White Fell, who switches between being a loving, beautiful young woman and a murderous beast. Housman, a campaigner for women's suffrage, used the structure to comment on the pressure felt by the increasingly liberated and independent New Woman to conform to social expectations. It's an extremely rare example of the werewolf being anything other than a fearsomely masculine marauder.

But what is a werewolf, after all? Is it a man who becomes a quadripedal wolf, indistinguishable from the wild animal? Or the bipedal, immensely strong, hairy man-beast currently favored by the movies? Or simply a man who, like Hermann Hesse's Harry Haller in *Steppenwolf* (1927), believes that a part of his nature is lupine? Transformations into animals are stock fare in classical myths, not least for Ovid, who

---

* Bram Stoker's initial draft of *Dracula* included a first chapter, discarded before publication, that describes Jonathan Harker's encounter with a Transylvanian werewolf.

writes of wolves that "counterfeit a human shape" and of a kind of man who grows into a wolf while "his jaws retain the grin and violence of his face." Pliny, as ever, believed all he heard, such as the tale of a man who transformed into a wolf after eating a child's entrails. The werewolf was a standard figure of Western folklore, akin to the quasi-feral vampire (another transfigured animal) that hunted by night. But there it was a quasi-natural creature: not an evil demon but a savage beast hardly different from a real wolf. The sixteenth-century German farmer Peter Stumpp, convicted as the Werewolf of Bedburg and brutally executed, was regarded as a bloodsucker who used black magic to transform himself into a child-devouring animal.

One of the modern innovations that make the werewolf a cousin of Hyde is its sexualization. The wolf becomes a sexual threat to young girls in the versions of Little Red Riding Hood written by Charles Perrault in the late seventeenth century and by the brothers Grimm in the early nineteenth. One scarcely needs the Freudian interpretation of Bruno Bettelheim to see from Perrault's own explanation of the moral where his tale was leading:

> From this story one learns that children, especially young lasses, pretty, courteous and well-bred, do very wrong to listen to strangers, And it is not an unheard thing if the Wolf is thereby provided with his dinner. I say Wolf, for all wolves are not of the same sort; there is one kind with an amenable disposition—neither noisy, nor hateful, nor angry, but tame, obliging and gentle, following the young maids in the streets, even into their homes. Alas! Who does not know that these gentle wolves are of all such creatures the most dangerous!

Perrault is clearly thinking of the smooth-talking libertine who will get his way with young girls by seduction or rape. In Angela Carter's retelling of the story, "The Company of Wolves" (1979), the wolf begins as a charming young man: "she'd never seen such a fine fellow before."

There's nothing fine or charming about Hyde, of course, and the modern werewolf tales echo Stevenson's in having no cautionary function for young ladies. Rather, they present an exculpatory snarl from the man: Look out ladies, I can't control my urges! The Wolfmen like Lon

Chaney who threatened screaming starlets were instantly understood as Hyde-like molesters hiding behind a supernatural presence, taking by force what Count Dracula took with charm. Suave Christopher Lee could never be Hyde or the Wolfman,* but Oliver Reed (*The Curse of the Werewolf*, 1961) was made for the role.

It's the dishonesty, more than anything else, that makes the werewolf's sexual assaults so ugly: the pretense, say, that poor Tony in *I Was a Teenage Werewolf* (1957) (Michael Landon, launching his career a long way from *Little House on the Prairie*) is transformed into a beast by the din of the school bell rather than, as we understand perfectly clearly, by lust for the young woman who is working out so voluptuously in the gym. What follows is presented as murder by strangulation by a frenzied creature; we all know that in reality, like March's Hyde assaulting Ivy, it is in fact violent rape by a hairy-handed youth. That's confirmed by Tony's horrified shame at his crime: "I know what I am ... what I've become. ... Don't let anyone see me like that ... !" By the time *Werewolf in a Girl's Dormitory* (1961) arrives, you want to say yes, we get the point.

The werewolf has, however, the perfect alibi for his transgressions: the moon made me do it.† So too does the audience excuse its voyeurism: it's just a horror film about monsters! Just as the bubbling potion and scenery-chewing were necessary components of the Jekyll-to-Hyde transformation, so it has become essential to show the werewolf transition, as though to reassure us that the man has been erased by the beast and, as such, can be absolved. After the tour de force of latex face-stretching in John Landis's *An American Werewolf in London* (1981), it was *de rigueur* to judge a werewolf movie by its transformation sequence.

---

* Lee appears in Hammer's *The Two Faces of Dr. Jekyll* (1960) but not in the title role. Here Jekyll's doppelgänger is not a monstrous Hyde but a handsome young man—bringing the plot closer to that of *The Picture of Dorian Gray*.

† In fact the agent of transformation in *I Was a Teenage Werewolf* is a drug given to Tony by an abusive therapist he is seeing for his antisocial behavior, who regards the boy as fodder for his experiments. This doctor believes, like some pseudo-Darwinian Rousseau, that humankind needs to return to its primitive state—to "unleash the savage instincts that lie hidden within."

For all its comic swagger, *American Werewolf* can be read as a parable of male sexual anxiety. The werewolf David, a mild-mannered soul, flees from his girlfriend Alex in fear of what he might do to her, and it is his failure to control his savage impulses when she confronts him in lupine form and declares her love that brings about his death. Thus did Landis manage to reassert a kind of honesty after all.

The other fairy-tale cousin of the Jekyll and Hyde myth is of course "Beauty and the Beast." It's the benign version: the Beast is fundamentally good, although he can look large, hairy, and scary to a girl in the process of negotiating her detachment from an oedipal father-daughter relationship. Be patient, the story says, and all will be well. We can only hope that daughters can distinguish the Beasts from the Hydes.

Stan Laurel as you've never seen him before: as "Mr. Pride" in a 1925 spoof of Stevenson's mythic novel.

Jekyll and Hyde has proved one of the most versatile of modern myths. It is the ancestor of the slasher genre, in which the assailant is now commonly a sexual deviant in the mode of a Norman Bates or Peeping Tom who makes a Ripper-like mess of his generally female victims. The appeal is complicated and unsettling: voyeuristic, certainly, and often darkening into the misogynistic—all gory fangs, talons, and knives, a penetrative rape fantasy served up as popcorn entertainment. But at least psychotic monsters like Freddy Krueger don't possess a respectable double to offer male audiences an alibi for their arousal.

At the same time, you can hardly watch Spencer Tracy's contortions or Fredric March's hairy jowls without appreciating the scope for slapstick. Stan Laurel starred in the 1925 silent spoof *Dr. Pyckle and Mr. Pryde*, in which the devilish Pryde's "crimes" include stealing a boy's ice cream and alarming a woman on the street by popping a paper bag behind her back. Besides Abbott and Costello's encounter with the doctor, there were cartoon capers by Tom and Jerry ("Dr. Jekyll and Mr. Mouse" (1947), where the agent of transformation is milk mixed with mothballs and insecticide) and Jerry Lewis's *The Nutty Professor* (1963), in which the nerdy Professor Kelp transforms himself, through the wonders of chemistry, into the suave Buddy Love, a ladykiller only in the romantic sense. Elsewhere Jekyll and Hyde go into space and get groovy in blaxploitation horror. There was little prospect, really, that we would be spared *Dr. Sexual and Mr. Hyde* (1971), in which the doctor performs sexual experiments on the female inmates of an asylum.

There was a lot kept hidden in that box that Hyde, a good decade before Freud, opened up: all those fears and longings waiting to escape and make us do things we wish we could disown. "I don't know what came over me," says our Jekyll. "I wasn't myself." But oh, if only I could be!

Chapter 5

THE BODY AND THE BLOOD

DRACULA (1897)

An exotic being or group of beings, typically outsiders able to pass unseen in society, achieves immortality by drinking the blood of others, either killing them or turning them into similar beings. Those who are transformed in this way become undead souls and can only be put to rest through their ritualized, gory murder.

Mary Shelley's *Frankenstein* was not the only hideous progeny of that dark and stormy night at the Villa Diodati on Lake Geneva. Byron's physician John William Polidori also produced a short chiller, which was published in a periodical in 1819 under the title *The Vampyre*, advertised (to Polidori's fury) as "A Tale by Lord Byron." The story tells of a mysterious and rather obviously Byronic nobleman named Lord Ruthven, who seduces the sister of the story's hero, a young gentleman by the name of Aubrey, and persuades her to marry him. On her wedding night she is found dead, drained of her blood by Ruthven.

Our monsters, like our gods, are what we make them; they are what we need them to be, and never is this more true than for the vampire. By turning an old legend of uncouth, feral bloodsuckers into a sexualized allegory about predatory aristocrats, Polidori created the template for a flood of nineteenth-century vampire tales that explored the social, sexual, and racial anxieties of the era.

Penned and published in the shadow of *Frankenstein*, *The Vampyre* has often been dismissed as a trifle. But it had immense influence in its day, even inspiring Leo Tolstoy to dabble in vampire stories. (His effort *The Vampire*, published pseudonymously in 1841, was poorly received, and he didn't bother publishing any of the others.*) Yet it is Bram Stoker's version—coming only at the tail end of the Victorian vampire craze—that is now regarded as the progenitor of one of the most popular modern myths of all. Why is that?

The question is all the more puzzling given that, while *Dracula* was fairly well reviewed, it didn't immediately become a hit. It is partly to the interest from the nascent movie industry that the novel owes its fame. Stoker's widow Florence fought bitterly against F. W. Murnau's unauthorized 1922 adaptation, *Nosferatu*, but the popularity of the book on stage and screen soon captured the attention of Hollywood. In 1931 Universal Studios cast Bela Lugosi, a Hungarian with matinee-idol looks who had already played the vampiric Count on Broadway, in the title role, and the rest is. . . . Well, it suffices to say that Dracula has now made more than two hundred cinematic appearances, and there's no sign of the pace slowing.

The critical literature on Dracula and vampires is vast, but still it can't quite pin down what made Stoker's novel so . . . deathless. *Dracula* is, even by the standards of most mythopoeic source texts, a mess. Its prose is overheated, its characters are wooden, and their actions defy credibil-

---

* "The Family of the Vourdalak," written by Tolstoy's cousin Alexei Tolstoy in the 1830s–1840s and published in 1884, was rather more successful. *Vourdalak* was a Slavic word for vampire. The story was the basis of Mario Bava's 1963 horror film *Black Sabbath*, starring Boris Karloff, from which the legendary English heavy-metal band took its name.

ity. Stoker's other works expose him as something of a hack, his tastes frequently chauvinistic and occasionally bizarre.

None of that, however, inhibits the potential of a story to birth a myth. In fact it often helps. If we analyze *Dracula* as literature, it will always disappoint. If we think of it as myth, everything makes sense.

Vampires are not what they used to be—and thank God for that. I can understand the wish to be bitten by Christopher Lee, Frank Langella, even hammy old Lugosi. But the vampires of folklore were repulsive, and you wouldn't want them in your bedroom.

They haunted eastern Europe, from Prussia and Silesia to Hungary and Serbia. Vampire epidemics—outbreaks of people returning from the dead as bestial blood-leeches—were recorded in those regions throughout the seventeenth and eighteenth centuries. The French botanist Joseph Pitton de Tournefort described a journey through Greece in 1701 in which he heard about an undead being that the locals called Vroucolacas, although he dismissed it as mere superstition. But rumors of an episode of vampirism in Austrian Serbia in the early 1730s led Charles VI, the Holy Roman Emperor and Archduke of Austria, to commission a report on the events. It related how a soldier who died of a broken neck after falling from a cart rose from his grave to prey on others, as well as attacking cattle. His body was disinterred and found to be undecayed, and was dispatched in what would later become the traditional manner: a stake through the heart (followed by incineration). But others had become infected with the vampire plague. Several more bodies were dug up and discovered also to be uncorrupted; they were decapitated and burned.

The Austrian report dismissed these beliefs as credulous delusion, but the details found their way, often luridly embellished, into popular accounts. The most influential was *Treatise on the Vampires of Hungary and Surrounding Regions* (1746) by the French Benedictine abbot Dom Augustin Calmet. The vampires in Calmet's accounts are ambiguous beings, more akin to ghosts and ghouls than flesh and blood; sometimes they take the shape of a dog, he said, and sometimes of a man.

When Calmet's book was published in Paris, Enlightenment philosophers including Voltaire sought to rationalize these ridiculous superstitions. Maybe the people weren't actually dead when buried? Or perhaps certain types of soil preserved dead bodies? Maybe the frenzied attacks of so-called vampires were actually due to rabies? Calmet himself made no strong claims of veracity: these were stories told by peasants, he said, without any solid proof, and he was surprised that local authorities had not put a stop to the grotesque business of mutilating disinterred bodies. He left it to better minds than his to resolve the mysterious matter, but wondered, "Supposing there be any reality in the fact of these apparitions of vampires, shall they be attributed to God, to angels, to the spirits of these ghosts, or to the devil?"

After another vampire epidemic in Silesia in the 1750s, an official report expressed outrage at the violation of sacred burials because of the idiocy of "men of low birth." Beheading and staking of corpses was horrible enough, but a French account of an outbreak in Moldavia in 1770 tells of yet more stomach-churning measures: it reports that corpses (apparently of people who had died of the plague) had their teeth pulled out and blood sucked from their gums.

Even then the potential of vampirism to serve as a social metaphor was appreciated. Jean-Jacques Rousseau, who was convinced that vampire tales were mere fables, pointed out how belief in the miraculous and supernatural mirrors the way the church asserts power over the minds of men. Voltaire accused not only the church but also businessmen and officials of being vampire-like in their exploitation of the common people:

> There was no talk of vampires in London, or even Paris. I admit that
> in these two cities there were speculators, tax officials and businessmen
> who sucked the blood of the people in broad daylight, but they were
> not dead (although they were corrupted enough). These true blood-
> suckers did not live in cemeteries, they preferred beautiful places. . . .
> Kings are not, properly speaking, vampires. The true vampires are the
> churchmen who eat at the expense of both the king *and* the people.

The metaphor gained in appeal as the age of capitalism gathered pace. "Capital is dead labour," wrote Karl Karx in the mid-nineteenth century, "which, vampire-like, lives only by sucking living labour, and lives the more, the more labour it sucks."

—

"For nearly a century," writes Gothic scholar Nick Groom, "the vampire was a medical, sociological and political reality—or disputed reality— of Eastern Europe, as well as a theological conundrum. Then it came to Britain and was transformed." It was entrained in the currents of Gothic sensibility, it inspired the Romantic imagination, and it was never the same again.

When it was first published in the *New Monthly Magazine* in April 1819, Polidori's *The Vampyre* was prefaced with an anonymous "Extract of a letter from Geneva," which relates the story's genesis at the Villa Diodati in terms that make clear the reader is to understand it as Lord Byron's contribution to those proceedings:

It appears that one evening Lord B., Mr. P. B. Shelly, the two ladies and the gentleman before alluded to, after having perused a German work, which was entitled Phantasmagoriana, began relating ghost stories. . . . It was afterwards proposed, in the course of conversation, that each of the company present should write a tale depending upon some supernatural agency, which was undertaken by Lord B., the physician, and Miss M. W. Godwin. My friend, the lady above referred to, had in her possession the outline of each of these stories; I obtained them as a great favour, and herewith forward them to you, as I was assured you would feel as much curiosity as myself, to peruse the debauches of so great a genius, and those immediately under his influence.

The letter had accompanied the manuscript when it was sent to the publisher Henry Colburn, who reasonably concluded that the story was Byron's—which, given the poet's reputation, he was only too happy to publish.

When Polidori saw the issue, he was furious. He wrote at once to Colburn insisting that the work was his. It was, he admitted, based on a "Fragment" of a tale Byron had begun to write in Geneva that summer (which in fact was published, in that incomplete form and without the poet's permission, also in 1819). But Byron had soon abandoned the effort, and Polidori had taken up the idea and produced his own completed version at the request of a Russian noblewoman called the Countess of Breuss, who had visited the group at the villa. He had sent her the manuscript soon after, and that was the last he saw or heard of it until it appeared in Colburn's magazine. Polidori imagined that Byron had stolen his story.

That was not true—Byron had no knowledge of the business—but it would have been in keeping with the poor relations that had developed between the two men after the summer in Switzerland. They had fallen out badly at the end of that escapade, and Byron had sacked the physician. Polidori had seemingly taken his revenge by given Byron a thinly disguised starring role in his vampire tale.*

Lord Ruthven is the archetypal aristocratic cad, and irresistible to women:

> In spite of the deadly hue of his face, which never gained a warmer
> tint, either from the blush of modesty, or from the strong emotion
> of passion, though its form and outline were beautiful, many of the
> female hunters after notoriety attempted to win his attentions, and
> gain, at least, some marks of what they might term affection . . . such
> was the apparent caution with which he spoke to the virtuous wife and
> innocent daughter, that few knew he ever addressed himself to females.
> He had, however, the reputation of a winning tongue.

If this is an act of revenge, it was constructed with tools borrowed explicitly for that purpose. Polidori lifted the name Ruthven from *Glenar-*

---

* "There were probably one or two dead bodies in that man's past" was Goethe's assessment of Lord Byron.

von (1816), the first novel by Byron's former lover Lady Caroline Lamb, where the poet is also given an unflattering portrait as Lord Glenarvon, Clarence de Ruthven. There was something about this predatory caricature that appealed: Cyprien Bérard appropriated him too as the blood-sucking roué of *Lord Ruthven ou les Vampires* (1820), and one of the characters in Alexandre Dumas's *The Count of Monte Cristo* (1844) is sometimes referred to by that name.

Byron did have a genuine interest in vampires, as his fragment attested. In his poem *The Giaour* (1813), the ghoul of the title was a kind of vampire particular to Greece and the Levant, based on the old Calmet-style beast that preyed on its kin:

> *But first, on earth as Vampire sent,*
> *Thy corse shall from its tomb be rent:*
> *Then ghastly haunt thy native place,*
> *And suck the blood of all thy race.*

In a characteristically Byronic touch, this creature sucks the blood of its sister, wife, and daughter: acts tantamount to incest.

Polidori's Ruthven is another kind of beast altogether. He gambles and has "sought for the centres of all fashionable vice," in which he indulges with icy glee. His charisma ensnares the young hero Aubrey, who becomes Ruthven's companion on travels through Europe. While in Greece, Ruthven is fatally shot by robbers, and on his deathbed implores Aubrey to promise that for a year and a day "you will not impart your knowledge of my crimes or death to any living being in any way, whatever may happen, or whatever you may see." This the young man does; the next day he finds that Ruthven's corpse has vanished. A robber tells him that he and his comrades took the body to lie on the peak of a mountain, but when he goes there to bury it, Aubrey finds nothing.

Back in England, Aubrey returns to the family home, only to discover during a ball that Lord Ruthven has reappeared and now has designs on his sister. Aubrey is confused and alarmed, but his former friend snarls in his ear, "Remember your oath!"

Aubrey is bound by his word—but even if he broke it, who would believe him? He is driven crazy by his predicament. When a physician attending him tells that Aubrey's sister is to be married the next day, he goes to offer his congratulations, and learns from the image in her locket that her husband-to-be is—well, you guessed it. Desperately he entreats her not to go through with it, but this is interpreted as resurgent madness and he is restrained.

The wedding is to happen (naturally) just a day before Aubrey's spell of promised silence is due to expire. He writes to his sister imploring her to delay the proceedings for just a few hours. But her physician, given the missive, decides it is best not to distract her with the ravings of a madman. The next day Aubrey hears the wedding preparations in the household and rushes out of his room to find Ruthven on the stair. Once again the vampire hisses at Aubrey, "Remember your oath"—and adds now the threat that, if he does not, "my bride to day, your sister is dishonoured." In his rage, Aubrey falls ill again, and his condition is now moribund. At the stroke of midnight, released from his promise, he tells all to his sister's guardians, and dies. They rush to her chamber, to find that Ruthven has fled and Miss Aubrey lies dead. The young woman has "glutted the thirst of a VAMPYRE!"

It's pretty pedestrian fare, to be frank, but its influence on the legend was immense. No longer the wild, skulking creatures of old tradition, vampires now walk among us with impunity, even with privilege, bewitching and enchanting us to feed their terrible passion for blood. Before *The Vampyre* there was nothing to suggest there was anything sexual about these undead beings. But Polidori made that aspect a key component of the story: the vampire's goal was the violation of young maidens and the corruption of their purity and innocence. Mixed into this carnal stew are hints of sexual boundaries being transgressed; not least, of incest once again. When Aubrey first returns in distress from Greece to his sister in England, their meeting sounds alarms to the modern ear:

> He hastened to the mansion of his fathers, and there, for a moment, appeared to lose, in the embraces and caresses of his sister, all memory of the past. If she before, by her infantine caresses, had gained his affection, now that the woman began to appear, she was still more attaching as a companion.

And when Aubrey implores his sister to end the relationship with Ruthven, his words are more like those of a lover trying to win the object of his desire from another suitor: "Oh, do not touch him—if your love for me is aught, do not go near him!"

The dissolute visitors at the Villa Diodati seemed determined, then, to transgress the conventions of early nineteenth-century Gothic. As Jane Austen made clear in her Gothic satire *Northanger Abbey*, the stories of the genre's doyenne Ann Radcliffe, such as *The Mysteries of Udolpho*, owed their success partly to the thrilling frisson they offered to polite young ladies. As fairy tales for impressionable grown-ups (or so Austen would have it), any hint of illicit material was buried far out of sight. But *Frankenstein* was not like that, and neither is *The Vampyre*. Something much more disturbing and less well-mannered is beginning to rise to the surface.

Frankenstein and *The Vampyre* were bound together not just by their genesis but by the way they catered to a public taste for chilling monster stories. By the 1820s, stage productions of the *Frankenstein*-based play *Presumption* were sometimes accompanied by an adaptation of Polidori's story called *The Vampire; or, The Bride of the Isles*, written by the English dramatist James Robinson Planché.* Set in Scotland (the opening scene is in Fingal's cave), it has Ruthven die after being struck by lightning, and is notable for introducing the "vampire trap," a trapdoor that permitted an actor to exit and enter the stage with supernatural alacrity. In London performances, Thomas Cooke played the lead roles in both plays: this was the original horror double bill. "In the wake of Polidori's tale," says Groom, "theatres were overrun with vampires"—mostly in comic performances lacking the dark themes of their counterparts in Gothic literature. By the later part of the century Shelley's and Polidori's stories had started to merge: a musical burlesque called *Frankenstein; or, The Vampire's Victim* appeared in London in 1887.

---

* The play was in fact based on a French adaptation of the Polidori story by Charles Nodier, produced in 1820.

Like the comic books of the 1950s and the "video nasties" of the 1970s, vampire stories were accused in the mid-nineteenth century of corrupting public morals. James Rymer's *Varney, the Vampire: The Feast of Blood* was just the kind of thing these complaints had in mind. It was the archetypal penny dreadful, sold in installments as weekly pamphlets in which the stories were meandering, barely coherent, poorly written, and deliciously gory. The central character is an English vampire named Sir Francis Varney, who was cursed with blood-craving immortality during the time of the English Civil War as punishment for killing his son. Now living in the early eighteenth century, he mostly torments an aristocratic family fallen on hard times called the Bannerworths, and their circle, targeting in particular maidenly young women like Flora Bannerworth (although it's partly her father's money he is after). The Bannerworths put up with this with remarkably good grace, regarding him more as a pest than anything else. "He is rather a troublesome acquaintance," one of them sighs—but "I think there are a good many people in the world a jolly sight worse vampyres than Varney." In an early dash of mythic metanarrative, a Count Polidori of Venice pops up in one episode, while elsewhere Varney gets electrically revived from the dead in an echo of *Frankenstein*. He ends up, after hundreds of pages, incinerating himself in Mount Vesuvius.

There's something of the werewolf in Varney: he is *transformed* into a vampire by the moon, and emits a "strange howling cry." The overlap was well established in folklore—in his *Book of Werewolves* (1865), the first detailed account of this ancient monster, the Reverend Sabine Baring-Gould pointed out that the word for "vampire" in the Slavic lands of Eastern Europe (recorded here with some approximation as "*vrkolak*") is the same as that used for "werewolf."

The *Varney* yarn unspooled in a series of more than a hundred pamphlets between 1845 and 1847, at which point they were collected into a massive single volume of 220 chapters that would have taxed even the most avid reader's stamina. *Varney* was absurd but immensely popular, and helped secure the vampire as a stock character of thriller fiction. Many nineteenth-century writers succumbed to its allure, the great Russian writers being particularly keen to dabble: as well as Leo and Alexei

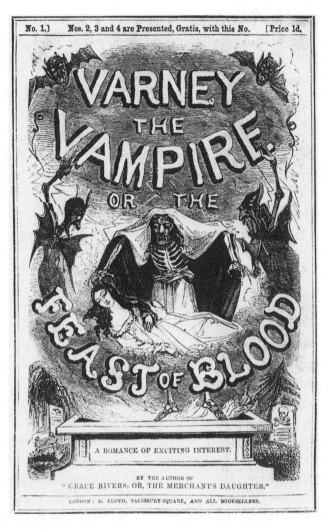

One of the "penny dreadful" title pages for Varney the Vampire *(1845).*

Tolstoy, vampire short stories were penned by Alexander Pushkin, Nikolai Gogol, Ivan Turgenev, and later the science-fiction writer Alexei Tolstoy (not to be confused with the aforementioned dramatist, he was only a remote relative of Leo). Vampire metaphors came easily to hand for writers of the period. Mr. Rochester's mad wife Bertha in *Jane Eyre*

(1847) is described as a red-eyed "goblin" who puts Jane in mind of "the foul German spectre—the Vampyre." (For Charlotte Brontë, the aristocratic vampire newly minted by Polidori had evidently not yet eclipsed the traditional folkloric version.)

It's appropriate that this trend was begun by a medical man, because Victorian vampires refract the old folklore tales through a medical lens. For one thing, they link vampirism with notions of contagion and infection, which became especially important with the advent of epidemiology and germ theory, and an understanding of the role of hygiene in public health, in the second half of the century. A medical perspective was also brought to bear on the prevailing symptom of Victorian vampirism: not savagery or bloodlust, but languor.

The victims of vampires fall into a delirious swoon that can sound rather alluring and undoubtedly had an illicit sexual flavor to it. Languor and exhaustion were, after all, precisely the consequences doctors warned about from masturbation, as well as being the familiar postcoital coda to *le petit mort*. Women were considered to be particularly prone to fatigue. Everyone knew that their nerves were weaker and could be more easily worn out, a condition that the American doctor George Beard called neurasthenia in the 1880s. What's more, menstruation, by depleting the body of blood, was thought to bring on symptoms akin to those of vampirism; as the British psychiatrist Henry Maudsley put it in 1870, "The monthly activity of the ovaries which marks the advent of puberty in women has a notable effect upon the mind and body; wherefore it may become an important cause of mental and physical derangement."

Languor and confusion are precisely what befall Laura, the heroine of the novella *Carmilla* (1873) by the Irish writer Joseph Sheridan Le Fanu, when she becomes the prey of the female vampire of the title. Laura is exiled with her father in a castle in remote Styria, where the mysterious Carmilla appears by her bedside, sometimes as a black cat, sometimes as a woman soaked in blood, even when all the doors are locked. She is eventually identified as a countess who died in the seventeenth century, whereupon an official deputation exhumes her coffin and finds the body looking fresh and warm. She is decapitated and staked.

Carmilla has a tooth that is "long, thin, pointed, like an awl, like a needle," and her nocturnal visits leave Laura with a "sense of exhaustion, as if I had passed through a long period of great mental exertion and danger." Little is left to the imagination about the sexual nature of Carmilla's predation: "Sometimes it was as if warm lips kissed me," Laura says, "and longer and more lovingly as they reached my throat, but there the caress fixed itself." Remarkably to our eyes now, the lesbian implications of *Carmilla* drew no comment in the nineteenth century. Or perhaps it is not really so remarkable: what cannot be accommodated in social mores is fated to remain invisible to it.

There had been a hint of same-sex passion in vampirism ever since Lord Ruthven seduced and flattered the young Aubrey into being his companion and confidante. By translating the relationship to a lesbian one, says cultural historian Nina Auerbach, *Carmilla* offered a safe scenario in which to situate male homoerotic fantasies—although this seems an overgenerous interpretation of what looks now decidedly like an all-too-familiar male heterosexual fantasy of lesbian sex. *Carmilla* has its roots in the Romantic tradition: Samuel Coleridge's 1816 unfinished poem *Christabel* tells of the titular heroine's encounter in a wood with a mysterious woman called Geraldine, who seems to have unearthly powers.* Geraldine takes Christabel back to her home and undresses to sleep by her side, whereupon we get an eroticized glimpse—no more—of something terrible:

> *Behold! Her bosom and half her side—*
> *A sight to dream of, not to tell!*

Carmilla was a key source for one of the truly great vampire films, Carl Dreyer's near-silent *Vampyr* (1932). But it was not until decades later

---

* Byron read *Christabel* to the company at the Villa Diodati shortly before they set to composing their ghost stories, causing Percy Shelley to have a vision of a women with eyes for nipples that sent him screaming from the room.

that lesbian vampires could be openly portrayed. I'm afraid you can probably guess how that went with Hammer's *The Vampire Lovers* in 1970 (sample dialogue: "I wish I felt tired—I never do, not at night. Just sort of . . . excited"). The relationship between Catherine Deneuve and Susan Sarandon in Tony Scott's *The Hunger* (1983) was, on the other hand, all art-house glamour and opulence. Auerbach argues that these relationships are all about intimacy, friendship, and sharing—which sounds all very *Twilight* (we'll come back to that), but I'm not sure the argument survives the journey through the "male gaze" it encounters.

Auerbach meanwhile dismisses heterosexual female vampires as "sexually orthodox" (which seems, by her standards, tantamount to saying not sexual at all). It's this kind of reasoning that leads her to label *Dracula* "less the culmination of a tradition than the destroyer of one"—by which she means that Stoker undermined the homoerotic "intimacy" of Ruthven and Aubrey or Carmilla and Laura. But this judgment seems to me to reveal the hazard of plotting out an agenda for a myth, such that one must proclaim it has "gone wrong" when it fails to comply. (Sherlockians and Bat-fans, I'll be looking at you too.) Everything Auerbach says about Stoker "paying extravagant lip service to heterosexual love" while "feeding on [male] hero worship" is right on the mark. But this is precisely one of the reasons we remember *Dracula* and not *Carmilla* or *The Vampyre*. If those earlier tales were truly metaphors for same-sex intimacy and friendship (and I don't think for a minute it's as simple as that), one reason *Dracula* is so rich is that it replaces this cozy picture with a seething cauldron of sexual anxiety. Stoker let the vampire possess him; perhaps he just made the mistake of inviting it in. Where before there had been an enticing scent of sexual allure, now there was lust and rape fantasy. Previously maidens had been menaced; now they were graphically violated. Where there had been demure bites on the neck, now there was gushing blood and writhing bodies. We can call the earlier Victorian vampires wicked, but now they are transgressive and invasive, deviant far beyond the routine sexual predator. Now vampires were able to penetrate to the soul.

Dracula, according to horror expert David Skal, is "one of the most obsessional texts of all time, a black hole of the imagination." Cultural critic Christopher Frayling says that it "was probably transgressing *something*—but the critics weren't quite sure exactly what." Exactly so— and probably because the author was not sure either.

On the face of things, Bram Stoker was an eminently respectable late-Victorian family man—and we know well enough today not to trust that persona an inch. For what he produced in *Dracula* was aptly described by English writer and critic Maurice Richardson as "a kind of incestuous, necrophilious, oral-anal-sadistic all-in wrestling match." Which means (I am not being entirely glib here) that it had something for everyone.

Stoker was born and raised in Dublin to a Protestant family clinging to the lower rungs of the middle classes. Respectability mattered to him; he had the judgmental morality of the socially precarious, for example proclaiming after a visit to America that the hobos and tramps there should be branded and sent to labor colonies to learn what hard work was. His rather shrill worship of "manliness" and his conventional views on gender roles look now like part of a determined act of self-deception.

Still, you can't fault his work ethic. Stoker never seemed to feel that one full-time occupation need preclude others, and so even while he was studying mathematics at Trinity College Dublin he took a job in the civil service, established himself as a theater critic, and in 1873 accepted the editorship of the *Irish Echo*, where he was essentially the only member of staff. (Somehow he still got his degree.)

In 1867 he saw a production of Sheridan's *The Rivals* starring the actor Henry Irving. Stoker was smitten, and when Irving, by then actor-manager at the Lyceum Theatre in London, returned to Dublin in 1876, Stoker's reviews of his production of *Hamlet* were so effusive that the acutely vain actor invited the young critic to dinner, at which he treated Stoker to a rendition of a melodramatic poem. It's tempting, given accounts of his acting technique, to imagine that it was a performance of hokey self-adoration, but Stoker attested that he was so moved he became reduced to "a violent fit of hysterics" (at that time a stereotypically womanly response—and he's not talking about laughter). His devotion was sealed. "Soul had looked into soul!" Stoker wrote. "From that hour began a friendship as profound, as close, as lasting as can be between two men." At the end of 1878 Irving asked him to come to London and be his front-of-house manager.

It's easy to see why Count Dracula is often said to be based on Irving, even though there is no compelling reason to think that is true. For the man truly was a sort of monster, mainly of egotism: he could be charming, but also cruel, haughty, and callous. And he drained Stoker for all he was worth, treating him like a servant if not indeed a slave, and exploiting his house manager's blind idolization to make outrageous demands

Bram Stoker (1847– 1912), photographed around 1906.

at all times of day and week. That Stoker not only bore it but sustained his adoration of Irving seems to speak of more than just hero worship; he had a masochistic infatuation with Irving that persisted until the actor died, a feeling stronger than any he showed for his wife and family. Stoker's friend, the Manx writer Hall Caine, attested that "I have never seen, nor do I expect to see, such absorption of one man's life in the life of another." Stoker's feeling for Irving, he added, was "the strongest love that man may feel for man."*

One might reasonably assume from this that Stoker was a repressed homosexual. While studying at Trinity in the early 1870s, he developed an infatuation with the American poet Walt Whitman, who is now widely considered to have had homoerotic (if not necessarily homosexual) relationships. Stoker wrote Whitman an impassioned letter, full of ambiguous remarks about his own sexuality: "How sweet a thing it is for a strong healthy man with a woman's eyes and a child's wishes to feel that he can speak to a man who can be if he wishes father, and brother and wife to his soul." A poem of Stoker's written in 1874, "Too Easily Won," speaks of the anguish of being rejected by another man: "His heart when he was sad & lone / Beat like an echo to mine own / But when he knew I loved him well / His ardour fell." Stoker met Whitman during an American tour with Irving in 1884, and the two men talked warmly.

But whether we can consider Stoker a gay man who couldn't come out is a complex matter. Only in his day were the boundaries of sexuality becoming fixed—indeed, it was then that the word "homosexual" was coined. In the same year *Dracula* was published, *Sexual Inversion* by Havelock Ellis and John Addington Symonds argued that same-sex desire was a common state of affairs with a long history, albeit an "inversion" of the norm. Until this tendency to medicalize sexuality, it didn't function as a label of identity, and the distinctions between mutual

* There was quite possibly a homoerotic element too in Stoker's closeness to Caine—highly strung and flamboyant, with a thinning mane and piercing stare, he was the "Hommy-Beg"—his Manx nickname—to whom Stoker dedicated *Dracula*.

affection and sexual desire between men received little scrutiny (although that is certainly not to say gay sexual relationships were socially accepted).

Homosexual preference wasn't seen as incompatible with heterosexual marriage, for the objectives of the two types of relationship were quite different. What a wife might bring to a man who loved men was almost as much a matter of aesthetic balance as of social respectability. Love didn't necessarily have to feature in that balanced equation. Just as Oscar Wilde's marriage can't simply be considered a socially conforming sham, neither can Stoker's. And the comparison could hardly be more apt, for both men once courted the same woman: the "Irish beauty" Florence Balcombe. She chose Stoker but got little joy from it; the marriage was said to be devoid of passion, and Bram only ever speaks of love and loving in the context of his male relationships. Poor Florence was left at home with their young son while her husband worked long nights at the Lyceum. At the end of her life she hinted at regrets that she hadn't opted for Wilde after all, in spite of what had befallen him.

The Stoker and Wilde families knew each other in Dublin, but Oscar's flamboyance seems to have made Bram wary of his acquaintance when the two men lived in London: they remained in uneasy contact, but Stoker never once mentioned Wilde in his letters. He must have watched with horror as Oscar's public disgrace unfolded in 1895 during the fateful trial against the Earl of Queensberry, provoked by Wilde's relationship with the earl's son Lord Alfred Douglas. After Wilde was convicted, his brother Willie wrote to Stoker saying, "Poor Oscar was *not* as bad as people thought him"—but it's not clear Bram ever replied.

In many ways Oscar Wilde's *The Picture of Dorian Gray* represents his more metaphorical take on the vampire myth: here again is the corrupt, Byronic aristocrat (Lord Henry Wotton) who saps the life-force from those around him. Wilde's book is far superior to *Dracula* in literary terms, and provides a more nuanced and imaginative perspective on vampirism. Wilde was also much more in conscious control of his material, especially its homoerotic subtext (and if *that* is what Auerbach's "intimacy and friendship" means to the psychic vampire, I think most of us would prefer to remain friendless). This is precisely why *Dorian Gray* has entered the modern literary canon but is not a modern myth.

◆

Even if we accept that Stoker's arrogant, domineering Count Dracula was not a surreptitious portrait of Irving, still he seemed to have hoped that the actor might play the role on stage. But that was never likely. After an informal reading of Stoker's stage treatment of the book at the Lyceum shortly after its publication, Irving is reported to have walked out muttering, "Dreadful!"

It was unkind, but perhaps not entirely unjust. Stoker seems an unlikely author of so enduring a book; the judgment of Henry Irving's grandson Laurence that Stoker was "inflated with literary pretensions" isn't far from the mark. Much of his output was little short of dreadful: lurid, potboiling fantasies and lame, sentimental romances such as *The Shoulder of Shasta* (1895) and *Miss Betty* (1898). Stoker's early literary efforts—short stories published here and there in magazines— often leaned toward the Gothic and fantastical. They revealed too some rather peculiar notions lodged in his imagination. His illustrated book of children's fairy stories, *Under the Sunset* (1881), contains grim tales of death, doom, and demons that reviewers suspected would mystify more than terrify most young readers, while the short story "The Dualitists" (1887), allegedly also written for children, makes you positively fearful about what he was repressing. Here two delinquent young boys argue about who has the best pocketknife (yes, I know) and torment other children by brandishing them. The boys capture two smaller children and carry them onto a roof. The children's alarmed parents try to shoot the pair of hooligans but succeed only in blowing off the heads of their own offspring. The boys throw the headless corpses onto the parents, killing them too—and the parents are posthumously convicted for the murder of their children, while the two boys are knighted for bravery. This abhorrent tale is delivered in a tone that insists it is all a jolly jape; it sounds like the ravings of a psychopath.

For all its popularity, *Dracula* didn't change critical opinion that Stoker was basically a hack; some reviewers saw it as just more of the same incredible and grotesque nonsense he'd produced before. The critic Ludovic Flow captures the spirit of Stoker's writing brilliantly:

> There is a semi-heroic, Everyman quality about his intense command of the mediocre—as if the commonplace had found a champion who could wear its colours with all the ceremony of greatness.

But *Dracula* is not famous for its prose or literary qualities, which are adequate at best and comically overblown or pedestrian in places. The novel's decent but rather dull heroes Jonathan Harker, Arthur Holmwood, and John Seward have voices that are barely distinguishable, and Stoker's attempts to introduce a diversity of accent in Dutch vampire expert Abraham Van Helsing and the Texan adventurer Quincey Morris ("We've told yarns by the camp-fire in the prairies . . .") are almost risible. It's hardly surprising that the book was considered anything but canonical until recent times—even in 1992 its inclusion in the Penguin Classics series led to raised eyebrows in the literary establishment. George Orwell included it in the category of a "good bad book," while horror-literature critic James Twitchell suggests that we must want to hear this story *really badly* if we're willing to put up with so much bad writing. "Like *Dr Jekyll* and *Frankenstein*," he says, "it is a mess."

Stoker seems to have set out to do no more than write a successful adventure-shocker, figuring that the popular vampire trope might hold wide appeal. Hall Caine attested that his friend had no great aspirations. Bram, he said, "took no vain view of his efforts as an author. . . . He wrote his books to sell." David Skal concludes, "In the final analysis, the most frightening thing about *Dracula* is the strong probability that it meant far less to Bram Stoker than it has come to mean for us."

Stoker had to write the book in the interstices of his frantic schedule.

He began it around 1890; during a wet family holiday in Whitby that summer he often abandoned his wife and son to their own devices while he explored local sites, sought out stories, and scribbled in the library. He wandered around the ruined abbey above the beach and learned of the shipwrecked vessel *Dmitri*, washed onto Whitby sands earlier in the century—which became the *Demeter* on which Dracula reaches England.

Like Mary Shelley and Robert Louis Stevenson, Stoker insisted that much of his mythic story came to him in a dream.* But the story is anything but a spontaneous outflowing of the subconscious. Stoker often jotted down fragments of the tale on whatever piece of paper came to hand, and wrestled for seven years with how to organize this material into a coherent narrative.† The long gestation has encouraged the view that Stoker researched *Dracula* meticulously, whereas in fact his exploration of vampire folklore was probably rather haphazard and cursory. While the name "Dracula" had been used by the fifteenth-century Wallachian ruler (Voivode) Vlad III Tepes, there's no basis for the common belief that Stoker based his count on this historical figure. He just liked the sound of the name, which he understood to mean "devil" (actually it means "Son of the dragon/devil," *Dracul* being the nickname of Vlad III's father Vlad II). And there's no sign Stoker knew of the reputedly cruel Vlad III's favored method of dispatching prisoners captured from his enemies the Turks: by impalement on stakes, earning him the historical sobriquet Vlad the Impaler.

For the initial chapters of the book set in Transylvania, Stoker took many of the local details (sometimes pasted into his text with just a light

* Bathetically, Stoker claimed this nightmare was induced by "a too-generous helping of dressed crab at supper."

† H. P. Lovecraft claimed to have known a woman who was almost contracted to edit the chaotic manuscript, had her price not been too high. There is an Icelandic translation from 1901 of a book credited to Stoker called *Powers of Darkness*, a version of *Dracula* in which the action mostly relates to Jonathan Harker's visit to Transylvania. David Skal thinks it might be an early draft that got to Iceland via Hall Caine.

paraphrase) from *The Land beyond the Forest*, an account of that region's folklore by E. D. Gerard (a pseudonymous composite of two sisters, Emily and Dorothea). Here he found the word *nosferatu* to designate a vampire, although its origin and true meaning are unclear. And in 1890 Stoker and Irving met with Ármin Vámbéry, a Hungarian traveler and specialist in Turkish culture, when the scholar visited London; at that stage Stoker was hoping that his vampire idea might become a theatrical vehicle for Irving. Vámbéry makes a brief appearance in *Dracula* as Van Helsing's friend "Arminius of Buda-Pesth University."

Despite this use of folkloric subject matter, *Dracula* has all the modernity of Stoker's contemporary H. G. Wells. It teems with references to the speedy new technologies of the telegram and railway and alludes to the latest developments in medical science.* Stoker took and borrowed where he pleased. The epistolary format of the book echoes that of *Frankenstein*, but a more immediate model may have been *The Woman in White* (1860) by the most successful thriller writer of the age, Wilkie Collins. Stoker takes this device further than Collins: his novel is a patchwork of voices in diaries and letters, giving it the feel of a bundle of papers thrown rather carelessly together. The shifting viewpoint makes its narrative hard to grasp, and once again the ambiguity acts to the benefit of the myth: intentions are unclear, narrators are unreliable, no one knows quite what is going on, and readers are free to make their own interpretations.

The book opens with entries in the journal of Jonathan Harker, a young man sent to Transylvania to complete the paperwork for Count Drac-

---

* The reference to Van Helsing as "one of the most advanced scientists of his day" is, however, somewhat unconvincing given how little use he makes of such knowledge, aside from performing a blood transfusion. It's not clear that Stoker really knew much about the "science of his day."

ula's purchase of an English estate called Carfax, near the village of
Purfleet in Essex. On his journey Harker encounters now hackneyed
omens of ill fortune ahead: villagers muttering about Satan, hell, and
vampires (*vrolok*); dogs howling in the hills; passengers making ges-
tures to ward off the evil eye; and so forth. He is taken to the Count's
castle in a carriage driven by a man with a long beard and a great black
hat (possibly Dracula in disguise) and is greeted there by his host: "a
tall old man, clean shaven save for a long white moustache, and clad in
black from head to foot, without a single speck of colour about him
anywhere." His hand, when it shakes Jonathan's, is icy cold, with a hairy
palm. As the cry of wolves echoes in the valley, Dracula utters the immortal words, "Listen to them—the children of the night. What music
they make!"

As he wanders around the castle alone the next morning, Harker discovers that all of its doors are bolted. There is no exit: he seems to be a
prisoner. The Count reappears after dark to tell Harker of his ancestry
among the Szekely, the tribe who fought the Turks, Magyars, Bulgars,
and other barbarian invaders. And one night as the young man looks out
of his window, opening onto the sheer cliff-face of the castle as it stands
above the shadowy valley, he sees Dracula emerge from a window just
below "and begin to crawl down the castle wall over that dreadful abyss,
face down, with his cloak spreading out around him like great wings . . .
just as a lizard moves along a wall." He begins to suspect he might never
leave this place.

Venturing into a remote suite of rooms in an effort to find an escape
route, Harker falls asleep and awakens in moonlight to find he has the
company of three young women. They approach him coquettishly as he
lies in a kind of swoon. "He is young and strong," one of them purrs.
"There are kisses for us all." But as they prepare to swoop and feast, the
Count appears and hurls them back in fury. "This man belongs to me!"
he thunders. "You never love!" they taunt him—but he tosses them
something after all: a bag from which comes "a gasp and a low wail, as of
a half-smothered child."

Just as Jonathan has given up hope of escape, the Count tells him

Count Dracula crawls down the wall of his castle on the cover of a 1919 edition of the book.

that he may return to England, where his fiancée Mina Murray awaits him anxiously. For Dracula too is leaving, having arranged for the local gypsies to ferry his own effects abroad. Harker sees these people at work, bringing great wooden boxes in their wagons.

Soon enough he discovers what these caskets are for. Searching for the Count, who he has never seen abroad in daylight, he finds a stairway that descends to a ruined chapel. Within the vaults, lying in one of the boxes on a pile of freshly dug earth, is Dracula:

> He was either dead or asleep, I could not say which—for the eyes were
> open and stony, but without the glassiness of death—and the cheeks
> had the warmth of life through all their pallor, and the lips were as red
> as ever. But there was no sign of movement, no pulse, no breath, no
> beating of the heart.

This, evidently, is how Dracula intends to travel to England, protected and preserved on the soil of his homeland. Harker flees from the sight, but the next morning, finding himself again trapped in the castle, he returns in desperation to the Count's resting vault. Horror-struck at the sight of the prone vampire with fresh blood trickling from his mouth, he determines to strike the creature dead with a shovel. But his resolve is broken when the Count's eyes open and lock onto his "with all their blaze of basilisk horror," and he retreats to Dracula's chamber, where he finds himself locked inside. He hears the gypsies nail down the lid of the box, and with the crack of a whip the cargo is carried away.

We're taken now to England, where Mina is exchanging letters with her friend Lucy Westenra in Whitby. Lucy is being courted by three suitors, all of them old friends: Lord Arthur Holmwood; Dr. John Seward, who runs a lunatic asylum at Carfax; and the American Quincey Morris. Whom should she choose? Mina travels to Whitby to help her friend resolve the dilemma, and while she is there a great storm throws up a shipwreck: the *Demeter*, a Russian schooner, whose only cargo is "a great number of wooden boxes filled with mould," bound for delivery to a Whitby solicitor. As the ship crashed ashore, a great dog had been seen to leap from it, and the beast is now suspected of having killed a coal merchant's mastiff—found with "its throat torn away, and its belly . . . slit open as if with a savage claw."

Lucy begins to behave oddly. She is often sleepy, yet has "more colour in her cheeks than usual," and she sleepwalks. A great bat is seen outside her window.* One night Mina glimpses her friend up in the ruined abbey above the harbor—and when she climbs the steps to reach her, she sees "something, long and black, bending over the half-reclining white figure." As Mina calls out, the thing looks up to show a "white face and red, gleaming eyes." But by the time she reaches Lucy, the young woman is alone, asleep and breathing "in long heavy gasps, as though striving to get her lungs full at every breath."

Lucy's state worsens: "the roses in her cheeks are fading, and she gets weaker and more languid day by day," Mina writes. Her anxious suitors convene to see what is to be done, and Seward writes to an "old friend and master," Professor Van Helsing of Amsterdam, "who knows as much about obscure diseases as any one in the world." When Lucy, now at her family home in Hillingham in London, takes a turn for the worse, Van Helsing diagnoses that "she will die for sheer want of blood to keep the heart's action as it should be," and he arranges for all three young men, as well as himself, to be donors for a transfusion. As they proceed, Van Helsing notices two red puncture marks on her throat, which were previously hidden beneath a black velvet band. He announces gravely that he must return to Amsterdam at once to consult his books.

The Dutchman knows about more than rare diseases. He brings Lucy a bundle of white flowers, from which he makes a wreath to hang about her neck. They are medicinal, he tells her—and when Lucy recognizes them as common garlic and laughs that he is playing a joke on her, he replies sternly, "I never jest!"

From Holland, Van Helsing sends a telegram to Seward warning him

---

* The association of vampires with bats goes way back. Bloodsucking bats do of course appear in nature: the pioneer of taxonomy Carl Linnaeus called one of the carnivorous South American species *Vampyrum spectrum* in 1758. A discussion of vampirism in the *Imperial Magazine* in 1819, inspired by Polidori's tale, was followed by a letter describing vampire bats—the correspondent exaggerates their wingspan as up to six feet, and claims that such creatures have been known to attack humans. Dracula's ability to turn into a giant bat is of course heralded in Harker's dreadful vision of his descending the castle wall with his cloak spread behind him.

to get to Hillingham at once, but the message is delayed and the doctor, soon joined by Van Helsing, arrives to find Lucy lying still and white next to the dead body of her mother, the wounds in her throat "horribly mangled." Still she is not dead, though there is hardly any life left in her. Despite the attentions of her men, she weakens fast, and two days later they gather round as her chosen fiancé, Holmwood, holds her dying hand. On the brink of death, Seward reports, she speaks "in a soft voluptuous voice, such as I had never heard from her lips: — 'Arthur! Oh, my love, I am so glad you have come! Kiss me!'" At this, Van Helsing springs into action. Hurling Holmwood backward, he cries, "Not for your life! Not for your living soul and hers!" Then Lucy takes her last breath, and Seward declares, "There is peace for her at last. It is the end!" — only for Van Helsing to tell him ominously, "Not so; alas! Not so. It is only the beginning!"

Once Lucy is in her coffin, the Dutchman shocks his young friend by announcing that he wants to perform an autopsy — but the like of which the doctor has never encountered before: "I want to cut off her head and take out her heart." He has deduced, of course, that Lucy has become a vampire.

He explains to Seward days later that he believes the now-undead Lucy is to blame for the children found in a weakened state around Hampstead with marks on their throats. Seward is incredulous, but at Van Helsing's insistence accompanies him that night to inspect Lucy's coffin, still lying in the churchyard at Hillingham. They find it empty.

Yet when the two men return to the tomb the next afternoon, there is Lucy, "more radiantly beautiful than ever. . . . The lips were red, nay redder than before; and on the cheeks was a delicate bloom." Van Helsing explains that they now have a terrible job to do. The next night, accompanied by Holmwood and Morris, they go again to the churchyard and witness Lucy feasting on the blood of a child. Warded off by a crucifix, Lucy passes like a specter through the closed door of her tomb. The men return the following night, and while Lucy still lies in her coffin, Arthur drives a stake through her heart. After the frenzy of the vampire's throes subsides, "There in the coffin lay no longer the foul Thing that we had so dreaded and grown to hate . . . but Lucy as we had seen her in her life, with her face of unequalled sweetness and purity."

Now the chase is on for the architect of this crime. Dracula has been making himself at home in his Carfax residence, while his acolyte Renfield, a madman in Seward's nearby asylum, raves about the master's return and eats flies. Jonathan Harker, having escaped after all from Translyvania and returned to England, tells the band of vampire-hunters about the events at the castle. "This vampire which is amongst us is of himself so strong in person as twenty men," Van Helsing explains to them in Stoker's tortured version of a foreigner speaking English. "He is of cunning more than mortal, for his cunning be the growth of ages." He is what they call in Eastern Europe a *nosferatu*.

The men discover that Dracula is having his boxes of earth—his safe places of rest during the daylight hours—installed far and wide in England, some of them in a mansion in Piccadilly that he has bought: a lair in the heart of society, perilously close to the monarchy itself. And now his prey is Mina. One night Seward, Holmwood, and Van Helsing burst into the Harkers' lodgings at the asylum to see Jonathan lying in a stupor and the Count bending over Mina's prone, blood-stained body. They repel him with crucifixes, and he vanishes into vapor. The only way to prevent Mina from suffering Lucy's fate is to destroy the Count—"even if," says Van Helsing, "we have to follow him into the jaws of Hell!"

They do not need to go quite so far. But when Dracula, his plans in England now foiled, makes his escape by a sea route back to Transylvania, the redoubtable crew aim to get there ahead of him, taking the Orient Express and tracking by telegram the Russian vessel on which he departed so as to ambush him before he can reach the castle. Van Helsing breaks into the castle and stakes the three vampire women who had so nearly attacked Jonathan ("it was butcher work . . . the plunging of writhing form, and lips of bloody foam"), and then the group ambushes Dracula's gypsy entourage on the mountain pass, just as the sun is setting. As the Count is uncovered lying in his box of earth, "the eyes saw the sinking sun, and the look of hate in them turned to triumph."

But oddly, there is no stake to finish him off; instead, Jonathan cuts his throat, while Morris, mortally wounded in the foray, plunges his bowie knife into the vampire king's heart.

First and foremost, "the vampire is an erotic creation," according to the Italian writer Ornella Volta: "The vampire can violate all taboos and achieve what is most forbidden." Those taboos surely include, *inter alia*, dominance and submission, rape, sadomasochism, bestiality, and homoeroticism. We might throw in masturbation and incest too: Christopher Frayling sees in the voluptuous nightly visitation of a being who leaves the victim in a swoon and depleted of vital fluids the imprint of erotic dreams and nocturnal emissions; while for Freudian psychoanalyst Ernest Jones, writing in 1910, the vampire expresses "infantile incestuous wishes that have been only imperfectly overcome." What makes *Dracula* so compelling and potent is that its sexual work is done largely unconsciously. If his great-nephew and biographer Daniel Farson is to be believed (which is by no means always the case), Bram Stoker "was unaware of the sexuality inherent in Dracula." Which is why David Skal is right to suggest that we can read the book today as "the sexual fever-dream of a middle-class Victorian man, a frightened dialogue between demonism and desire."

It seems unlikely that Stoker ever knew quite what he had written. There is every sign that he regarded *Dracula* as neither more nor less than what he intended it to be: a rather conservative shocker that celebrated manly virtues in the face of rising female assertion and independence. It is a tale in which maidens are given chivalrous protection by a band of valorous young blades, assisted by a wise and still virile father-figure. The final score: one lost, one saved. But the moral calculus is plain, for Mina Harker is redeemed by her demure meekness, whereas the confident modern woman Lucy Westenra ends up . . . well, where Lucy ends up, in terms of the sexual politics, is what is truly shocking about *Dracula*.

Lucy is a girl who likes to play the field. She entertains all three suitors on the same day, one after the other. Goodness knows what we are supposed to make of her remark that Dr. Seward's wooing involves "playing with a lancet in a way that made me nearly scream," while she gives the hapless Quincey Morris a kiss just because he asks for it, with no intention of taking things any further. Yet it is "calm, resolute" Arthur Holmwood who wins her affections, although you have to suspect she would have grown bored of him before very long.

"Why can't they let a girl marry three men, or as many as want her?" she asks Mina lewdly. All three men *do* eventually become her sexual partners—not literally, but it doesn't take much imagination to understand what is being enacted when Lucy, weakened by loss of blood from the Count's nocturnal visitations, needs a transfusion. The three men, and then the fatherly Van Helsing himself, take turns mingling their vital fluids with hers—and to drive home the metaphor, Van Helsing jokes to Seward that "this so sweet maid is a polyandrist"—making him (married to a wife with dementia) a "bigamist."

Yet even these ministrations aren't enough to save Lucy from her fall: she becomes a bride of Dracula. Stoker was hardly the kind of man knowingly to employ Freudian symbolism, so we can only surmise that there was a powerful unconscious at work when he wrote Lucy's account to Mina of the dream she had during one of the Count's visits:

> My soul seemed to go out from my body and float about the air. I seem
> to remember that once the West Lighthouse was right under me, and
> then there was a sort of agonizing feeling, as if I were in an earthquake,
> and I came back and found you shaking my body.

It gets worse. The scene in which the men seek out the coffin in which the vampiric Lucy sleeps and drive a stake through her heart enacts a routine familiar to modern vampire buffs. But Stoker's description is chilling, because there really can be no doubt that what we are really witnessing is a gang rape, as if the suitors are taking revenge for their earlier rejection:

The body shook and quivered and twisted in wild contortions ... But Arthur never faltered. He looked like a figure of Thor as his untrembling arm rose and fell, driving deeper and deeper the mercy-bearing stake, whilst the blood from the pierced heart welled and spurted up around it.

"The sexual implications of the scene," says critic Elaine Showalter, "are embarrassingly clear." Apparently not, however, to Stoker, who was prudish about sex and complained of "works of shameful lubricity" that were "corrupting the nation." Never one to miss a chance of laboring the point, he shows us the true monstrosity of a woman like Lucy: feeding on children in her crypt and casting their bodies callously aside, she desecrates all that is pure and holy in motherhood.

In short, Lucy is being punished for a trait deeply disturbing to the Victorian male: a passionate, even animal, sexual appetite. This misogynistic violence and fear is papered over with a tissue of respectability— for surely all the men are trying to do is protect the ladies' virtue. Lucy's bloody fate is for her own good.

The disquieting prospect of female erotic desire was dramatically portrayed in Philip Burne-Jones's painting *The Vampire*, produced in the year of *Dracula*'s publication. (Stoker knew Burne-Jones and sent him a copy of the book.) Here the woman mounts her supine male victim with haughty entitlement, her hair wild and loose and a triumphant, sensuous smile on her face. A poem written by Rudyard Kipling after he viewed the work made the point clear: "The fool was stripped to his foolish hide / Which she might have seen when she threw him aside ..."

Women were getting on top, and it made men nervous. The assertiveness of the New Woman was regarded as unnatural, emasculating, "mannish," and indeed positively vampiric. "I suppose the 'New Woman' won't condescend in future to accept [marriage proposals]," Mina writes disapprovingly in her diary apropos Lucy; "she will do the proposing herself." When in the 1910s Marie Stopes began to write frankly about female sexuality, she was chastised by a certain "Lord X" (her own anonymizing designation), who told her in a letter that she was teaching

Philip Burne-Jones, The Vampire *(1897)*

women things "only a prostitute" should know. "Once you give a woman a taste for these things," wrote the outraged peer, "they become vampires, and you have let loose vampires in decent men's homes."

But decent men could not, it seems, suppress a shudder of excitement at that prospect. The highly sexualized mixture of terror and arousal coalesces whenever female vampires appear in *Dracula*. There's no denying that Stoker's hyperventilating prose is alarmingly effective when he describes Jonathan Harker's encounter with Dracula's brides in his Transylvanian castle:

> The fair girl went on her knees, and bent over me, fairly gloating. There was a deliberate voluptuousness which was both thrilling and repulsive, and as she arched her neck she actually licked her lips like an animal, till I could see in the moonlight the moisture shining on the scarlet lips and on the red tongue as it lapped the sharp white teeth . . . Then the skin of my throat began to tingle as one's flesh does when the hand that is to tickle it approaches nearer—nearer. I could feel the soft, shivering touch of the lips on the supersensitive skin of my throat, and the hard dents of two sharp teeth, just touching and pausing there. I closed my eyes in a languorous ecstasy and waited—waited with beating heart.

All this focus on throats, lips, supersensitive skin, makes it clear that the sex we're really talking about here is oral—with all the anxieties *that* invoked for a late-Victorian man with likely repressed homosexual urges.

Sex and moral deterioration terrified Stoker ever more in his later life. In 1908 he criticized the sort of novel he deemed to be "foully conceived" and argued that censorship of such filth was needed to prevent degradation of humankind: "A close analysis will show that the only emotions which in the long run harm are those arising from sex impulses, and when we have realized this we have put a finger on the actual point of danger." Stoker's suppressed sexual anxieties erupted with full force in his late novel *The Lair of the White Worm* (1911), the titular creature of which can manifest itself as a beautiful aristocratic woman. At times the book is frankly repulsive, while at the same time a textbook case of Freudian symbolism: the worm's lair is a "fetid hole" that, when blown apart with dynamite, becomes "a horrible repulsive slime in which were great red masses of rent and torn flesh and fat." This theme of foul corruption may have been appallingly relevant for Stoker himself, who was now probably in the final stages of syphilis, and his book is, in David Skal's words, a work of "almost hallucinatory misogyny." The

Times Literary Supplement dismissed it at the time with the comment that "No one asks for probability or even possibility, but coherence is a necessity."

Lucy, I like to think, got the last laugh in *Dracula*: her vampish gusto makes her a far more attractive character today than the insipid Mina. The sexual liberation that terrified Stoker is now part of the vampire's appeal. It's true that, seen from the twenty-first century, the Hammer films of the 1950s and 1960s that feature female vampires can look like crude vehicles that simply exploit a more permissive context for maximum titillation and commercial potential. But there are "subliminal surprises in [this] waste of staked bimbos," as Nina Auerbach deliciously attests. To an American like Auerbach, trammeled by the stifling conformity that imposed "family values" on women "with a ferocity alien even to most Victorians," these British films were a "sadistic fairyland" that "open[ed] windows beyond the family." Despite all the leering and ogling at heaving bosoms, those romping female vampires were bursting the bonds of domesticity. There's something rather desperate and doomed in the efforts of Peter Cushing's buttoned-up Van Helsing to pin these women down with stakes. They're having too much fun, and they have glimpsed a future in which Kate Beckinsale cracks heads and kicks ass as the vampire heroine of the *Underworld* movie series. It is, I suppose, progress of a sort.

Given what we know of Stoker, we might expect the subliminal eroticism of *Dracula* to be slanted toward male same-sex relations. It doesn't disappoint: whether or not, as Auerbach claims, *Dracula* "was fed by Wilde's fall," the book positively throbs with frantically suppressed homoeroticism. From the outset, the Count can barely contain his urge to feast on Jonathan Harker; he dismisses his "brides" as they prepare to violate the young man with the command "This man belongs to me!" Friedrich Wilhelm Murnau, the director of the iconic 1922 film adaptation *Nosferatu*, was gay himself and seems alert to Stoker's unconscious subtext when his Dracula figure Graf Orlok licks blood from the finger of his bewildered guest after he, alarmed by the count's apparent invisibility in his mirror, has cut himself shaving.*

While Dracula has sometimes been presented as a model of queer identity, as such it hardly seems a positive portrayal. He embodies everything perceived to be "bad" about homosexuality—all that Stoker sensed, feared, and loathed in himself. Like vampirism, homosexuality was at that time becoming a deplorable "condition" from which one needed to be rescued.

The heroes of the novel do their best to show how. Contrasting with the vile lechery of the Count as he looms over Harker, and the young man's fascinated disgust as he discovers the vampire's blood-gorged body, the camaraderie among the band of men who vanquish the vampire models Stoker's own solution to his predicament. As Auerbach says, the book "abounds in overwrought protestations of friendship among the men, who breathlessly testify to each other's manhood." Over them all presides the Whitmanesque figure of Van Helsing, who assures Arthur Holmwood that "I have grown to love you—yes, my dear boy, to love you." For all his prejudice and dissembling, one can't help feel for the anguished Stoker here, seemingly so desperate to neutralize and normalize his feelings.

The homoerotic allure of the vampire had been explored, with far less inhibition and repression, three years before *Dracula* was published, by the Slavic aristocrat Eric Stenbock. His short story "A True Story of a Vampire" records the recollection by an old woman called Carmela who lives in a castle in Styria—an obvious reference to the homoerotic *Carmilla*—of an episode from her youth. A mysterious Hungarian guest called Count Vardalek arrives at the castle and enthralls Carmela's young brother Gabriel. We can guess well enough from his appearance what the Count has in mind: "He was rather tall with fair wavy hair, rather long, which accentuated a certain effeminacy about his smooth face."

---

* We should be cautious about ascribing too much explicit intention to Murnau, however, for he used a screenwriter.

Gabriel pines when Vardalek, now welcomed into the family home, has to go away on trips to Trieste. "Vardalek always returned looking much older, wan, and weary. Gabriel would rush to meet him, and kiss him on the mouth. Then he gave a slight shiver: and after a little while began to look quite young again."

But Gabriel, previously so healthy, starts to succumb to a mysterious illness. On his deathbed, "Gabriel stretched out his arms spasmodically, and put them round Vardalek's neck. This was the only movement he had made, for some time. Vardalek bent down and kissed him on the lips." When the boy dies, the Count leaves, never to be seen by the family again.

A poet as well as an author of fantastical tales, Stenbock was more Wildean than Wilde, and made no bones about it. He conducted many homosexual relationships while studying at Oxford, and he was said to have gone everywhere with a life-size doll that he called his son. He was said to be eccentric, morbid, and perverse — qualities he put to good use in his vampire story.

The gay subtext of *Dracula* was pursued in *The House of the Vampire* (1907) by the writer George Sylvester Viereck. A complex and controversial figure, Viereck met with Adolf Hitler in Germany in the 1930s and was imprisoned in the United States in 1942 because of his Nazi sympathies (specifically, for failing to register as a supporter of National Socialism). His book relates the story of Reginald Clarke, a dissipated Henry Wotton figure who draws young men into a web of corruption. He represents a popular character type of the early twentieth century: a *psychic* vampire, who absorbs the energy and creativity of his victims. "Your vampires suck blood," cries one of his victims, "but Reginald, if vampire he be, preys upon the soul." Viereck's description of Clarke's fate leaves little doubt that the character was modeled on Wilde himself:*

---

* As a young man, Viereck met Lord Alfred Douglas for lunch in New York shortly after Wilde's death, and the two men remained in correspondence until Douglas's death in 1946.

Many years later, when the vultures of misfortune had swooped down
upon him, and his name was no longer mentioned without a sneer,
he was still remembered in New York drawing rooms as the man who
brought to perfection the art of talking.

Clarke's vampirism is portrayed here not as a vile perversion but as the
right of a superior being to draw the life force from his inferiors. True
to his fascistic leanings, Viereck wrote, "My vampire is the Overman of
Nietzsche. He is justified in the pilfering of other men's brains." He is
a revival of the Byronic vampire, a creature of entitlement, taste, and
refinement.

It's puzzling why gay men like Stenbock, Wilde, and Viereck chose
to reiterate the widespread association of homosexuality with the dese-
cration of youthful innocence. Yet the theme of corruption being spread
via body fluids during illicit, decadent sexualized embraces, would, in
Stoker's time, have resonated too with fears about syphilis. Science-
fiction writer Brian Aldiss considers *Dracula* "the great nineteenth-
century syphilis novel," although Oscar Wilde's biographer Richard
Ellman argues that this was also the real subject of *The Picture of Dorian
Gray*. The disease announced itself as lesions on the body, much like
the bodily symptoms of vampirism for which the characters in *Dracula*
search their skin. It has been suggested that Stoker died from syphilis
contracted from a prostitute; this seems unlikely, although that is his
fate in Aldiss's metafictional *Dracula Unbound* (1991).

Aldiss's book was published when an entirely new sexually transmit-
ted disease had grown to epidemic proportions: AIDS, as an allegory
of which vampire mythology seems peculiarly tailored. There is the infection by blood, the enervated state of the victims, and also the suggestion pushed by homophobic media reports that AIDS is associated
with perverse sexual behavior. As is often the way with modern myths,
there was already a new version available to explore the metaphor: Anne
Rice's Vampire Chronicles, beginning with *Interview with a Vampire*
(1976), and followed by *The Vampire Lestat* (1985) and *The Queen of the
Damned* (1988). The critical reception and public response was much

more positive for the latter two novels than for the first, and perhaps that reflected how society had changed—not just because the strength of the gay rights movement by the mid-1980s had created an environment more receptive to the themes, but because the precarious existence of Rice's vampires seemed to speak to the devastating effects of AIDS in the gay community. "I happened to encounter *Interview with a Vampire* at the height of the AIDS epidemic," writes author Audrey Niffenegger,

> and it seems in retrospect to be a prescient book. Anne Rice could not have known how her creation would resonate with a world in which blood itself was dangerous, in which male homosexuality and death became closely entwined.

Although Rice was comfortable with such interpretations, she denied having any intention of making her Vampire Chronicles a gay allegory. But the homoeroticism is plain to see; the vampires Louis and Lestat live for years like a gay couple with their young adoptive daughter Claudia. That analogy arguably does gay parents no favors, however, since the prospect of Claudia maturing into an adult woman within the body of a perpetually young girl makes for some of the most disturbing—and, it must be said, mythopoeic—material in *Interview with the Vampire*. "You're spoiled because you're an only child," Lestat tells her, to which she languidly responds, "I suppose we could people the world with vampires, the three of us." By daring to voice such things, Rice gives the vampire myth a genuinely contemporary infusion, the horror element here being far less significant than the capacity to provoke taboo-busting unease.

<hr />

Viereck's psychic vampires stalked late-Victorian and Edwardian sensibilities. Arthur Conan Doyle's 1894 short story "The Parasite" tells of a "man of science," Professor Gilroy, who falls under the mesmeric spell of an elderly female psychic named Miss Penclosa, who becomes able to command his body and mind, and to make him her lover against his will:

I am for the moment at the beck and call of this creature with the crutch. I must come when she wills it. I must do as she wills. Worst of all, I must feel as she wills. . . . She can project herself into my body and take command of it. She has a parasite soul; yes, she is a parasite, a monstrous parasite.

The root of the anxiety here is the perceived power of women to make men go bad. As we'll see later, Conan Doyle came up with the antidote.

That "mesmerism" was still in popular currency at all, let alone with a medically trained (if notoriously credulous) man like Conan Doyle, testifies to a long-standing fear of this kind of psychic possession and domination. Proposed in the eighteenth century by the German doctor Franz Anton Mesmer, mesmerism was a form of hypnotism that invoked the spurious principle of "animal magnetism," a quasi-scientific theory according to which mental and physical health could be altered by directing the flow of a magnetic fluid in the body. Mesmer himself claimed to have restored the sight of a blind girl in Austria who played the piano so marvelously that Mozart was said to have composed a piano concerto for her. Mesmer's healing powers were discredited by a French royal commission in 1784, but his ideas merged during the nineteenth century with a burgeoning interest in both parapsychology and hypnotism — the latter became a standard technique of psychoanalysis. The charismatic figure who dominates others by casting a hypnotic spell was exemplified by Svengali in George du Maurier's 1895 novel *Trilby*, where the link to sexual seduction is made explicit: Svengali maliciously exploits a young Irish girl while granting her the "immortality" of fame.

There is something of this nature too in Dracula. Jonathan Harker is under the spell of the Count's commanding presence from the moment they meet in Transylvania, while poor creatures like Renfield are easy psychic prey. Bela Lugosi seems intent throughout the 1931 movie on transfixing his victims with his stare, and by all accounts many female (and perhaps not only female) viewers were happy to submit. He is, Auerbach says, "more mesmerist than biter." Like Svengali (who is generally viewed now as a stereotype of the menacing Jew), this allure comes in part from his status as outsider.

The vestiges of mesmerism merged in the late nineteenth century with a resurgent mysticism and the enthusiasm for spiritualism, lending credence to ideas about mind control, telekinesis, and spirit beings that exist on an "astral plane." Theosophical writings of that time are saturated in such notions, and the 1904 book *Modern Vampirism: Its Dangers and How to Avoid Them* by A. Osborne Eaves is a typical tract of this sort. Instead of regarding vampires as supernatural beings, Eaves wraps them up in mystical pseudoscience, their threat being less to our soul or vital fluids than to our psyche. Vampires are undead beings that wander the astral plane and "must feed on the emanations arising from blood and alcohol; public houses and slaughter-houses are thronged with these unhappy creatures, which hang about and feed thus." They have the power to seize our will and compel us to commit crimes.

There is a moral message in all this: "It is only possible," Eaves declares, "for a man to become a Vampire by leading a really wicked and utterly selfish life." You can escape the creatures' depredations with "purity of thought," avoiding "sensuality" as well as alcohol ("which retains evil magnetism"). In other words, the threat of vampirism is here a tool for enforcing prescribed social behavior: no sex or drink. Eaves even quotes *Dracula* in support of his claims, as if it were an authoritative tract on the phenomenon and not a potboiler novel. It's characteristic of this literature, too, to find the argument bolstered with a strong dose of sub-Darwinian racism: people of Hungary and Serbia, says Eaves, have "a strain of fourth-race blood in them, and the true Vampire belongs to this race." The noble British, meanwhile, "are of the fifth great root-race, and have outgrown these beings of previous evolution." Vampires, like werewolves and the savage Mr. Hyde, represent evolutionary regression.

The psychic vampire often appears in the form of the seductive femme fatale or "vamp" of pulp fiction. "The Girl With the Hungry Eyes," a 1949 short story by the fantasy writer Fritz Leiber, later adapted into a film and TV series, features a vampiric pinup who entraps her victims via the billboards on which her image appears—a clever allegory on the mesmeric, soul-draining sexual allure flaunted by the advertising industry. "There are vampires and vampires, and not all of them suck blood," says the narrator. This collusion of vampires and advertising came full circle when the horror series *True Blood*, which ran on HBO from 2008 to 2014, licensed its vampire tropes in a series of tie-ins advertising (in highly sexualized fashion) Gillette razors and Harley-Davidson motorcycles.

———— ◆ ————

Lord Ruthven—urbane and demure, aristocratic, insistent that a gentleman do the right thing—is a very British vampire. Stoker's Count is different. He is suspiciously foreign—specifically, he is Eastern European. The Briton of the late nineteenth century knew what that meant, for *they* were over here already with their harsh tongue and strange ways, threatening to pollute the Anglo-Saxon bloodline. Between 1867 and 1883 around twelve thousand of the immigrants who arrived in Britain from Eastern Europe were refugees from the Jewish pogroms; most settled in the East End of London. Newspaper headlines—"Still They Come"—whipped up xenophobic fervor, leading to civil unrest, protests, and marches by far-right groups. Eventually the British government passed the first legislation in peacetime Britain to curb immigration, the Aliens Act of 1905.

Dracula played to the subliminal invasion fears of Stoker's audience. Like the refugees, the Count has been forced from his native country—"barren of people" on whom to feed, as Van Helsing explains, whereas Londoners offer sustenance "like the multitude of standing corn." "So he came to London," says Mina, "to invade a new land." Might he not then spawn a whole race of vampires, "a new and ever widening circle of semi-demons to batten on the helpless," as Jonathan Harker puts it? Didn't the Count tell Harker, after all, that he comes from a "conquering race"?

So it's surely not incidental that the Count sets up one of his London residences in Whitechapel in the East End, nor that when he flees the country he depends on the assistance of a Jew, who is doubtless arranging all kinds of traffic in illegal immigration. Stoker indulges all the usual anti-Semitic stereotyping here, calling the man "a Hebrew of rather the Adelphi Theatre type, with a nose like a sheep, and a fez," and adding that, like Judas, for a fee he gives Harker and colleagues the information they will need to follow Dracula's movements. They are two of a kind, the vampire Count and the Jew: hoarders of gold and agents of internationalist capital, infiltrating the nation by buying up property and setting up business.

This xenophobic invasion rhetoric often wore a quasi-medical mask in the late nineteenth century. It was couched in terms of infection and contagion—the same language used in the new germ theory of disease developed by Louis Pasteur and Robert Koch. Whereas previously disease had been considered specific to a location—the "miasma" theory ascribed it to bad air (*mal aria*), of the sort one might find in fens and swamps—now the early microbiologists were saying that agents of disease could be passed between individuals. The blame, by implication, fell upon those who carried the disease—all too often, foreigners.* In this respect, germ theory provided a modern scientific alibi for the old prejudice that diseases came from other nations and peoples: the Spanish disease, the French pox. The question is essentially that posed by David Punter in his survey of Gothic literature: "to what extent can one be 'infected' and still remain British?"

This is the same species of question explored in John Ajvide Lindqvist's 2004 vampire novel *Let the Right One In*, better known from the 2008 Swedish movie and the 2010 American remake *Let Me In*. Who were the right immigrants to "let in" to Sweden was a question being furiously debated when Lindqvist's book was published: the country had imposed tough entry restrictions and, like the rest of Scandinavia, took a strongly assimilationist attitude toward incomers. The vampire Eli looks pale enough to be Nordic (even too much so?), but her identity is decidedly ambiguous: is she a girl or a hermaphrodite? That's the problem with vampires today: you can't spot them easily. They "might as well be

---

* In the febrile atmosphere of post-Brexit Britain, these enduring prejudices were apt to leak into the most unexpected places. During a seemingly anodyne BBC radio discussion about pets, one "dog expert" was asked about the effects of Brexit and volunteered the view that it might stem the influx of disease-carrying dogs from Eastern Europe. Perhaps she was thinking of ones like the "fierce brute" seen jumping from the wreckage of the *Demeter* shipwrecked in Whitby Bay?

our next door neighbor," writes literary academic Jules Zanger, "as Dracula, by origin, appearance, caste, and speech, could never pretend to be."

This sense of plaguelike infestation and contagion is one of the ways Stoker locates *Dracula* within the currents of the science of his day. Vampirism, like other diseases, may be contracted through the blood, and can be cleared by transfusion. When Jonathan Harker, Van Helsing, and their companions go looking for Dracula's casket in his house at Carfax, they are confronted by a horde of rats: "their moving dark bodies and glittering, baleful eyes made the place look like a bank of earth set with fireflies." Werner Herzog returns to that image, so evocative of ancient fears of the plague, in his 1979 remake of *Nosferatu*, in which the arrival of the Count in Bremen is heralded by rats swarming through the street as if following a demonic Pied Piper.

In Stoker's time there was an existential element to these fears of infection and illness. In his book *Degeneration*, translated into English in 1895, the physician Max Nordau (born in Pest in today's Hungary) warned that degeneration in art and morality, along with the stresses of an urban industrial culture, were corrupting Western society, sapping its vigor and causing both physical and spiritual decline. Here too the language was medical, but the symptoms—languor and sexual degeneracy—were all but indistinguishable from those experienced by Dracula's victims. Stoker was familiar with Nordau's ideas. "The Count is a criminal and of criminal type," Van Helsing explains. "Nordau and Lombroso would so classify him"—referring here also to the Italian criminologist Cesare Lombroso, to whom Nordau dedicated his book.

Degeneration was, as we saw in the previous chapter, a specter awakened by Darwin's *Descent of Man* (1877), which linked humans to more primitive forms of life. Darwin's ideas were crudely recycled in the supposed hierarchy of races, with northern European whites placed highest on the evolutionary tree. Darwinism also led to a belief that individuals could regress to a more bestial state (page 150) and that this, like the "primitive" character of non-European people, would be apparent from their physiognomy. These ideas, fueled by prejudice about race and class, were promoted by Lombroso, and Stoker absorbed them. In *The Lair of the White Worm*, the servant of the novel's villain is a grotesque stereotype of an African: "a negroid of the lowest type, hideously ugly, with the animal instincts developed as in the lowest brutes . . . in fact, so brutal as to be hardly human."

Degeneration theories also inform the portrayal of Renfield. British psychiatrist Henry Maudsley believed that mental illness represented a regression along the evolutionary scale: a decline not only in powers of reasoning but also in morality. As Renfield turns ever more crazy in Seward's institution, he begins to eat insects like a "savage": an object lesson of what lies in store for us if the vampire bites.

The threat Dracula poses is the oldest racist slur in the book: he will rape white women and turn them into whores. Lucy succumbs, but Mina's recovery is confirmed at the end of the book when she becomes the acme of decent womanhood: a mother.* Yet the Count, for all his strength and ferocity, is ultimately less advanced and is defeated because his foes are both spiritually and technologically superior—the latter evident in how the final pursuit and ambush of the fleeing Dracula relies on railway timetables, ship tracking, and telegrams.

With his prominent nose and exaggerated brow, Murnau's Dracula/ Orlok is not so much the criminal type as the stereotypical anti-Semitic caricature. Vampire tales became the perfect allegory for conspiracy theories about cabals who walk unseen among us: when *Nosferatu* was made, the anti-Semitic *Protocols of the Elders of Zion*, a fabricated text alleging to describe plans for Jewish world domination, had begun circulating. The trope of a vampiric Illuminati remains popular today. American writer Dan Simmons's best-selling horror novel *Carrion Comforts* (1989) describes an amoral sect of psychic vampires who use their mesmeric power—called The Ability—to direct the course of world history, orchestrating events ranging from the Holocaust to the murder of John Lennon, in the manner of the best (worst) conspiracy theories. And in David

---

* I like to imagine that Jonathan and Mina's baby son, named Quincey after their brave Texan companion, represents the first of those demon children on which the camera zooms at the end of the horror film: is the curse still alive within him? But I admit it is highly unlikely that Stoker had anything of the sort in mind.

Mitchell's 2014 novel *The Bone Clocks* (and the spin-off *Slade House*), a team of resistance fighters called the Horologists wage eternal battle against a secret society of soul-devouring psychic vampires called the Anchorites, against a backdrop of climate-induced civilizational collapse.

The *Athenaeum* magazine loftily proclaimed *Dracula* "wanting . . . in the higher literary sense." Other reviewers failed to see the point of implanting this old superstition in the heart of English society. "A vampire with a cheque-book, a solicitor, and a balance at the bank," wrote Andrew Lang, "is not a plausible kind of creature." How wrong he was; such things were precisely what made vampires *modern*.

But the reviews of *Dracula* were mostly enthusiastic: the *Daily Mail* compared it favorably with *Frankenstein*, *Wuthering Heights*, and Poe's *The Fall of the House of Usher*. The book was popular enough that Stoker soon had to worry about pirate editions. He was never to repeat this success, though not for want of trying. *The Jewel of Seven Stars* (1903) was an orientalist mummy yarn that even the interest in Howard Carter's discovery of the tomb of Queen Hatshepsut in the Valley of the Kings in the 1890s could not rescue from its mundane melodrama. *The Lady of the Shroud* (1909) was a late Gothic potboiler set in the vampire territory of Dalmatia. Of *The Lair of the White Worm*, we've heard quite enough already. When Stoker died in 1912, his widow Florence was left with a sizable estate. But the couple had a long habit of living in properties that were beyond their means, and soon her circumstances were straitened. She sold her late husband's books and papers at auction for less than she might have hoped and kept a close eye on the rights to his works in order to stay solvent. When Murnau's *Nosferatu* appeared, advertised as "freely adapted" (without permission) from Stoker's novel, Florence started legal proceedings to suppress it. The film company Prana went bust, however, before she could recover a penny for copyright infringement.

It's fortunate that Florence Stoker was unable to enforce her wish to have all copies of the film destroyed, for the movie is now recognized as a

Count Orlok goes on the rampage in F. W. Murnau's Nosferatu *(1922).*

classic of German expressionist cinema. It has itself inspired at least two inventive reinterpretations: Werner Herzog's aforementioned 1979 remake and E. Elias Merhige's inventive metafictional *Shadow of the Vampire* (2000), in which the actor Max Schreck,* who played Orlok (and is here played by Willem Dafoe), turns out to be a real vampire.

Murnau had rather crudely tried to duck the issue of copyright by changing the names and locations. Orlok seeks to move from Transylvania to the fictional Wisborg in Germany, while his servant Renfield becomes Herr Knock, the employer of Hutter (Harker), who is in league with the vampire from the start. Orlok is more akin to the feral vampires of folk legend than to Stoker's lordly count, his features bestial and rat-like: a devil from a children's fairy tale. The French critic Roger Dadoun memorably describes him as "an agglomeration of points." Murnau's use of theatrical trickery is still unnerving today, as for example when Orlok rises from a prone position in his coffin propelled by a hidden mechanism. The movie abandoned all of Stoker's modernism to produce

---

* His (real) name means "terror" in German.

images drenched in primal fear. When prints reached the United States, the *New York Post* declared it "no momentary horror, bringing shrieks from suburban ladies in the balcony, but a pestilential horror coming from a fear of things only rarely seen."

It was Murnau who introduced the idea—absent from any traditional accounts—that sunlight will kill a vampire. In *Dracula*, the setting of the sun is merely the cue for a vampire to awaken—but the destructive power of the sun's rays invoked in *Nosferatu* is perhaps extrapolated from the way Dracula's body "crumbled into dust" when he is killed by Harker and Quincey Morris as the sun sets behind the Carpathians. The idea that a vampire can't survive in daylight doubtless proved such a popular addition to the myth because it emphasizes the isolation of the vampire from normal society, forcing these creatures to live at the margins—like the ultraviolent vampire "family" in Katherine Bigelow's film *Near Dark* (1987).

Florence Stoker finally accepted that the only way to monetize her late husband's sole successful book was to permit adaptations. In 1924 she sanctioned a stage play written by the actor-director Hamilton Deane, who had once been in Henry Irving's touring company (Deane's family, of Irish descent, knew the Stokers). With just a tiny budget to draw on, the production set the action almost entirely in an English drawing room (Dr. Seward's sanatorium); of Transylvania there was no glimpse. This setting explains the tuxedo worn by Bela Lugosi in Tod Browning's 1931 film based on the play. His high-collared cape, meanwhile—now a stock part of the Dracula fancy-dress costume, faithfully observed in spoofs like *Hotel Transylvania*—was introduced for a stage effect in which the Count, played in Deane's production by Raymond Huntley, could vanish down a trapdoor without his head being visible once he turned his back In another piece of trickery, a dummy was substituted for the actor in the climactic killing so that the body would seem to disintegrate.

The play updates the Victorian setting: Dracula flies (by airplane!) to England and Jonathan Harker travels by motorcar. Oddly, Mina and Lucy swap roles: Mina is killed and staked as a vampire (all decorously offstage), while Lucy is Jonathan's fiancée. It's a static affair, with a gratuitous love subplot: a Noel Coward melodrama with fake fangs. But despite the critical panning the play received in London, the American publisher and theater impresario Horace Liveright saw some potential and reached an agreement with Florence Stoker for the play to be adapted by writer John Balderston for Broadway, where it opened in October 1927. Huntley had asked for too high a fee to reprise his role, and so it was given instead to an unknown Hungarian actor named Béla Blaskó, who took the stage surname Lugosi.

The following year, Universal became interested in turning the play into a film, for which another deal was agreed with Florence Stoker. Dracula was originally going to be played by Lon Chaney, who had already made a vampire movie with Browning for the Metro studio*—but he could not escape his Metro contract, so Lugosi was retained in the role for a paltry fee, having made rather too plain his desperation for the part.

Lugosi's mannered, static performance seems camp today—Skal describes it as "Valentino gone slightly rancid," while Nina Auerbach says he is "so dependent on commanding attitudes and penetrating stares that he practically turns into a monument." He spoke little English and allegedly recited his lines phonetically, without necessarily understanding what he was saying; whether or not that is true, his stilted delivery makes it believable. All the same, this figure in elegant evening dress with slicked-back hair made a very different impression from the white-locked and slightly fetid Dracula of Stoker's book, and ensured that every Dracula since—from Christopher Lee to Frank Langella† to Gary Oldman—has had to ooze sex appeal.

---

* It appeared as *London after Midnight*, with Chaney looking more like a camp music-hall version of *Chitty Chitty Bang Bang*'s Child Catcher—a weird character who became the inspiration for the silhouette-demon of *The Babadook* (2014). Browning remade it in 1935 as *Mark of the Vampire*, with Lionel Barrymore in the title role.

† Langella took the role in a revival of the Deane-Balderston play for Broadway in 1977, and kept it for the filmed version by John Badham that same year.

Enter Bela Lugosi as the most famous screen Dracula.

In its opening scenes, Browning's film surpassed anything the stage could offer. Dracula's castle in the Carpathians was thrillingly rendered in a lavish Hollywood set, the sweeping stone staircase festooned with cobwebs and raked with the light of a sinking sun in scenes that evoke a cathedral-like religiosity.* While Universal squeezed as much use from this set as possible by recycling some of it for Whale's *Frankenstein* later that year, it consumed much of the budget and only serves to emphasize

---

* Undermined somewhat by the armadillos that, bizarrely, scuttle amid the shadows.

the cramped style of the second act, where Dracula reverts to menacing the British contingent in a stuffy drawing room. Browning altered the plot to make Dracula's visitor in Transylvania not Jonathan Harker but a bizarre, mincing Renfield, played with his usual zest by Dwight Frye. As in the Deane-Balderston play, "John" Harker is no hero but a blundering upper-class idiot, excitable and dim. "You're so—like a changed girl," he moons to the vampire-bitten Mina (now, as Seward's daughter, restored as his fiancée). "You look wonderful!" As Dracula threatens his beloved, his only response is that they should "call the cops."

Lugosi became a cult figure, of course—but he was ultimately another of Dracula's victims. He never escaped the association with his screen debut, and his cachet diminished with every subsequent appearance. By 1948, eight years before his death, he was camping it up as Dracula in *Abbott and Costello Meet Frankenstein*, after which he could find only low-budget roles and spiraled into poverty and drug addiction. He famously reprised the caped vampire in his last film, Ed Wood's *Plan 9 From Outer Space* (1959), which has been generously described as "the worst film ever made." The silent footage of the former star was shot by Wood three years earlier for a different (unmade) movie, but he inserted it both as a desperate bid to include a "big" name on the bill and in tribute to his late friend.

Tony Scott's 1983 vampire film *The Hunger* opens with Pete Murphy of the Goth-rock group Bauhaus singing "Bela Lugosi's dead." It's a stylish acknowledgement that vampire stories are no longer versions of *Dracula* but of the Dracula myth—which has its roots as much in Hollywood as in Stoker's novel. To see, in *The Hunger*, the impossibly beautiful, pale, and bloodsucking David Bowie age into a wrinkled creature and die (even as his elegantly wasted young acolyte Murphy inherits his vocal stylings) is now to witness a poignant indication of how uncannily well Stoker's tale created a template of androgynous, sexualized decadence to serve the coming century and beyond.

When *Dracula* reached celluloid, the message found its medium.

The iconography of *Frankenstein* was forged in film, but *Dracula*, more than any other modern myth, seems in retrospect to have been created with cinema in mind: Stoker's epistolary patchwork was just waiting for the cross-cut to dissolve space and time. Two years before the book was published, the Lumière brothers staged the public premiere in Paris of their Cinématographe, showing not their famous short film of a train rushing toward the audience (to which the hysterical response of viewers has probably been exaggerated) but a more mundane scene of workers leaving a factory. When he witnessed one of these spectacles in 1896, Maxim Gorky wrote:

> It is terrifying to see, but it is the movement of shadows, only of shadows. Curses and ghosts, the evil spirits that have cast entire cities into eternal sleep, come to mind and you feel as though Merlin's vicious trick is being enacted before you.

Francis Ford Coppola includes the Lumières' screening in his *Bram Stoker's Dracula* (1992), and the sepia fades and fast-motion sequences of that movie step outside the narrative to tell us that the medium, Derrida's "art of ghosts," is now a part of the story too. Coppola is commenting on cinema even as he creates it.

From the outset, motion pictures were understood to offer a kind of immortality, creating not just the frozen static likeness of the photographic portrait but eternal youth with a semblance of animation. And that was not merely a foolish superstition of credulous *fin de siècle* audiences: film really does reconfigure time in our minds, the present changing the past. It's impossible now to watch Bowie's ravaged old corpse in *The Hunger* without thinking of his gaunt face, preparing for death, in the video for his 2016 song "Lazarus," or his pajama-clad body withdrawing into the cupboard-coffin and closing the lid on the light. The Thin White Duke was always the archetypal Wildean vampire.

It's no exaggeration to say that cinema's perpetual fascination with its power to confer immortality and, conversely, its public exposure of the tragic decay of youthful beauty and vitality, originates at the nexus of *Dracula* and *The Picture of Dorian Gray*. Whether it is Bette Davis as the faded movie star preying vampishly on her wheelchair-bound sister Blanche (Joan Crawford) in *Whatever Happened to Baby Jane?* (1962) or Marlon Brando's grotesque Kurtz inviting us to remember the gorgeously priapic rebel of his youth as he mutters "The horror!" in Coppola's *Apocalypse Now* (1979), the movies self-consciously enact the corruption of the flesh. Cinema today is only enhancing its capacity to resurrect the dead. If it is coincidence then it is an uncanny one that the actors most strikingly rejuvenated or brought back from the dead by digital manipulation have all appeared in roles that probe the boundaries of life, lifelessness, and death: the screen's most famous Van Helsing, Peter Cushing (*Rogue One*, 2016), Frankenstein's creature Robert De Niro (*The Irishman*, 2019), and the self-aware replicant Sean Young (*Blade Runner 2049*, 2017).

There have been over three thousand vampire movies, and countless works of vampire fiction. Many of them, of course, are quite dreadful, because many movies and novels of any kind are—but it should be clear by now that "artistic quality" is no measure of the mythic significance of a retelling. Sure, we might see in the bargain-basement production values, camp sexism, and scenery-chewing of the Hammer films something of the "prurient debasement" that writer Clive Leatherdale condemns in *Dracula: The Novel and the Legend* (1985), but this demotic character can be candid too. Nearly all of the tales of Dracula and his vampire brood have something to tell us, although admittedly it can sometimes be rather wearisome listening to what that is.

One of the best-known, Stephen King's *Salem's Lot* (1975), develops the nightmare articulated by Jonathan Harker of a "vampire invasion"—transferred now (as so often in King's work) to small-town America. "The thing that really scared me," King said, "was not the vampires, but the town in the daytime, the town that was empty, knowing that there were things in closets, that there were people tucked under beds, under the concrete pilings of all those trailers." Salem's Lot in Maine is terrorized by an Old World vampire named Kurt Barlow and his charming accomplice and frontman Richard Straker. Beginning with a young boy called Danny Glick, the duo turn the town's inhabitants one by one into creatures like themselves, who must stay hidden during the daylight hours because sunlight burns their flesh. Before long, half the town is sleeping away the day under the floorboards and roaming the roads for blood after sunset.

Salem's Lot makes shrewd use of the canonical features of the horror genre: everyone in town knows from the books and movies that vampires must be staked, and Danny's friend Mark keeps the vampire child at bay with the plastic cross seized from his Aurora glow-in-the-dark plastic model of Frankenstein's creature (yes, *that* one!). When the hero, Ben, is told that folklore dictates the vampire's bite marks vanish after death, he replies that he knows it already because he's watched the Hammer films.

The vampire epidemic that consumes the town seems to feed less on blood than on the simmering rancor and bile of American society: the jealousies, sexual tensions, ugly prejudices, and domestic abuses that constitute the dark underbelly of a nation ill at ease with itself. Like vampires, these social evils stay hidden behind closed doors but seep out after dark. All the traditional deterrents to the vampire are impotent here: garlic, crosses, and holy water won't save you, just as pious moralities are no defense against the nastiness that festers in communities like Salem's Lot. The settlement itself ends up literally a ghost town, and we're left to guess if ultimately the epidemic spreads further afield—but perhaps that hardly matters, for doesn't this kind of blight already afflict every small town in the United States? King has said that the book, written during the Watergate hearings when it seemed Richard Nixon's corruption had spread its infection throughout the US government, was partly also a subconscious political allegory.

The Dracula myth surely now includes the notion, unsanctioned in Stoker's book, that the Count is not simply ancestrally related to the original, historical Dracula, Vlad Tepes "the Impaler," but literally *is* him, preserved by his vampiric immortality. The notorious Vlad is barely mentioned in the novel, and never by name, but the link to the Wallachian Voivode was played up in the fanciful book *In Search of Dracula*

(1972) by Raymond McNally and Radu Florescu. This connection was incorporated into the TV movie *Bram Stoker's Dracula* the following year, with Jack Palance in the title role.* The conceit here is that the Transylvanian count visited by Jonathan Harker is the fifteenth-century Dracula himself, who has found a way to conquer death. Seeing Lucy standing next to Mina in a photograph Harker carries, Palance's Dracula recognizes her as an exact likeness of his former queen, and he comes to England intent on reuniting with her. Deadly he may be, but he is also a tragic, romantic, and thoroughly humanized figure.

Thus a retelling that presented itself as a return to the "authentic" story in fact recast it fundamentally: we are asked to find sympathy for the devil. (Poor Jonathan Harker, meanwhile, goes bad, becoming a vampire and getting staked in a pit of spikes.) Frank Langella shows the Count at his most alluring in John Badham's *Dracula* (1979): subtitled "A love story," it styles Dracula "the greatest lover who ever lived" (with hair like John Travolta's in *Saturday Night Fever*, how could he not be?). Mina and Lucy are here the daughters of Van Helsing and Seward, chafing under the yoke of strict paternal control, and Dracula seems more like a liberator than a predator.

The Vlad Tepes backstory was included in Coppola's *Bram Stoker's Dracula*, where now it is Mina who personifies the long-lost wife of the Count (Gary Oldman as a goateed Edwardian dandy). Mina is not merely a doppelgänger—she has glimpses of her former life as Queen Elisabeta, whose grief-stricken suicide while Vlad Dracula was fighting the Turks drove him to renounce God and swear vengeance. The film's tagline archly mimics the title of a corny rock ballad: "Love never dies."

Dracula's apocryphal origin story took on a life of its own in the 2014 movie *Dracula Untold*, set firmly in the fifteenth century, where Luke Evans's hunky Vlad the Impaler makes a pact with a wild vampire in the mountains of Wallachia so that he can acquire the power to defeat his

---

* The film's screenplay was by Richard Matheson, whose vampire novel *I Am Legend* (1954) shaped the emerging zombie myth (see chapter 9).

enemies. The film ends with the now immortal Dracula in modern-day London, once again meeting a woman who looks just like his fifteenth-century wife Mirena.

Humanizing the monster is one of the most obvious trends in the horror narratives of the past several decades. It is more surprising for Dracula than for Frankenstein's creature, who was after all treated with some sympathy by Mary Shelley herself. Perhaps this process reflects the gradual internalization of the vampire: a subconscious acceptance that, once the creature was transformed from werewolf-like beast to the all-too-human Ruthven, Varney, and Dracula, it could perform more effectively as a projection of our own desires. If the vampire is going to be not the invading alien but an aspect of ourselves, we want it to look its best.

Developing this sense of identification further, modern-day vampires tend to live in communities, often diseased or decaying and symbolizing the marginalized and downtrodden. "They're really one more oppressed minority, aren't they?" says a character in Aldiss's *Dracula Unbound*. Barbara Hambly's *Those Who Hunt the Night* (1998) present the undead as victims, threatened by a mysterious killer who the vampire hero must track down in order to preserve their race. The Ina, a vampire species, are a nocturnal minority group in Octavia Butler's *Fledgling* (2005); the young heroine Shori is made doubly an outsider because, in contrast to all other Ina, she is genetically modified to have dark skin. And regardless of whether Anne Rice meant her Vampire Chronicles to supply an allegory for gay oppression and homophobia, setting the tales in pre–Civil War Louisiana means that there is no avoiding the theme of slavery. It's complicated, though: Louis de Pointe du Lac, the subject of the "interview with a vampire," is an ex–plantation owner who preys on his slaves, and it is they who first come to see him for what he is, forcing him to flee with his companion Lestat. The 2010 horror novel *Abraham Lincoln, Vampire Hunter* by Seth Grahame-Smith (made into a movie by Tim Burton), in which the president fights vampire slave owners of the American South, might not be the most sophisticated take on Karl Marx's metaphor of vampiric exploitation, but you can see where it is coming from.

Marginalization and alienation are familiar territory for young vampires. The friendship between lonely, bullied twelve-year-old Oskar and his female vampire protector Eli is at the emotional core of *Let the Right One In*. Joel Schumacher's *The Lost Boys* (1987) introduced the gang of cool but aimless teenage vampire rebels, paving the way for the young-adult romance of Stephenie Meyer's *Twilight* series, where the floppy-haired, wan-faced, impossibly good-looking bloodsuckers are preserved that way forever—and isn't your eternal soul worth trading for Robert Pattison's cheekbones?

The *Twilight* books recast the vampire mythology as fairy tale. That means removing from Stoker's stew all that is disturbing and perverse, and creating instead a coded language for sexual initiation. Stephen King has said, "It's very clear that [Meyer is] writing to a whole generation of girls and opening up kind of a safe joining of love and sex in those books. It's exciting and it's thrilling and it's not particularly threatening because it's not overtly sexual." That's no criticism in itself (although King's comments on Meyer's literary skills certainly were). But it reflects the way popular culture will always, in dealing with modern myths, include versions diluted to the point of being more or less bloodless. The *Twilight* franchise is a gentle palliative for adolescent sexual anxiety, and there will always be a place for such things. But it makes *Twilight* pretty much the opposite of *Dracula*.

The flipside of the same coin, Joss Whedon's cult TV series *Buffy the Vampire Slayer*, with its adult-proof, postmodern code of rules and references, supplies another vehicle for exploring teenage issues, albeit with a strong dose of self-satire. As young people know, fighting monsters is just something you have to get on with, and can even be sort of fun if you're in the right gang—but high school is hell.

The trope of vampire as romantic social outcast is wittily inverted by Kim Newman in *Anno Dracula* (1992), the first in his series of vampire alternative-history mash-ups, in which famous characters from history (such as Queen Victoria and the "Bloody" Red Baron) rub shoulders with (meta)fictional heroes (like an undead British secret agent called Bond). The arrival of Dracula in Victorian England, far from being foiled by Van Helsing's crew ("Van Helsing is dead. His head remains on its spike"), succeeds in spreading vampirism throughout society until it becomes a respectable social norm—all the right people are vampires now. They lord it over the remaining "warm" humans, while Dracula has declared himself the queen's consort and keeps her in chains. As Newman implies, vampires are only sexy and romantic when they are a minority. Let their contagion spread too far, and they'll become as boringly conservative as the society they have invaded.

Here, as in so many of the modern vampire tales, the Dracula figure is no longer an outsider arrived from a distant land. No: vampires have always lived, walked, and fed among us, often as part of a debauched, clandestine society with its own hierarchies and codes and loyalties. This slightly paranoid trope harks back to the heretical and demonic sects that Europeans were always discovering and persecuting in the Middle Ages—but modern vampires do not lurk in the primeval woods up in the mountains, snaring the unwary traveler. Like Dracula they have entered our cities, our apartments, kitchens, motel bars. They ride in limousines and challenge our comfortable scientific certainties, offering a counternarrative to the modern world. Once the feral demons of folklore became modernized, they were ready to illuminate modern themes: changes in sexual expression and gender roles, the impact of medicine and technology, fears about contagious disease and foreign invasion.

Still, there's no denying that one aspect of the vampire mythology is surely very old indeed. By overcoming mortality, they force us to confront it. "One of the things that's terrifying is that to prove that we're human we have to die and rot in the grave," says Gothic expert Nick Groom. "Vampires present us with this dreadful prospect that our humanity lies in our death and putrefaction."

And blood is the key. By drinking blood, the vampire both mimics and desecrates the Christian sacrament, thereby staving off the corruption of the flesh. (It's surely no coincidence that only this mythical monster seems vulnerable to the crucifix and holy water.) Because modern science has rejuvenated this old anxiety about death, with its blood transfusions, tissue and organ replacement, stem cell treatments, and the ultimate whispered promise of a transhuman technological immortality, we still need the vampire as much as ever—to remind us that eternal existence might not be a blessing, that our life has a beginning and an end, and that this is a precious part of what makes us human.

❦

Chapter 6

WHO SHALL DWELL IN THESE WORLDS

THE WAR OF THE WORLDS (1897)

Aliens invade Earth. They are physically repulsive, but their
technology is vastly superior to ours and they begin to lay waste
to the planet, destroying cities and killing thousands. Through pluck,
ingenuity, or sheer luck, humanity survives, the alien hordes are
overcome, and we must rebuild civilization from the ruins.

===

Several centuries from now, according to a scenario imagined by the
Planetary Society, a person on Mars will stumble across a tiny plastic disk:

This person may be encased in an elaborate space suit, or perhaps
breathing freely in the newly oxygenated atmosphere of the former
Red Planet. The surrounding landscape might be that of a harsh red
desert, as it is today, or it might be green rolling hills, the result of

intense terra-forming. We do not know. All we know is that when this
future Martian picks up this tame-looking disk, he or she will hold in
their hands a message from our world, addressed to theirs.

She (let's say) is perplexed at first about what this artifact is. But using
her advanced technology to analyze it, she realizes that the microscopic,
light-diffracting pits that cover its metallic surface encode information
in binary form, just as digital computers do. She figures out that it holds
a message, and with patience, technical skill, and ingenuity she decodes
what it says.

This future Martian learns that the disk bears the title "Visions of
Mars" in a now obsolete language of Earth, and that it was prepared and
donated by the Planetary Society—which, historical records tell her,
was a nonprofit organization to promote space exploration founded on
Earth in the year 1980 by the planetary scientist Carl Sagan and others.
The disk was carried to Mars on board the robotic Phoenix spacecraft,
which landed on the planet in 2008, long before it was terraformed into
a habitable state. On it she finds digital images of Sagan himself, as well
as an ancient writer called Arthur C. Clarke and other Earthling vision-
aries of the fledgling technologies of space exploration. She will learn
what the terrestrial astronomers of the nineteenth and twentieth centuries knew and thought about her planet.

And she finds a story written by one Herbert G. Wells. It doesn't flatter her. This Martian—the descendant of settlers from Earth—discovers
than in the late nineteenth century, dwellers on Mars were imagined to
be utterly alien, horrific to look at, and terrible in thought and intent.
She reads *The War of the Worlds*.

This, at any rate, is what the Planetary Society hopes, and who can
say it will turn out otherwise? Ogilvy, the ill-fated astronomer in Wells's
book, might complacently state, "The chances of anything man-like on
Mars are a million to one," but it seems entirely probable that humans
will walk there one day. Whether they will establish colonies and even
transform the face of the planet is another matter altogether, but entrepreneur Elon Musk has his spaceflight company SpaceX working on it
right now, and he wants to see colonists set off within a decade.

Perhaps the DVD onboard Phoenix might be regarded like the stone tablets and papyrus scrolls of antiquity. Sagan will be hailed as a priest and prophet, and Wells's story will be seen for what it truly is: a myth.

Will the Martians understand what the myth is about? Do we?

The War of the Worlds provides the template for alien-invasion fantasies of every stamp, from John Wyndham's apocalyptic novel *The Kraken Wakes* (1953) to Tim Burton's movie spoof *Mars Attacks!* (1996) and the post-invasion thriller *A Quiet Place* (2018). It did not, of course, invent the notion of malevolent nonhuman monsters, but in pre-modern times these creatures dwelt in far-away lands, their sightings reported in tall tales by intrepid explorers. They were fought by heroes; they were not invaders, did not threaten meek citizens, families, cities, the whole world. And they were supernatural or bestial throwbacks, not representatives of an advanced civilization. Their threat was personal, not existential. It was a new kind of narrative of the Other that Wells provided, and we can't stop looking into the "cool and unsympathetic" glare of these beings and fantasizing about the catastrophe and ruin that they may visit upon us.

—◆—

The final decades of the nineteenth century spawned several modern myths, yet history offers us little sign of the psychic upheaval that, in retrospect, we might infer to have nourished them. While the ideas of Darwin, Nietzsche, and Freud were reconstituting the way we perceive ourselves, daily life scarcely looks from this vantage point to have been on the brink of serious rupture. The technologies of telecommunication and mass media—the radio and telephone—had not acquired enough market penetration to seem transformative. Railways, on the other hand, had already become familiar enough to be prosaic. Even science appeared increasingly to be a fixed and rather comprehensive body of knowledge. The physicist William Thomson (Lord Kelvin) is commonly credited as saying, as the century drew to a close, that physics was complete except for increasing accuracy of measurement. The attribution seems to be apocryphal, but the complacency it imputes is not.

By contrast, the first two decades of the twentieth century were marked by widespread apprehension of a social and moral crisis, by revolutions both political (in Russia) and scientific (quantum theory, atomic and nuclear theory, relativity), and of course by a global conflict that marked a new era of total war. Yet in that period began no new myths that I can discern.

This is not as strange as it might seem. Myths are not made in times of conflict and revolution. They come from the stress and unease that precedes or follows a seismic shock, not from the shock itself. Neither are they responses to new discoveries; rather, they are an accompaniment, possibly even a contributory factor, to discovery. If an anxiety can be clearly articulated ("war has been declared"), it becomes poor material for mythopoesis. What you get instead is allegory—which has a social function too, but is a wholly different beast.

No, the ennui of the *fin de siècle* was created not by a sense of dislocation but by a perception that it was imminent. The Industrial Revolution had happened, technological modernity was here, and yet an uncanny normality prevailed. It couldn't last. Something catastrophic was surely approaching. "It is hard to resist the conclusion," says the literary critic Bernard Bergonzi, "that a certain collective death-wish pervaded the national consciousness at the time."

Herbert George Wells, an aspiring English writer of modest means and origins, responded to that anticipation by writing the first compelling vision of the quotidian shattered by apocalypse.

"To some extent," writes critic Robert Philmus, "all science fiction draws upon the metaphors inherent in current ideas and transforms them into myth." Wells was the first great exponent of science fiction as a distinct genre, and that made him the most deliberate mythmaker of the late Victorian age. Every one of the early scientific romances that made his name—*The Time Machine* (1895), *The Island of Doctor Moreau* (1896), *The Invisible Man* (1897), *The War of the Worlds* (1898), *The First Men in the Moon* (1901)—touches on myth, and all of them are familiar in rough outline even to many people who have not read them. When the writer Will Self confessed that *The War of the Worlds* was "one of those books that I felt I'd read long before I actually did so," he articulated one of the generic features of modern myths.

H. G. Wells (1866–1946), modern mythmaker extraordinaire, photographed in 1920 by George Charles Beresford.

Science fiction wasn't Wells's invention alone, of course. His greatest rival was the French author Jules Verne, whose fantastical visions such as *Journey to the Center of the Earth* (1864) and *Twenty Thousand Leagues under the Sea* (1870) stirred the imagination of a bourgeois audience since before Wells was even born. But Wells's myths have proved more enduring than Verne's, and ironically this is because he cared less about the science in his fiction. Verne wanted to make his technologies plausible according to the science of his day, while Wells admitted that a little "scientific patter" was generally all he employed to help the reader accept the fantastic hypothesis on which the tale was hung. He wanted to "hold the reader to the end by art and illusion and not by proof and argument":

> For the writer of fantastic stories to help the reader to play the game
> properly, he . . . must trick him into an unwary concession to some

plausible assumption and get on with his story while the illusion holds. . . . Hitherto, except in exploration fantasies, the fantastic element was brought in by magic. . . . But by the end of the last century it had become difficult to squeeze even a momentary belief out of magic any longer. It occurred to me that instead of the usual interview with the devil or a magician, an ingenious use of scientific patter might with advantage be substituted. That was no great discovery. I simply brought the fetish stuff up to date, and made it as near actual theory as possible.

Wells argued that his aims were different to Verne's. "His work dealt almost always with actual possibilities of invention and discovery," he wrote. "The interest he invoked was a practical one. . . . But these stories of mine [the early scientific romances] do not pretend to deal with possible things; they are exercises of the imagination in a quite different field." Wells saw very clearly which field that was: "They belong to a class of writing which includes . . . the story of Frankenstein."

"Hard SF," of which Verne might be regarded as the progenitor, strives for scientific rigor. It displays rather little mythogenic potential (and has rarely anticipated the technologies of the future either), whereas the fantastical "scientific romances" of Wells are brimming with it. "They tell a story symbolic of processes that are somehow inherent in all human destinies," said Jorge Luis Borges. "I think they will be incorporated . . . into the general memory of the species and even transcend the fame of their creator or the extinction of the language in which they were written." Who can say that they will not be read by future Martians?

⸻

Born in Bromley, Kent, in 1866, Herbert was the fourth child of a struggling shopkeeper. As a young man he went to an undistinguished private school before taking up an apprenticeship at a draper's in Southsea, an outlying district of the down-at-heel naval city of Portsmouth on England's south coast, where he spent nigh on four years in joyless drudgery. If he sometimes seemed determined in his later life to assume the stuffy, chauvinistic mantle of the respectable bourgeois Victorian gentleman, it's not hard to understand why: who wouldn't wish to escape a background in which poverty was just a few missed bills away?

His social ascent began when the young Wells secured a teaching post at a grammar school in Midhurst, Kent, and was able to enroll in 1884 on a course to pursue scientific studies at what was soon to become the Royal College of Science (and later, Imperial College) in South Kensington, London. There he heard lectures on biology by Thomas Henry Huxley, Darwin's staunch advocate and grandfather of the eminent biologist Julian Huxley, with whom Wells would publish the popular text *The Science of Life* in 1929–1930.

The young Wells longed to be a writer—a *real* writer, taken seriously by the critics. He would have nothing to support him except his own wits, craft, and graft (plus, like most male writers of the time, a wife to cook his meals and attend to his household). He married his cousin Isabel Mary Wells in 1891, but they divorced three years later when Wells began an affair with one of his former students, Amy Catherine Robbins (whom he called Jane). They were married in 1895, by which time they were already living together in a rented house in Woking, Surrey, their unwed cohabitation having occasioned some eyebrow-raising and curtain-twitching in the neighborhood. During this brief stay—Wells was always on the move, and the couple relocated to Kingston-upon-Thames in the summer of 1896—his output was prodigious. In his small, comfortable, archetypally provincial house, just by the rail station, he worked on *The Time Machine*, *The Wonderful Visit* (also published in 1895), *The Wheels of Chance* (a realist novel about cycling, one of Wells's passions), *The Island of Doctor Moreau*, *The War of the Worlds*, *When the Sleeper Wakes*, and *Love and Mr. Lewisham*.

Was this stuff, produced at such a frantic pace, any good? It depends on what you're looking for. Wellesian characters are never as vivid as those of Dickens; their inner life is not what you'd call complex. His stories tend to start at the beginning and relate events until they finish. The prose is plain.

He has certainly suffered more critical condescension than many

other famous middlebrow writers. Even in his own lifetime, says critic
Bryan Appleyard, he was seen by the literary set "as some sort of gullible
upstart":

> Well-born leftists in the Fabian Society regarded his popular fame and
> success with disdain, largely because he was born the son of a house-
> keeper and a shopkeeper. George Bernard Shaw loved to mock his pre-
> tensions and deride his talents and even his own literary patron, W. E.
> Henley, pleaded with him, just at the moment when he was producing
> some of the greatest books of his age, "to do better—far better and to
> begin with you must begin by taking yourself more seriously."

Yet to Borges and Joseph Conrad (both subtle stylists in their own
ways), the literary shortcomings Wells might have had in the conventional sense were beside the point. They understood that art doesn't need
a refined surface to reach deep levels of human experience. In speaking
of what sustains Wells's best books, Borges is also talking about the characteristics of modern myth in general:

> Work that endures is always capable of an infinite and plastic ambigu-
> ity; it is all things to all men, like the Apostle; it is a mirror that reflects
> the reader's own features and it is also a map of the world.

What makes this possible in a story, he adds, is, in essence, a distinct *lack*
of craftsmanship. The work "must be ambiguous in an evanescent and
modest way, almost in spite of the author; he must appear to be ignorant
of all symbolism." Or, better still, he may *really be* ignorant of it.

In short, it doesn't pay to take too much care when (inadvertently)
crafting a myth. Wells, although capable of writing truly fine prose, more
or less guaranteed by his absurd rate of productivity that he would not
take that care. He continued to produce a book a year (sometimes two)
almost nonstop from the publication of *The Time Machine* until five
years before his death in 1946. And it shows. Many of his books are
sloppily plotted and riddled with flaws and inconsistencies, to the point

where his editors implored him to take more time and stop spreading himself so thin.

What marked Wells out, however, was his vivid, even visionary invention, as well as a keen eye for weaving social commentary and well-informed scientific speculation into the fantastical. These attributes were apparent from the outset with *The Time Machine*. When the unnamed Time Traveller, arriving in the far future of humankind, meets the fey, feeble Eloi and the demonic, subterranean Morlocks who prey on them, we can discern both a bourgeois fear of the underclass and Darwinian speculation about divergent evolution in the human species. Wells prods the anxieties of his respectable readers about social and moral decline—a concern that, as we saw earlier, was widespread in the late nineteenth century and which contributed to the enthusiasm for eugenics among progressive thinkers (Wells included) in the decades that followed.

But it is perhaps not this battle between refined ego and savage id—the aesthetes and the brutes—that makes *The Time Machine* so memorable. Where the novel acquires mythic dimensions is in the final descriptions of the world's far future under a fading, swollen red sun. On a bleak beach fringed by an icy sea, the Time Traveller glimpses what remains of animal life on the planet: a vague round thing, "the size of a football perhaps," trailing tentacles. Was it our own evolutionary fate to become like those awful things that were about to come crawling from Mars onto the golf courses of Surrey?

The Time Machine was widely praised, and it turned Wells from a cash-strapped, struggling writer into a rising literary star. "He is more likely to make a reputation than any young writer of his standing," said the *Pall Mall Gazette*. Six thousand copies sold within weeks of publication, which would qualify as a hit even today.

The Island of Doctor Moreau was (setting aside the minor work *The Wonderful Visit*) a less successful second act, at least in its critical reception. This story of a rogue surgeon on a remote island who operates on animals to give them humanlike forms was too dark and brutal, simply too nasty, for late-Victorian tastes. Even now the grisliness of the book has a shocking power; Wells puts just enough humanity into the trans-

formed Beast People to make you pity them even while they remain uncanny and revolting. The novel ends with a theme that was to recur not
just in *The War of the Worlds* but in Conrad's *Heart of Darkness* (1899),
which can be read almost as a metaphysical *Doctor Moreau*. When the
protagonist, Edward Prendick, escapes from Moreau's island and returns
to civilization, his view of the world has changed beyond recovery. The
things he has witnessed aren't some awful aberration but reflect the
darkness lurking in all humankind:

> I know this is an illusion, that these seeming men and women about
> me are indeed men and women, men and women for ever, perfectly
> reasonable creatures, full of human desires and tender solicitude,
> emancipated from instinct, and the slaves of no fantastic Law—beings
> altogether different from the Beast Folk. Yet I shrink from them, from
> their curious glances, their inquiries and assistance, and long to be
> away from them and alone.

Here is all the horror of a decaying age of empire, mixed in with a dash
of Freud and Darwin. No longer are Europeans blessed with Crusoe's
conviction of a right to dominate, a natural hierarchy among peoples
and races. *Doctor Moreau* tells a new version of the Island myth, and it
is ugly. We seethe with those same impulses that we projected onto the
subjugated "savages," and which civilization teaches us to suppress. Indeed, imperialism is no less savage and bloodthirsty, but simply presents
itself with more decorum: the animals wear masks.

Those thoughts were still in Wells's mind as he cast around for a new
story that would restore his nascent reputation after the cool reception
of *Doctor Moreau*.

The War of the Worlds capitalized on a craze for all things Martian. That
there might be life, even intelligent life, on the Red Planet was a conviction shared by many scientists at that time. Dark, linear features had
been reported on the planet's surface by the Italian astronomer Giovanni

Schiaparelli in 1877; he referred to them neutrally as "channels," but when the Italian word *canali* was mistranslated into English as "canals," some assumed that they were purposefully built artifacts. In his book *Mars* (1895), the American astronomer Percival Lowell argued that these structures were the remnants of a dying civilization.

Schiaparelli's channels turned out to be a classic example of our tendency to project form and order onto things barely glimpsed: the habit of mind called pareidolia. The drawings he adduced in his book bear little relation to the true appearance of the Martian surface, which exhibits no such linear streaks. As it happens, Mars does carry "channels," sometimes more like meandering canyons, which planetary scientists today recognize as a sign that natural watercourses once threaded the now-dry landscape. Mars is thought to have had a warmer, wet climate not unlike Earth's early in its existence—but the surface water, which probably comprised great, shallow lakes and perhaps even a kind of ocean, dried around two billion years ago as the planet cooled. Some water remains, mixed with frozen carbon dioxide (dry ice), in the polar ice caps; some has escaped the planet's gravity (weaker than Earth's) and been lost into space. As water is deemed an essential ingredient for life as we know it, it is entirely possible that microbial life once existed on Mars, in which case there could be "fossil" traces in the planet's surface geology. But we need not fear a Martian invasion.

Wells's audience did not know any of this, of course, and neither did he.

Lowell's speculations were developed by the German writer and philosopher Kurd Lasswitz in his 1897 science-fiction novel *Two Planets*. The book tells of the discovery of a Martian base in the arctic and our subsequent interactions with the Martians—who are more or less humanlike, highly advanced, and peaceful. Our planetary neighbors are also relatively benign and collaborative in George Griffith's *Olga Romanoff* (1894), where Martian astronomers try to warn the earth about the impending catastrophic impact of a comet. This image of aliens as akin to a foreign race was a traditional view, harking back to the proto–science fiction of the seventeenth century (see page 63): the lunar beings encountered in Francis Godwin's fantasy *The Man in the Moone* (1638) and

Cyrano de Bergerac's *The States and Empires of the Moon* (1657) are, save for their gigantic stature, little different from people one could find in any European court. (That goes too for the remarkable dream of space travel related by the German astronomer Johannes Kepler in his *Somnium*, written in 1608, which supplies the epigram to Wells's book.*) One could trade with them, discuss philosophy, share a meal. If humans clashed with such aliens, it was rather like a conventional conflict between nations, not an existential struggle for survival.

Wells offered a very different perspective. He imagined a highly advanced race, desperate and dying, on the cooling, drying planet that Lowell had described. "To carry warfare sunward," says the (nameless) narrator of *The War of the Worlds*, "is, indeed, their only escape from the destruction that generation after generation creeps upon them." And so the Martians had long plotted the conquest of our planet. As Wells puts it in the book's famous opening passage:

> Yet across the gulf of space, minds that are to our minds as ours are
> to those of the beasts that perish, intellects vast and cool and unsympathetic, regarded this earth with envious eyes, and slowly and surely
> drew their plans against us.

Wells had the edge on his rivals because of the depth of his scientific knowledge. He flaunts it a little clumsily in the first pages of the book, suggesting he scarcely need remind the reader that Mars "revolves about the sun at a mean distance of 140,000,000 miles." (You knew that, right?) But he marshals recent understanding of Mars effectively to make his narrative seem plausible. We might have anticipated the dangers, the narrator tells us, if we'd heeded Schiaparelli more. And he mentions a 1894 report in the eminent science journal *Nature* describing an ominous "strange light" seen on the planet's surface, which he suggests might

---

*   *"But who shall dwell in these Worlds if they be inhabited?...*
        And are we or they Lords of the World?...
        And how are all things made for man?"

have been created in the process of casting the "huge gun" from which cylinders bearing the Martians were fired toward Earth.

Wells had close connections with *Nature*, where his college friend Richard Gregory was assistant editor. He was often invited to contribute to the journal; his first article, also in 1894, called for popular writing on science to attain high standards of accessibility. Gregory's review of *The War of the Worlds* in *Nature* in 1898 commented on how scientifically literate fiction like this might "create sympathy with the aims and observations of men of science."

Readers of the novel could see at once that this was a very different sort of story from the romantic melodrama of George Griffith. It seemed all too plausible. And the invaders were genuinely alien; communication and understanding between us and them seemed not just impossible but unimaginable. These space invaders have become archetypal: repulsive, tentacled, crawling things, inscrutable save for their cold, murderous resolve.

The book employs a reportage format. Apart from some retrospective accounts of events that the narrator has collected (mostly from the exploits of his younger brother, a medical student in London), we see only what he sees, with all the attendant confusion and terror. Wells might have taken this technique from Daniel Defoe's *Journal of the Plague Year* (1722), a fictionalized eyewitness account of the Great Plague in London in 1665—perhaps the closest equivalent in the early Enlightenment to the catastrophic threat to civilization itself posed by the Martians. It turned out to be a device perfectly attuned to the emergence of mass media in the twentieth century, in which events could be reported in real time through the narrow window of the reporter's lens.

When the first of the Martians' cylindrical craft crashes into Horsell Common in Surrey, it is regarded as little more than a curiosity. The narrator visits the site the next morning with the astronomer Ogilvy, and they are the only spectators. Even the headline in the London evening papers (deliciously bathetic now to the British reader: "A Message Received from Mars: Remarkable Story from Woking") provokes no more than idle chatter. When the lid of the cylinder unscrews and octopus-like arms emerge, a delegation of officials, including Ogilvy, is sent into the impact pit to parley. That's when the intentions of the visitors become clear.

Out of the pit rises a "humped shape," and from it flashes a beam that incinerates everything in its path—including the hapless deputation. This heat-ray is by no means the worst of the Martians' weapons of war, for out from the pit strides "a monstrous tripod, higher than many houses . . . a walking engine of glittering metal . . . articulate ropes of steel dangling from it, and the clattering tumult of its passage mingling with the riot of the thunder." The British army forces are powerless before this colossus, and when more tripods emerge from the other cylinders that have fallen in the environs beyond the western fringes of London, there seems to be no hope for humankind. The machines disgorge clouds of toxic "black smoke" made of an "unknown element"* that kills all who breathe it. A chilling harbinger of the gas warfare of the First World War, this stuff disperses only when the Martians spray it with water to turn the thick vapor into a harmless powder. Meanwhile, Martian vegetation seeded from the spacecraft begins to colonize the English countryside, sending out fast-growing, blood-red tendrils that infect and smother all terrestrial plants. The English Home Counties are transformed into an alien landscape.

The narrator, having evacuated his wife to the nearby town of Leatherhead (which soon looks no safer than Woking), finds himself on the run from the Martians. He ends up trapped in the basement of a ruined house in Sheen, near Richmond-upon-Thames, with a miserable curate whose mind is disintegrating in a welter of self-pitying, eschatological religiosity: he believes the end of days has come. When another Martian cylinder crashes nearby, the two men must hide from the robotlike "handling machines" that allow the piloting Martians to conduct their operations without suffering the debilitating effects of the stronger

---

* In a sign of how well apprised Wells was of the science of the day, he later writes that scientific analysis of the black residue suggests that this element might combine with argon to produce a poisonous compound. Argon had itself been discovered only three years earlier, so an aura of mystique still surrounded it.

Mars attacks: An illustration from a 1906 Belgian edition of The War of the Worlds, *drawn by Henrique Alvim Corrêa.*

terrestrial gravity. Eventually the Curate succumbs to madness and strides, heedless and noisy, from the scullery, convinced that his time of divine judgment has come. The narrator knocks him unconscious, only to beat a fast retreat from a handling machine that drags off the Curate's body to nourish its Martian controller.

At length the narrator perceives that the world beyond the scullery has gone quiet. When a dog, unmolested, comes snuffing and scratching, he is emboldened to crawl out and see what has become of the world. He wanders toward central London, destruction all around him, seeing scarcely any humans—and those mostly blind drunk or mad. When he reaches South Kensington he hears the awful "Ulla, ulla" cry of the Martian war machines, and follows the sound to its source on Primrose Hill. Resigned to his death, he marches toward the titan, only to notice carrion birds circling the high hood above the tripod legs. He soon sees that they are pecking at the remains of a dead Martian.

You probably know the denouement: the Martians have been slain by their vulnerability to terrestrial germs. "Directly these invaders arrived, directly they drank and fed," explains the narrator, "our microscopic allies began to work their overthrow."* Little by little, people return to stricken London, and life begins again with barely plausible alacrity. (What must seem most unlikely of all to Londoners today is that the trains are one of the first public services to resume.)

This "War of the Worlds" was then, after all, not much more than a war of west London and its suburbs. It's testament to Wells's skill that from so apparently parochial a setting he conjures so universal and apocalyptic a myth.

The War of the Worlds was serialized in two magazines, Pearson's and Cosmopolitan, simultaneously between April and December 1897—an indication of Wells's wiliness and persistence in making deals with publishers. It was published in book form by William Heinemann at the start of 1898. Most reviews were positive. "It is one of the books which it is imperatively necessary to sit up and finish," said the Spectator. A few found it too horrible, while the response of the historian Basil Williams in the Athenaeum displayed exactly the kind of snobbery guaranteed to

---

* Wouldn't a superior Martian intelligence have anticipated problems with the microbes of a foreign planet? They failed to imagine such threats, Wells tells us (in contradiction of all biological and evolutionary logic), because there are no bacteria on Mars.

The denouement of The War of the Worlds, *as depicted by Henrique Alvim Corrêa in 1906.*

infuriate Wells: there is, Williams announced, "too much of the young man from Clapham attitude." (To the non-British reader: that is as clear a statement as you could then adduce that the book was *infra dig*, not really serious or mature stuff.) Still, why should he care? As George Orwell later said, Wells "*knew* that the future was not going to be what respectable people imagined."

Few English books at the turn of the century, even when they concern alien invasion, could escape the vexed topic of social class. One of the many paradoxes of Wells is that he deplored the complacency of the English middle classes while seeming to hanker after their lifestyle. He arraigned convention and conformity and advocated social reform even while evincing a distrust of the "lower classes." His claim that *The War of the Worlds* was intended as "an assault on human self-satisfaction" is a barb aimed at the middle classes—on what the Artilleryman in the novel scornfully describes as "lives insured and a bit invested for fear of accidents. And on Sundays—fear of the hereafter." Yet Wells himself did not live so far from that life, and bourgeois sentimentality creeps into his writing, for example, in the lame reunion of the narrator with his wife at the end of the book. (Was ever a wife evoked with more indifference? Did Wells's partner Jane notice, one wonders?)

It's tempting to see hypocrisy here. But that was Wells for you. An advocate of sexual liberty, he displayed notorious sexual incontinence—his affairs with women including the birth-control activist Margaret Sanger and the novelist Rebecca West were allegedly conducted by arrangement with Jane—while outwardly playing the part of the late-Victorian patriarch. From a cozy home in Surrey he denounced the complacency of English society and deplored the vulgarity of the encroaching suburbs. Yet these contradictions supply some of the productive, unconscious tension that fueled his mythmaking. Griffin, the maniacal protagonist of *The Invisible Man*, could represent an attack on both the conventional society that brands him an outcast (personified by the smug superiority of the doctor Kemp who he seeks out to persecute) and the superstitious ignorance of the working-class rustics of Iping in Sussex where he wreaks havoc.

Wells also had an ambivalent attitude toward science and technology. *Doctor Moreau* and *The Invisible Man* take a rather dim view, offering almost caricatures of mad scientists who have lost their moral compass. Yet some of Wells's later works depict scientific utopias that look naïve and boosterish today. When such utopian visions were devastatingly satirized by Aldous Huxley (the brother of Wells's friend Julian) in *Brave New World* (1932), Wells read the book as a betrayal, calling it "incredibly dismal" and "a sour grimace at human hope." It is because these inner conflicts undermine simplistic allegorical readings that Wells's stories have the space to expand into myth.

The War of the Worlds is permeated with these tensions about class and change. The destructive force of the Martians in the region west of London is related with something approaching glee (those familiar with the modern urban sprawl between Woking and Richmond might be inclined to share it). As he wrote to a friend:

I'm doing the dearest little serial for Pearson's new magazine, in which
I completely wreck and destroy Woking—killing my neighbours in
painful and eccentric ways—then proceed via Kingston and Richmond to London, which I sack, selecting South Kensington for feats
of peculiar atrocity.

Much of the power of the novel comes from the shockingly ruinous
outcome: cheerful, prosaic suburbia ripped apart and turned into a
bleak wasteland. There is no hint of joy and liberation in it. As with
The Time Machine, Wells evokes this end-of-time desolation to haunting
effect:

> When I had last seen this part of Sheen in the daylight it had been a
> straggling street of comfortable white and red houses, interspersed
> with abundant shady trees. Now I stood on a mound of smashed
> brickwork, clay, and gravel, over which spread a multitude of red
> cactus-shaped plants, knee-high, without a solitary terrestrial
> growth to dispute their footing. The trees near me were dead and
> brown, but further a network of red threads scaled the still liv-
> ing stems.

We can't help now reading this as foreshadowing both the shattered
battlefields of the Great War and the smoking wreckage of the London
Blitz. (After the Martian rampage, the iconic St Paul's Cathedral bears
a gaping hole in its dome.) But the visions (as well as their geographical
location) more closely resemble the surreal transformations wrought by
flood, crystallized air, and sheer imagination in the novels of J. G. Ballard: *The Drowned World*, *The Crystal World*, and *The Unlimited Dream
Company*. This landscape is not so much the consequence of total war
as the outcome of a Freudian Ragnarok and the ingression of another
world altogether. Wreathed in red weed, Earth has not been conquered
by Mars but has *become* Mars: "I found about me the landscape, weird
and lurid, of another planet."

There's more fertile tension in the political dimensions of *The War of the Worlds*. Wells intended it as a comment on imperialism, noting how pitifully unable to repel the might of the British armed forces were the indigenous peoples they subdued. In one account he cited a conversation with his brother Frank:

> We were walking together through some particularly peaceful Surrey scenery. "Suppose some beings from another planet were to drop out of the sky suddenly," said he, "and begin laying about them here!" Perhaps we had been talking of the discovery of Tasmania by the Europeans—a very frightful disaster for the native Tasmanians! I forget. But that was the point of departure.

The British had colonized Tasmania in the early nineteenth century, establishing settlements for whaling, sheep farming, and trade while eventually wiping out entirely the Aboriginal population. What if we, convinced of our cultural and technological supremacy, were to face an invader equally superior and ruthless, Wells asked. "The Tasmanians, in spite of their human likeness, were entirely swept out of existence in a war of extermination waged by European immigrants, in the space of fifty years," the narrator says:

> Are we such apostles of mercy as to complain if the Martians warred in the same spirit? . . . Surely, if we have learned nothing else, this war has taught us pity—pity for those witless souls that suffer our dominion.

On another occasion Wells stated that the conversation with Frank happened instead on Primrose Hill in London (where the Martians make their final camp as disease overwhelms them), and that they were discussing the "bombardment of some village in the South Seas." "How would it be with us," Frank had allegedly said, "if some creatures of a vastly superior power came down upon us and behaved like a drunken man-o'-war's crew let loose among some gentle savages?"

On that account, *The War of the Worlds* might seem to be a depiction of humankind—the British in particular—getting their just desserts. But although Wells seems to have been no disciple of imperialism, the idea that we should view the Martians as agents come to teach us a moral lesson evaporates when we encounter them. They prove to be so loathsome—leathery heads creeping on tentacles to feed on the blood of living human prey (yes, Martian vampires!)—that one feels their destruction is itself a moral imperative, like the eradication of some especially foul and lethal vermin. As later adaptations of the tale made abundantly clear, the aliens are without question the bad guys now.

Even in Wells's own time, his cool appraisal of imperialism was liable to be misunderstood. Speaking on the aptness of British rule in Uganda in 1907, Winston Churchill characterized the British as "as remote from, and in all that constitutes fitness to direct, as superior to the Baganda [the country's majority ethnic group] as Mr Wells's Martians would have been to us." It's almost as though Churchill imagined Martian governance would have been a good thing, sorting out human disarray and corruption.

Despite its critique of imperialism, *The War of the Worlds* is very much a part of the "invasion literature" genre that arose in the *fin de siècle*. From the 1880s until the outbreak of the First World War, there was widespread fear that Britain had let its military power decay, leaving the isles vulnerable to invasion by our old European foes: the French and the Germans. Noting how easily Prussia had defeated the French in the Franco-Prussian War of 1870, some commentators asserted that the British armed forces had become complacent about the threat posed by Germany (or another European power). It was time, they said, for a thorough overhaul of the military.

Scaremongering titles that warned of the consequences of continued neglect included George Chesney's *The Battle of Dorking*, Philip Howard Colomb's *The Great War of 189–* (1895), Beckles Willson's "The Siege of Portsmouth" (1895), and William Le Queux's *The Great War in England in 1897* (1894). These tales of military conquest or defense of the realm were very popular. When *The Battle of Dorking*, first serialized in *Blackwood's Magazine*, was republished as a novella, it sold eighty thousand copies in a month.

The authors of these saber-rattling fables were often military men:

Chesney an army general, Colomb a naval admiral. Le Queux was a journalist, diplomat, and aviation enthusiast who shared their jingoistic paranoia, writing of "the absolute necessity which lies upon England to maintain her defences in an adequate state of efficiency." His sequel, *The Invasion of 1910* (1906), described a German invasion of Britain in which London is occupied but eventually liberated by a plucky resistance movement. It was commissioned and serialized by the *Daily Mail** and was a great success.

George Griffith had already connected science fiction to invasion fantasies in *The Angel of the Revolution* (1893), the precursor to the aforementioned *Olga Romanoff*, in which humans make contact with Martians. *The Angel of the Revolution* is a rambunctious epic, written while Griffith was an editor at *Pearson's*, which told of a global conflict instigated by a secret brotherhood of utopian anarchists. The armed forces fight with Verne-style airship fleets and submarines: an image Wells would use in *The War in the Air* (1908) and *The Shape of Things to Come* and, decades later, Michael Moorcock would deploy in his steampunk trilogy beginning with *Warlord of the Air* (1971). Griffith himself developed this scenario in *The Great Pirate Syndicate* (1899), in which a British airship fleet blockades Europe and forces it to accept an open trade market.

In 1920 Wells distanced his warning in *The War of the Worlds* from all this jingoistic huffing about the Germans, French, and Russians. It was not the threat of war against England that people had been complacent about in the 1890s, he said, but the nature of war itself. Its potential for destruction had become intolerable, even apocalyptic:

> In those days I was writing short stories, and the particular sort of
> short story that amused me most to do was the vivid realization of
> some disregarded possibility in such a way as to comment on the false

---

* Lest anyone imagine an unbroken tradition in the Europhobia evident in the *Daily Mail* today, let's not forget its support for Hitler's Nazi regime in the 1930s.

securities and fatuous self-satisfaction of the everyday life—as we
knew it then. Because in those days the conviction that history had
settled down to a sort of jog-trot comedy was very widespread indeed.
Tragedy, people thought, had gone out of human life forever. A few of
us were trying to point out the obvious possibilities of flying, of great
guns, of poison gas, and so forth in presently making life uncomfortable if some sort of world peace was not assured, but the books we
wrote were regarded as the silliest of imaginative gymnastics.

"Well," he concluded bitterly, "the world knows better now."

The danger of catastrophic military conflict preoccupied Wells
throughout his career. *The War in the Air* describes a global conflict using flying machines that destroys major cities, including London, Paris,
and Berlin, and eventually brings about the collapse of civilization. More
fearsome still is the vision laid out in *The World Set Free* (1914), in which
scientists figure out how to unleash the awesome energies then recently
discovered within the atomic nucleus. Wells's prediction of "atomic
bombs" was in the Hungarian physicist Leo Szilard's mind when in 1933
he realized how nuclear fission might be triggered in a runaway process.
It took just twelve years to make and detonate such a device in wartime,
obliterating cities at a stroke. By 1945, the scale of destruction depicted
in *The War of the Worlds* and *The War in the Air* was not just feasible but
realized. Their end-of-the-world scenario was no longer merely a part of
myth, but a dark cloud that loomed over the world.

Wells was acknowledged as a mythmaker even during his lifetime. The
critic Edward Shanks, writing on *Doctor Moreau* in 1923, could have
been speaking of all Wells's scientific romances:

> Whereas an allegory bears a single and definite interpretation, a myth
> does not, but can be interpreted in many ways, none of them quite
> consistent, all of them more alive and fruitful than the rigid allegorical
> correspondence.

Another contemporary reviewer identified what it is about Wells's writing that gives it these qualities: "precision in the unessential and vagueness in the essential." That was indeed a vital ingredient of Wells's genius. His competitors in the new genre of fantastical fiction were apt to paint in garish primaries on a massive canvas, the result being all too obviously a contrivance for which suspension of disbelief required heavy ropes. Wells knew better: "I had realized that the more impossible the story I had to tell, the more ordinary must be the setting."

The War of the Worlds is set among Wells's real-life surroundings in Woking, often reconnoitered on his beloved bicycle. A sense of place is related with the diligence of a local guidebook: the cylinders carrying the Martians and their contrivances fall on Horsell Common (between Woking and Ottershaw), the village of Pyrford, and Addlestone Golf Links. The names invite us to imagine picnic blankets laid with sandwiches and flasks of tea on a slightly chilly weekend in May, but what we find instead are slimy octopoid extraterrestrials.

Of the essentials in Wells's tales, though, we hear little. How exactly does the Time Machine work? (Something, Wells hinted, to do with time as the fourth dimension, an idea discussed by scientists even before Albert Einstein introduced four-dimensional spacetime into his theory of relativity.) Just how does Moreau "humanize" animals by surgery? What are the mysterious "rays" that turn Griffin invisible (and why don't they work on his clothes)? Show me this "cavorite" that supposedly suppresses gravity in *The First Men in the Moon*, Jules Verne is said to have scoffed. But as we saw earlier, Wells made no pretense about what he was up to: to "get on with [the] story while the illusion still holds."

It's with Wells that we see the first major divergence of the mythic and realist modes of writing—and the beginnings of a hierarchy that distinguishes literary and genre fiction. Wells himself wrote both sorts, but the fact that he felt he would not be taken seriously if he were known only for his "scientific romances" attests to the significance of the distinction even then. Although he seemed to regard science fiction as a developmental phase out of which he had to mature, rather few of his social-realist novels retain any currency today; *Kipps* and *Tono-Bungay* might still be read, but has anyone even heard of *Mr Blettsworthy on*

Rampole Island? Even the less successful of his science-fictional works—
The War in the Air (1908), *Men Like Gods* (1923), *The Shape of Things
to Come* (1933)—attract more praise and attention now than his many
worthy but dated efforts at social commentary.

While Poe and Stevenson, and even Wells to a degree, were discussed
in their time alongside Dickens and Balzac, by the early twentieth century the kind of novels that I am categorizing in the mythic mode were
a species apart. Arthur Conan Doyle too fretted, with good reason, that
his "detective stories" were disdained by serious critics (although his
feeling that his true calling was chivalrous romances in the style of Walter Scott says a great deal about the era in which his sensibilities were
rooted).

Wells was a master at combining realism and fantasy. Joseph Conrad,
an early fan, called him the "Realist of the Fantastic," and Well's own
claim to "domesticate the impossible hypothesis" perfectly captures what
he was up to. But the characteristics that would make it increasingly hard
for the mythic mode to be taken seriously by literary commentators are
already apparent in Wells's scientific romances, in particular their shal-
lowness of psychology and character. We don't really care about the narrator of *The War of the Worlds*, even though we can feel his terror and be-
wilderment. Like the Time Traveller of *The Time Machine*, he is a cipher
who enables and relates the action, not even worthy of a name. Griffin
in *The Invisible Man* is too unpleasant and unstable from the outset for
us to feel any dismay at his corruption by the power of invisibility. Prendick in *Doctor Moreau* has nothing memorable about him, no powers of
insight beyond a capacity for terror and revulsion at what he witnesses.

If you come to these works in search of the humanity of a Hardy,
Lawrence, or Woolf, you'll be disappointed. If you think, however, that
literature has nothing else to offer than that, you're missing out on the
powerful cultural force of myth.

The Martians did not prevail; they got only as far as Primrose Hill before
British germs destroyed them. But they would be back.

Literally so in an execrable 1898 "sequel" to Wells's book, *Edison's Conquest of Mars*, commissioned from the undistinguished hack Garrett P. Serviss by two American newspapers after their serialization of *The War of the Worlds* was met with an enthusiastic response. In this dire novel, a few Martians survive and return to their planet, after detonating an explosion that, in the first of many such sci-fi catastrophes, destroys New York. To preempt a second assault, the leaders of the world collaborate to send a force to Mars that includes inventor Thomas Edison and the scientists Lord Kelvin and Wilhelm Roentgen, the discoverer of X-rays. The Martians are wiped out in precisely the kind of aggressive display Wells had sought to satirize.

This, it seemed, was what America wanted: blustering validation of a "manifest destiny" to overcome any challenge to US values, to conquer even vastly superior (but slimy, morally repugnant) civilizations. And this would have depressed Wells no end. Literary scholar Robert Crossley summarizes Serviss's effort aptly: "As fiction it is hackwork; as a commentary on *The War of the Worlds* it is astonishingly impervious to Wells's anti-imperialist motive; but as an example of how cultural and national values get drawn into the Martian myth it is both instructive and appalling."

It was bad enough what became of the story even in Wells's lifetime.* He was furious that the American newspaper serialization placed the action in the United States, and he wrote to the *Boston Post*, which ran the story in 1897–1898, to complain about the changes it made. But he was considerably more perturbed by the hysterical response to the infamous 1938 radio broadcast of a dramatization by Orson Welles—which, characteristically for Welles, largely rewrote as well as relocated the story.

The radio broadcast became legendary in its own right, although the real extent to which the nation was gulled into thinking an alien invasion was under way has been questioned. There was surely some sensa-

---

* A theatrical adaptation of *The War of the Worlds* for the Edwardian London stage was considered by Wells's friend J. M. Barrie, author of *Peter Pan*. It's an image to conjure with—but sadly, Wells persuaded Barrie against it.

tionalism in the press, and Welles himself didn't hesitate to exaggerate the response to burnish his reputation. The panic, he claimed in 1955,

> was in fact nationwide! . . . there wasn't a phone you could get to . . . really, anywhere in the States, the highways were jammed with cars going one way or another, those people who were in the cities were going to the hills, and those people who were in the hills were going to the cities.

He spun stories of sailors on shore leave being recalled to duty to defend the country from the Martian invaders. In retrospect Welles seems to be almost sketching the screenplay for a cinematic portrayal of the episode. The reality was a little different.

In the 1930s Welles had become the highest-paid star of American radio, and in 1937 the Columbia Broadcasting System (CBS) offered him a contract to produce a nine-week run of weekly radio plays on subjects of his choice. For this he gathered about him a small repertory company and, in partnership with theater producer John Houseman, commandeered the run-down Mercury Theatre on Broadway for rehearsals. The broadcasts themselves went out live from the CBS studio. The first production of *Mercury Theatre on the Air* was an adaptation of *Dracula*.

The "War of the Worlds" episode was broadcast on 30 October 1938, perfectly timed for Halloween. Like other of the Mercury programs, it was thrown together at the last minute, its creator showing how his creativity thrived on crisis. The adaptation was, in truth, rather brilliantly done, adopting the form of live coverage repeatedly interrupted by news flashes and thus translating rather faithfully the reportage style of the novel. Some have said that the show seemed almost calculated to mislead the listener to believe what they were hearing was genuine, but one would have to be rather credulous to suppose that all of the events related could be taking place in real time.

Newspaper headlines the next day made a meal of the alleged panic. "Attack from Mars in radio play puts thousands in fear," the *New York Herald Tribune* announced, an assessment echoed by the *New York Daily News*: "Fake radio 'war' stirs terror through US." According to the *New*

York Times, "Throughout New York families left their homes, some to flee to near-by parks." Some apparently thought the invaders were the Germans or the Japanese.

A scholarly book on the events by Princeton psychologist Hadley Cantril, *The Invasion from Mars: A Study in the Psychology of Panic* (1940), fed this legend of mass hysteria, accepting the exaggerated claims at face value. Cantril suggested that "at least a million Americans became frightened," and painted a picture of chaos across the nation:

> Some ran to rescue loved ones. Others telephoned farewells or warnings, hurried to inform neighbors, sought information from newspapers or radio stations,* summoned ambulances and police cars.

But was this fake news about fake news? Today's fashionable skepticism likes to imply that the story of nationwide panic was all manufactured post hoc. The truth, however, is where it usually is: somewhere in between. Some people, perhaps tuning in midway through the broadcast, after the introduction that made its fictional nature plain, did seem to be alarmed. "I ran out into the street with scores of others, and found people running in all directions," said one witness. Public alarm was sufficient for the police to be sent to the CBS studio, where a company executive allegedly gave an order (unheeded) to interrupt the broadcast because of growing rumors about the effect it was having. A rival radio station sent out reassurance: "Mr. and Mrs. America, there is no cause for alarm! America has not fallen. I repeat, America has not fallen!"

It surely added to the impact of Welles's production that it was perfectly attuned to the anxieties of the time. War loomed in Europe, where Hitler had just invaded the Sudetenland, and the threat of aerial warfare and mass destruction made possible by new military technologies was apparent, even three years before the Japanese strike on Pearl Harbor

---

* That other radio stations were still broadcasting as if nothing was amiss was one sign from which alert listeners could have drawn reassurance that the world was not ending.

How Orson Welles's broadcast of The War of the Worlds *was reported in the* New York Times *and the* Boston Globe.

gave form to American fears of a sudden "alien" attack from the sky. Reports of a hysterical response to the broadcast even reached the fascist leaders of Europe; Mussolini called it "a sign of the decadence and cowardice of American democracy," while the German Nazi press suspected a "terror-psychosis" spread to serve Jewish interests.

To some American commentators, meanwhile, the scare was a sign of how easily an ignorant public could be manipulated. Dorothy Thompson wrote in the *New York Tribune* that the Mercury Theatre production

> demonstrated beyond question of a doubt, the appalling dangers
> and enormous effectiveness of popular and theatrical demagoguery.
> They have cast a brilliant and cruel light upon the failure of popular
> education. They have shown up the incredible stupidity, lack of nerve
> and ignorance of thousands. They have proved how easy it is to start a
> mass delusion. They have uncovered the primeval fears lying under the
> thinnest surface of the so-called civilized man.

In retrospect we might instead see the affair as an early indication of how primed our consumerist culture is for conspiracy theories spread by mass media. It also demonstrated the illusionistic power of broadcast media in particular. As the statement issued by Welles and CBS in the wake of the episode said, "We can only suppose that the special nature of radio, which is often heard in fragments, or in parts disconnected from the whole, has led to this misunderstanding."

We'd be wise not to place too much trust in what Welles later said about his motives, but he touched a nerve when he claimed:

> We were fed up with the way in which everything that came over
> this new magic box, the radio, was being swallowed. . . . So in a way
> our broadcast was an assault on the credibility of that machine; we
> wanted people to understand that they shouldn't take any opinion pre-
> digested, and they shouldn't swallow everything that came through the
> tap, whether it was radio or not.

As cultural scholar Jeffrey Sconce has suggested, the "Wellesian panic" became its own myth for the same reason that all myths are born: because they feed a need. And what we needed here, says Sconce, was a cautionary tale about the power of the mass media. Welles provided a new narrative for *The War of the Worlds*, in which it becomes a story not just about invasion and apocalypse but about how fears of catastrophe infect our daily life. These anxieties lead to a breakdown in trust: radio and television no longer present the voice of authority but must be listened to with cautious skepticism. In Welles's remark about "everything that came through the tap" is a presentiment of Americans' growing rejection of anything (including the water supply!) that represented authority and government. It's all a plot! This mix of confusion, superstition and conspiracy theory, fed by consumerist excess and media saturation, in the face of threats beyond our control is captured with disconcerting relish in Don Delillo's 1985 novel *White Noise*—but even that did not anticipate the surreal and tragicomic mix of paranoia, reality denial, and social media–fueled fantasy that unfolded in Donald Trump's America during the Covid-19 pandemic of 2020.

H. G. Wells had given his permission for the Mercury Theatre production, thinking it would simply be a reading of the book. Evidently he did not know Orson Welles. Not, at any rate, at that point; but when the two men met in Texas during Wells's tour of the US in 1940, all was forgiven: Welles's charm apparently won the day and the two got

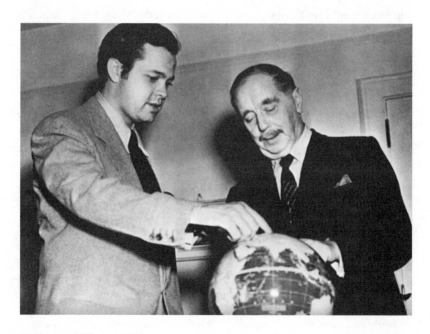

Wells meets Welles in 1940.

on royally. Seeming to forget his earlier threats of legal action over the liberties the American had taken with his book, Wells said that meeting "my little namesake here, Orson," was "the best thing that has happened so far" on his trip. "I find him the most delightful carrier," Wells went on. "He carries my name with an extra 'e', which I hope he will drop sooner or later. I see no sense in it."

The Welles scare only served to make *The War of the Worlds* more famous than ever. The book was promptly reissued by the New York publishers Grosset and Dunlap, who put some of the newspaper headlines reporting the panic on the front cover and sold it as "The Book that terrified the Nation over the Air!"

Before concluding that the impact of Welles's radio broadcast revealed how uniquely gullible and paranoid Americans are, bear in mind that unrest was reported also in Brazil, Chile, Ecuador, and Portugal following later radio broadcasts of the story. In Ecuador there were allegedly riots in which the broadcasting station was burned down and people killed, and a report to the British Foreign Office asserted that many people tried to escape the Martians by spending the night on a local mountain.*

Such accounts should be treated with caution, but paranoia is precisely what the ominous opening words of Wells's novel invite us to indulge:

> No one would have believed in the last years of the nineteenth century that this world was being watched keenly and closely by intelligences greater than man's and yet as mortal as his own; that as men busied

---

* And for British readers tempted to feel smug about all this: audiences in the UK were allegedly panicked by a reportage-style radio show in 1926 about a revolution in the country. Here too, though, some transatlantic Schadenfreude is justified; the *New York Times* trumpeted on that occasion, "Such a thing as that could not happen in our country."

themselves about their various concerns they were scrutinised and
studied, perhaps almost as narrowly as a man with a microscope might
scrutinise the transient creatures that swarm and multiply in a drop
of water.

We are being watched. But not by Big Brother, nor by some prurient Peeping Tom: the fantasy is neither Orwellian nor Freudian. The
watcher seems closer to Milton's all-seeing Satan, looking down at the
globe from a perch beyond this world. These observers are of another
order of being: "minds that are to our minds as ours are to those of the
beasts that perish, intellects vast and cool and unsympathetic."

As I suggested earlier, what sets *The War of the Worlds* apart from
previous fantasies about denizens of other planets, and from contemporary tales of invasion and empire, is that the attackers are so ineluctab-
bly, horrifically different from us. And yet they are not monsters in the
traditional sense, which serve an allegorical purpose (as the word tells
us, being derived from the Latin *monstrare*, to show). They are merely
another race of intelligent beings, a product of nature rather than the
supernatural. Very probably this idea would not have been thinkable
until Darwin argued for the plasticity of biological form and reasoned
that humankind itself originated from creatures utterly alien in shape
and mind.

While the traditional monster makes us feel terror, Wells's creatures
invoke the subtly different feeling of horror: they give you the creeps.
Terror, says critic James Twitchell, is a temporary, real-life sensation in
response to a present danger—but horror is dreamlike and permanent,
a sensation of mythic proportion. Terror can be overcome: we can run
away from the homicidal maniac. But horror is always there, always
glimpsed, never fully seen.

Wells intuited this, and despite his detailed, quasi-scientific account
of Martian physiology and ecology, there remains an elusive quality to
the description:

They were, I now saw, the most unearthly creatures it is possible to
conceive. They were huge round bodies—or, rather, heads—about

four feet in diameter, each body having in front of it a face. This face
had no nostrils—indeed, the Martians do not seem to have had any
sense of smell—but it had a pair of very large, dark-coloured eyes, and
just beneath this a kind of fleshy beak. . . . In a group round the mouth
were sixteen slender, almost whip-like tentacles, arranged in two
bunches of eight each.

The War of the Worlds altered the whole notion of the extraterrestrial
being, giving it a nonhuman form as foul and inscrutable as it was malevolent. Wells speaks to an inherited fear of things squirming and slimy,
many-legged, bloodthirsty. Here are the progenitors of H. P. Lovecraft's
cosmic demon Cthulhu, of John Wyndham's Triffids and Krakens, of the
toadlike Jabba the Hut from the *Star Wars* saga. It wasn't until Dennis
Villeneuve's 2016 movie *Arrival* (based on the 1998 novella *Story of Your
Life* by Ted Chiang) that cephalopod beasts like this were permitted
to exhibit civilized behavior—and even then the human characters evidently struggled against an instinctive distaste for the squidlike forms.

Cut loose from the norms of bipedal urbanity, these new aliens acquired a disturbingly protean aspect. Insectoid metamorphosis became
common: the larval chest-burster of *Alien* transforms into a mucus-
slimed humanoid, half reptile and half ant. Most distressing of all is the
possibility that, like Kafka on rewind, these creatures might not just in-
fest but *become* us, even as their minds remain "cool and unsympathetic."
A horror that the crustacean alien will colonize our bodies is played to
great effect in Neill Blomkamp's 2009 movie *District 9*, where the aliens
are disparagingly called "prawns."

The Wellsian motif of the alien spaceship that has smashed into
Earth's crust features in John Campbell's 1938 novella *Who Goes There?*,*
where such a craft remains frozen in antarctic ice for twenty million
years until (don't they watch the movies, for goodness' sake?) a group

---

* Campbell wrote the story, first published in *Astounding Science Fiction*, under the pen name
Don A. Stuart.

Henrique Alvim Corrêa's vision of the Martians in 1906.

of scientists try to thaw it. The alien that emerges can take on the form and thoughts of any living thing it devours, beginning with a sled dog but soon enough. . . . You doubtless know the plot, for Campbell's story was the basis of the 1951 monster movie *The Thing From Another World*, directed by Christian Nyby for Howard Hawks's Winchester Pictures Corporation—with its closing line that supplied the perfect slogan for postwar UFO mania: "Keep watching the skies!"

The story is better known today from John Carpenter's iconic 1982 remake *The Thing*. That film's remarkable special effects, created by the pre-CGI master of visceral transformation Rob Bottin, illustrated with unforgettable style the contemporary analogue of the invasion fantasy:

now aliens invade not just our world but our minds and bodies. We are no longer consumed but *subsumed* by the creature. To see bodies split open like fruit and heads grow spider legs as the Thing transmutes is to witness the darkest imaginings of Bosch given flesh, now not by a race of demons in a flaming hell but by aliens in the icy wastes where that original alchemist of the human body, Victor Frankenstein, met his end. As Susan Sontag commented in her 1965 essay on science fiction, "The Imagination of Disaster," the alien invaders "practice a crime which is worse than murder. They do not simply kill the person. They obliterate him." Terror is a fear of death—but this is true horror.

This body horror too is foreshadowed by Wells. His Martians don't eat us; they don't eat anything, for they lack a digestive system. "Instead, they took the fresh, living blood of other creatures, and *injected* it into their own veins . . . blood obtained from a still living animal, in most cases from a human being, was run directly by means of a little pipette into the recipient canal." Here is an echo both of the reanimating forces that inject a spark of being into the body parts of Frankenstein's creature, and the tortured persistence of flesh beyond death, sustained by blood, portrayed in *Dracula*. This is more than Cold War paranoia; it is a very ancient horror of the merging of man and beast, biologized and technologized for the modern age.

The alien spacecraft in Campbell's *Who Goes There?* is also indebted to Wells: a featureless, vaguely cylindrical structure, "like a submarine without a conning tower." That cigar shape evolved into the cross-sectional view of a more familiar form: it was universally acknowledged by the 1950s that aliens built their craft as spinning disks, called "flying saucers." (Odd, in retrospect, that these ominous devices should be named after such mundane domestic items.)

The term appeared first in newspaper accounts of a report by an amateur pilot in Idaho in 1947 of unidentified flying objects arrayed in echelon formation, like geese from Mars. They flew, the pilot said, "like a saucer if you skip it across the water." By 1950 we had the movie

The Flying Saucer and a TV series, *Flying Disc Man from Mars*, followed soon by *Earth vs the Flying Saucers* (1956), based on the 1953 nonfiction book *Flying Saucers from Outer Space* by Donald Keyhoe. The legacy of *The War of the Worlds* is abundantly clear in these charmingly terrible concoctions. The poster for the latter movie (with stop-motion special effects by Ray Harryhausen) shows the disk-shaped craft hovering over the land like Wells's tripods, obliterating all in their path with death rays. Those tripods themselves became the looming, vaguely saucerish craft in George Pal's 1953 movie adaptation of Wells's book.

On the one hand, these films knew their place: alongside the horror pantheon in B-movie land. *Earth vs The Flying Saucers* was released in a double bill with Columbia's *The Werewolf*, while *The Flying Saucer* appeared with a rerelease of the 1941 zombie feature *Man-Made Monster*, now titled (for the post-Hiroshima age) *The Atomic Monster*. For monster horror was now fully fused with science fiction, and nuclear physics

The flying saucer is the spacecraft of choice for Cold War–era aliens.

The Martian tripods becoming hovering flying saucers in George Pal's 1953 adaptation of The War of the Worlds.

rather than Jekyll's potions was the favorite agent of transformation. (In fact the film's zombies were animated, Frankenstein-style, by electricity; but in 1950 atomic energy was a much better symbol of science run amok.) Even *The Werewolf* was now a product not of the supernatural but of science *manqué*, the creature brought into being by a serum made with an extract of wolf's blood.

On the other hand, the merging of the flying saucer–style UFO with Wells's coldly observant aliens created a new cultural phenomenon that attained mythical status in its own right. Sex may have begun in Great Britain in 1963,* but aliens had been at it in America for at least two years by then: Barney and Betty Hill of New Hampshire became the first recorded alien abductees in 1961. Driving home from vacation, they saw strange lights in the night sky and arrived with weird, half-formed memories and a sense they had mentally suppressed some disturbing events

---

* At least for the poet Philip Larkin, whose "Annus Mirabilis" falls "Between the end of the 'Chatterley' ban / And the Beatles' first LP."

en route. That Barney felt compelled to examine his genitals in the bath-
room and Betty's dress was torn and smeared with a strange pink powder
offered disturbing clues; so did Betty's subsequent dreams of being examined inside a disk-shaped craft and having a needle thrust agonizingly
into her navel. She took out a book on UFOs from her local library—a
book written by none other than the aforementioned Donald Keyhoe, a
Marine Corps aviator and author of *Flying Saucers from Outer Space* and
its best-selling prequel *The Flying Saucers Are Real* (1950).

Keyhoe, who also published horror and fantasy short stories in pulp
magazines such as *Weird Tales*, was a key figure in the early days of the
UFO and alien abduction craze. He alleged that, like Well's Martians,
aliens of unspecified origin had been observing Earth for centuries. By
invoking the atomic-bomb explosions of 1945 as the impetus for their
stepping up surveillance and other activity, he yoked this narrative to
postwar nuclear foreboding. Keyhoe introduced the idea that the US
military knew about this activity but covered it up for fear of public
panic. The military itself, never short of odd characters akin to those in
Stanley Kubrick's *Doctor Strangelove*, was not uniformly dismissive of
Keyhoe's ideas; an Air Force press secretary called them "responsible" and
"accurate." Betty and Barney Hill, meanwhile, were model abductees:
respectable, modest, their imaginations steeped in Freudian repression.
When Betty wrote in 1975, "My concern is not whether a person believes or not; my concern is the number of abductions that are taking
place now," it is hard to doubt the sincerity of the belief that had taken
over her life.

Belief in UFOs and alien abduction is now classified as a genuine
and specific psychological condition—although for religious scholar
David J. Halperin, who surveys its history in his book *Intimate Alien*, it
is rather one of the key "UFO mythic themes." Its cluster of images and
tropes reflects the repression and forebodings of the era in which the
reports began: a fear of transgression and violation, a sense that something has gone wrong in our past and is now just beyond the grasp of
memory, a feeling that behind our daily life are strings being pulled by
unseen, omnipotent agents with designs on our bodies and our ways of
life, a notion that *the truth is out there*, in the twilight zone, in the X files.

Wells's alien invasion myth created a vehicle for these feelings, which could be readily adapted as requirements changed. During the era of Cold War détente and sexual liberation in the 1970s, then, visiting aliens appeared more benevolent, less sexually transgressive and repulsive. At last we were ready for *Close Encounters of the Third Kind* bringing, not the threat of Armageddon, but psychedelic disco lights flashing in the sky, or for a cute and friendly *E.T.* peddling to the moon.

But not for long. Not now that the nuclear end of days lies stored a hundred times over in silos, operated by red buttons in the hands of narcissists and megalomaniacs. Not now that we have seen cities laid waste by firestorms and mushroom clouds, skyscrapers explode into fireballs and topple to rubble, not now that the hands of the Doomsday Clock sit at two minutes to midnight.

Gigantic spaceships looming over famous cities—one of the instantly recognized motifs of the alien invasion story—don't feature in *The War of the Worlds*. But it's not hard to see the genesis of this image in the towering Martian tripods, symbols of a higher intelligence that has become our oppressor. The great airships of George Griffith's *The Angel of the Revolution* played a part in that imagery too: the barrage balloons floating over London during World War II triggered the idea of hovering alien craft for Arthur C. Clarke's 1953 sci-fi classic *Childhood's End*.

We think we know what comes next: death rays of searing fire that blast apart the White House or Big Ben. (Ever since *Edison's Conquest of Mars*, aliens have shown a strange obsession with razing Manhattan.) But that's not how it is in *Childhood's End*: the invasion is benign, or so it seems at first. Humanity is ruled peacefully by alien Overlords, who remain hidden but supervise human international affairs for our own good, ending the military space race. The reader knows, of course, that it won't stay that way for long, although the Overlords are no mere imperialists in waiting—it's way more complicated than that. *Childhood's End* is a particularly influential elaboration of the myth of *The War of*

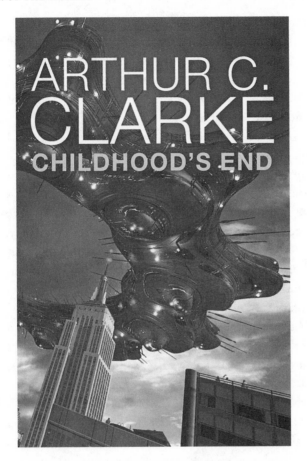

The image of spaceships looming over cities featured in Arthur C. Clarke's 1953 novel Childhood's End.

the Worlds, offering a different vision of what it might be like to be subjugated by a superior extraterrestrial power.

The scope for variation and elaboration is endless. The great ship casting a shadow over Johannesburg in *District 9* may look threatening, but the arthropod aliens here are the underdogs, marooned on Earth and treated as unwelcome migrants in a sharp comment on both contemporary geopolitics and the segregationist past of South Africa. The alien overlords in the 1980s American TV series *V*, meanwhile, are wolves in human clothing: more precisely, they are carnivorous reptiles who can dislocate their jaws to swallow large rodents (and worse) alive. This cult series began as a high-minded attempt to adapt Sinclair Lewis's antifascist novel *It Can't Happen Here* (1935)*—but indeed it wouldn't happen here, the American producers were told, unless the fascists became aliens who eat people. For how else would the average Americans recognize them?

The hovering black ovoids of *Arrival* recapitulate the appearance as well as the mission of the craft in *Childhood's End*, but first we have to figure out how to talk with these Wellsian aliens. It's a very twenty-first-century conceit that Chiang's original story explores: can we learn to communicate with one another across cultures, or will our mutual incomprehension turn to violence? *Independence Day* (1996), in contrast, filmed shortly after the first Gulf war, suggests that war is inevitable and only military firepower (plus good old American grit) will win the day. You might charitably call it homage to the B-movie; as film critic Roger Ebert of the *Chicago Sun-Times* put it, "'Independence Day' is not just an inheritor of the 1950s flying saucer genre, it's a virtual retread—right down to the panic in the streets, as terrified extras flee toward the camera and the skyscrapers frame a horrible sight behind them." Calling the movie "timid when it comes to imagination," Ebert lamented the cartoonish aliens and the "cornball ray-beams that look designed by the same artists who painted the covers of Amazing Stories magazine in the 1940s."

Steven Spielberg has never disguised his enthusiasm for aliens, and he had been a fan of *The War of the Worlds* since childhood. But although he later said he had long dreamed of putting it on the screen, he feared that *Independence Day* had queered his pitch—until shortly after the turn of the millennium, when Wells's book offered itself as a vessel for

---

* Remnants of that origin survive in the swastika-like insignia of the aliens (V for "Visitors"), along with their youth movement and SS-style uniforms. The minority they persecute here are the scientists, whose activities and movements are tightly restricted and monitored.

a different narrative. "After 9/11 it began to make more sense to me," Spielberg said.

David Koepp, the scriptwriter of Spielberg's 2005 adaptation, averred that "the genius of the story is that everyone can fit their personal fears around it, that someone out there—the fascists, the communists, the Americans—is coming to get them." It was still too awkward at that time for him to add "Islamist terrorists" to the list, but there's no mistaking the inspiration for the film. "Is it terrorists?" asks one character as the invasion begins, and soon enough an airplane is shot out of the sky.

Yet the movie is not really a political allegory about an assault on American freedom from an outside civilization. The aliens (now of unspecified extraterrestrial origin, for we know Mars too well) are not terrorists or would-be conquerors from another culture. This ultimately rather smart film is really about what it means to survive a catastrophe that dismantles all conceptions of normality. Spielberg's *War of the Worlds* is, like so many of his films, a portrait of a family under stress trying not to fall apart, at the center of it a flawed but loving father. It manages mostly (as Spielberg doesn't always) to avoid the sentimentality that could so easily derail such a plot. Perhaps the movie indeed amounts to "pretty expensive family therapy," as the *New York Times* critic rather derisively put it. But by following the principle that, as Koepp said, "we will not see anything unless Ray [played by Tom Cruise] sees it, and we'll not know anything unless Ray knows it," Spielberg managed to keep faith with Wells's first-person text and produced an adaptation that devotees of the book could respect. The imagery is genuinely scary, even if we've seen folks hiding from the gaze of prowling predators a little too often in Spielberg's oeuvre. But what unsettles most of all is that apocalyptic images of collapsing masonry and mountainous rubble piles had, after 9/11, become real again.

In the immediate aftermath of the terrorist attacks on New York's World Trade Center, disaster movies seemed the only credible points of reference for comprehending the scenes we had witnessed on our screens. At the end of 2001, American film critic James Hoberman wrote of "the déjà vu of crowds fleeing Godzilla through Lower Manhattan canyons, the wondrously exploding skyscrapers and bellicose rhetoric of

Independence Day, the romantic pathos of *Titanic*, the wounded innocence of *Pearl Harbor*, the cosmic insanity of *Deep Impact*, the sense of
a world directed by Roland Emmerich for the benefit of Rupert Mur-
doch." On September 11, said Hoberman, "the dream became reality."

There was then a sense of culpability in Hollywood. Many were
thinking what veteran director Robert Altman said out loud: "The movies set the pattern, and these people [the terrorists] have copied the movies. Nobody would have thought to commit an atrocity like that unless
they'd seen it in a movie. . . . I just believe we created this atmosphere and
taught them how to do it."

The response of the moviemakers was to suppress its own imagery,
as if guilt might be exorcised in the editing suite. Release of the Arnold
Schwarzenegger blockbuster *Collateral Damage*, in which the hero's
family dies in a skyscraper blown apart by terrorists, was postponed. The
trailer of *Spider Man* was doctored to cut scenes showing the World
Trade Center, and scenes for *Men in Black 2* were reshot for the same
reason. An adaptation of Wells's *The Time Machine* by DreamWorks was
changed to avoid showing New York City bombarded with fragments of
a disintegrating moon. "There are just some movies that you can't make
from here on in," said one of the company's producers.

In that febrile atmosphere, Spielberg's scriptwriters for *The War of
the Worlds* confessed to deliberately avoiding images of "Manhattan getting the crap kicked out of it." Perhaps that was simply respecting proper
boundaries of taste, but the impulse that Wells's book activates is not to
be ameliorated by a spot of geographical self-censorship. For the orgy of
destruction that he visits on Surrey and the west London suburbs is an
unbridled expression of what Freud characterized as the death drive: our
collective Thanatos complex.

More than a vehicle for invasion paranoia, then, *The War of the Worlds*
became the progenitor of an "end of the world" mythology, a technological Ragnarok in which, as Susan Sontag has written, we can experience
the vicarious "fantasy of living through one's death and more, the death

of cities, the destruction of humanity itself." Or, as the advertising slogan for *Independence Day* put it with trademark bluntness, "the end of the world as we know it." Actually the end of the world in that film was one we know all too well, for the overblown visual language of *Independence Day* is now stock fare in apocalyptic disaster movies. The darkening of skies over Manhattan by a menace from space, followed by the blazing destruction of iconic buildings, comes from that impressively terrible 1979 film *Meteor* and its countless successors, such as *Deep Impact* and *Armageddon* (both 1998).

In all of those films the worst disaster is averted, however, while *The War of the Worlds* gives a glimpse of our planet wholly ravaged by apocalypse. After the two worlds wars, and especially after we had seen the blasted plains, punctuated by ruins, where hours earlier Hiroshima and Nagasaki once stood, Wells's vision looks terrifyingly prescient.

He had painted a comparable picture in *The Time Machine*, where the Time Traveller travels into the far future "more than thirty million years hence," when the sun has swollen into a dim red giant and humankind has vanished. The scene has a desolate poetry:

> The red beach, save for its livid green liverworts and lichens, seemed lifeless. And now it was flecked with white. A bitter cold assailed me. . . . There were fringes of ice along the sea margin, with drifting masses further out; but the main expanse of that salt ocean, all bloody under the eternal subset, was still unfrozen. . . . From the edge of the sea came a ripple and a whisper. Beyond those lifeless sounds the world was silent. Silent? It would be hard to convey the stillness of it.

In that moment there is a solar eclipse, turning the sky "absolutely black," and the narrator attests that "a horror of this great darkness came upon me."

This is so much more chilling—because so much more plausible—than the sound and fury of attacks by Hollywood aliens, just as the becalmed, surreally red-festooned landscape at the end of *The War of the Worlds* carries the story still further into mythic territory. In that scene,

Wells prepared us to comprehend the century of total war in the terms
that it truly warranted: as a potential end of everything.

Those bleak final pages distance the book from the conservative fantasies of the *fin de siècle* invasion literature. This is a story, not about
defending the past, but about a radical transformation of society. "Cities,
nations, civilisation, progress—it's all over," says the Artilleryman. "That
game's up. We're beat."

According to Bernard Bergonzi, "*The War of the Worlds* can be read
as an expression of the traditional eschatological preoccupation with the
end of the world, which has been the source of so much religious imagery." That, after all, is how the deranged Curate interprets the Martian
invasion: as the end of days, with the Martians acting as "God's minis-
ters." Even when the danger has apparently passed and the aliens are rot-
ting in their high steel tombs, we are not quite sure if the end has already
happened and what remains is the shrouded illusion of an afterlife:

> I go to London and see the busy multitudes in Fleet Street and the
> Strand, and it comes across my mind that they are but the ghosts of the
> past, haunting the streets that I have seen silent and wretched, going
> to and fro, phantasms in a dead city, the mockery of life in a galva-
> nized body.*

In Pal's 1953 movie the Martians obliterate Los Angeles, as if Hollywood is now toying with self-destruction. Little more than a decade
after Pearl Harbor, it was the west coast, not Manhattan, that felt most
exposed and vulnerable—but by then, with thermonuclear bomb test-
ing in the Pacific, there was already a perception that the stakes of war
had changed irrevocably. Technologies for vaporizing cities and burning people into shadows and atoms being a relatively new invention, it's
hardly surprising that we have to psychically process them over and over
again, whether through the lunatic liberation of Slim Pickens riding a

---

* This, you might appreciate, is an allusion to *Frankenstein*.

hydrogen bomb like a rodeo steer in *Doctor Strangelove* or the dreaming Sarah Connor staring into mushroom clouds through a chain-link fence in *Terminator 2.*

We keep returning for more punishment: our appetite for Thanatos shows no sign of diminishing. Far from expelling the fear, Hollywood has become an industry devoted to its relentless and systematic manufacture. As the cinema critic Robert Muller writes:

> Film producers . . . seem to have come to the conclusion that, after Belsen and Hiroshima, in an era of ever-improving hydrogen bombs, the only manufactured fear likely to satisfy and calm us is one which comes from *out of this world.*

Susan Sontag agrees with this view of what she calls the "dispassionate, aesthetic view of destruction and violence" that typifies science fiction films, at least up to the 1960s when she was writing. They "normalize what is psychologically unbearable, thereby inuring us to it," she claims.

These are, however, strangely conflicted assessments. How are those images of wanton and wholesale destruction of civilization supposed to be calming and inuring—as Muller suggests, to "take our mind off" the very real threat of nuclear annihilation? No, this manufacture of fear is like the (equally lucrative) manufacture of horror: it erects a framework on which we can hang and contemplate our fears. That's why we have myths.

It's no wonder that when these projections of nightmare become real—when we see actual skyscrapers bloom into orange flame before sliding to the ground in an avalanche of glass and steel—we can find no better recourse than to try to root reality in the movies. The world is upside down; our imaginations have leaked into it and seem almost to have *caused* events. Myths are not supposed to body forth in this manner.

The world did not end for the first time in *The War of the Worlds.* Mary Shelley's *The Last Man*, published in 1826, baffled even those who anticipated that the author of *Frankenstein* might produce another tale of Gothic angst. Set in the late twenty-first century, it tells of a plague that devastates the human population of the Earth and shatters civilization. The story is narrated by Lionel Verney, the sole eventual survivor of a small group of European nobles who try to escape the affliction by journeying through Europe, and whose survival indeed makes him the last man on Earth. "One by one," he says, "we should dwindle into nothingness." The book is a meditation on death, which some critics believe to have been inspired by the drowning of Percy Shelley in 1822. As Verney declares:

> Surely death is not death, and humanity is not extinct; but merely passed into other shapes, unsubjected to our perceptions. Death is a vast portal, an high road to life: let us hasten to pass; let us exist no more in this living death, but die that we may live!

A plague that wipes out humanity, leaving scattered survivors to a precarious and perhaps futile existence, is another familiar, even clichéd staple of science fiction—and has come scarily back in vogue during the Covid-19 pandemic. But Mary Shelley's evocation of it in *The Last Man* shows, as *Frankenstein* had done already, that her sensibility was out of step with her times: the early Victorians lacked any reason to suffer from nightmares of global destruction and extinction. Reviewers didn't know where to start; one could do no more than blast the work as "sickening," another spoke of "diseased imagination" and "polluted taste." In fact Shelley's vision was quasi-religious: might not viewers of the English Romantic painter John Martin's lurid, mid-nineteenth-century eschatological landscapes have enjoyed the same kind of visceral thrill as today's multiplex audiences watching Manhattan get trashed once again by comets, tidal waves, or aliens?

Even Shelley's *The Last Man* was not the first end-of-the-world novel, however: the French writer Jean-Baptiste Cousin de Grainville's 1805 book had the same title (in French), although there is no indication that Shelley knew of it. The possible destruction of our planet by a comet impact became another popular doomsday scenario, featuring in Edgar

Victorian dreams of apocalypse: The Great Day of His Wrath *(1851 – 1853) by John Martin.*

Allan Poe's "The Conversation of Eros and Charmion" (1839), Camille Flammarion's *The Last Days of the World* (1893), and George Griffith's short story "The Great Crellin Comet" (1897). (Griffith anticipated *Armageddon* and *Deep Impact* in having the disaster averted by firing a projectile at the comet.) Flammarion's story was made into a film by Abel Gance in 1931, in which a quasi-fascist government tries to impose order while the citizens break out into mass orgy. In the event the comet misses, although much of the world has been destroyed.

The end of everything is in itself an old mythical image, of course: Martin's image came straight out of the book of Revelation, the word "apocalypse" itself originally referring to the revelation of terrible truths about human nature, existence, and sin. But it is notably rare to see the actual end of the world on screen today—and when we do, the sensation is quite different from that evoked by alien apocalypse: not so much terrifying as simultaneously deeply sad and ecstatic. So it tends to be only in the art house, where apocalypses are symbolic, that this happens: both Lars von Trier's *Melancholia* (2011) and Don McKellar's *Last Night*

(1998) make the end of the world a kind of brilliant, tragic epiphany, utterly compelling and undeniably sacred.

And unlike the aftermath of the end of days or the twilight of the gods, in the modern post-apocalypse there is no glorious, fresh new beginning—for everything ends *except us*. We are left eking out a pitiful, dreary existence in the ruins, survivors after the Fall, with nothing to look forward to except eternity, wondering if we are after all not the lucky ones but the opposite. Wells shows us this apocalypse as if through an eye pressed to the keyhole. In telling his story through the perspective of an ordinary man caught up in a devastating Armageddon that he can barely understand, Wells supplied a fruitful way to explore these existential anxieties. His narrator and Curate, holed up in a basement in Horsham while the world is reduced to ruin out of sight above, are in the same predicament as Hamm and Clov in Samuel Beckett's *Endgame*, looking out through a high window onto nothing: "Finished, it's finished, nearly finished, it must be nearly finished."

The emerging mythology of the post-apocalypse is starting to acquire a discernible shape, modulated now also by the threat of climate-induced collapse. Nevil Shute's *On the Beach* (1957) painted the end of civilization in realist tones, but it needed the fantastic to hit its stride, in works as diverse as the Mad Max movies, Robert Zelazny's *Damnation Alley*, Margaret Atwood's *Oryx and Crake*, David Mitchell's *Cloud Atlas*, and Russell Hoban's *Riddley Walker*.

No novel achieves a biblical sense of desolation more acutely than Cormac McCarthy's *The Road* (2006), in which the vaguely specified apocalypse might be climate-induced. What gives it the same mythical dimension as the red wilderness of Wells's Martian conquest is the isolation of the narrative perspective, allowing it to work on the metaphorical level of alienation from civilized society as well as the literal scenario of disaster. The threat of climate-induced collapse and extinction hung too over the 2019 serialized adaptation of *The War of the Worlds* by the BBC, set faithfully in period but otherwise taking considerable license with Wells's story. Here the aftermath is no passing moment before regular life begins to assert itself again; the survivors of the Martian attack wander for years in the ruins of a blasted land as red and barren as Mars itself, where no plant will grow and the sky is shrouded in a pall that evokes the suffocating ocher smogs of New Delhi and Beijing.

Why do we have this hunger to witness the end of the world? The ancient Greeks might have recognized it as catharsis: a purging and purification through the release of extreme emotion. What we are witnessing here is indeed classic tragedy—but now, in an age so powerful and so deranged as to make it conceivable, the tragedy of the world. As film theorist André Bazin said in the 1950s, the alien-invasion narrative supplies the oddly cathartic experience of "living through one's death" in an orgy of mayhem and destruction that he dubbed the Nero complex. What for Wells may have been an act of revenge against the stifling conformity of his society has become a means of confronting and grappling with our own obliteration.

<hr />

Anxieties about authenticity still hover over Wellsophiles' assessments of *War of the Worlds* adaptations. Do we risk "losing" Wells to Welles, to Spielberg, even (God forbid) to Jeff Wayne's massively popular musical extravaganza with its sonorous Richard Burton narration? Is it a travesty when Horsell Common and Woking become Trenton, New Jersey—especially given how important place was for Wells?

The answer, of course, is that Wells had to relinquish any say over such matters precisely to the extent that he succeeded in making his most memorable and momentous of myths.

The War of the Worlds justifies Ursula Le Guin's claim that "science fiction is the mythology of the modern world." But it also bears out her suggestion that this is only a half-truth so long as we regard myth as "an attempt to explain, in rational terms, facts not yet rationally understood." That is a fair summary of much of Wells's own motivation, in this book and others. But as we watch the Martians lay waste to our civilization, we get a glimpse of what Le Guin says is another function of myth: to take over when the intellect fails. It is, she says, "a non-intellectual mode of apprehension."

Nowhere is that tension between the ostensible operation of the intellect and the intervention of something deeper and atavistic more apparent than in the works of Wells's contemporary and friend, whose creation of an arch-rationalist ended up by revealing the limits of the intellect in navigating the currents of modernity. Here I mean Sir Arthur Conan Doyle.

❦

Chapter 7

REASON WEARS A DEERSTALKER

THE SHERLOCK HOLMES STORIES (1887–1927)

A hyper-rational person, accompanied by a bumbling
Everyman, displays an almost preternatural (but wholly explicable)
ability to penetrate mysteries and solve crimes. He faces an
implacable, equally brilliant foe who might or might not prove
to be his match in a final confrontation.

═══

Science, at the end of the nineteenth century, looked like the savior of civilization. Clear-sighted, resolutely male, and immune to the siren call of emotion and superstition, it would make the world a finer place. Sherlock Holmes personifies that vision as seen through a lens of British imperialistic dominance. He is not the most clubbable of souls, not exactly someone you'd want as a friend. But he is incorruptible, dependable, honest to a fault—and the incarnation of reason and logic. If anyone can foil the degenerates of the underworld, surely it is he?

The contrast between the rigorously rational Holmes and the notoriously gullible Arthur Conan Doyle, who was ready to believe in spirits and fairies on the flimsiest of evidence, has often been remarked. But there is nothing truly incompatible here. Holmes is precisely the kind of rationalist that a man like Conan Doyle—who was always part mystic and devoted his later life to the promotion of spiritualism—would imagine. His deductions *exceed* reason, and his successes often rely on absurd coincidences and improbabilities, not to mention simple inconsistencies of plot. Holmes does not so much bring order to chaos as represent the banishment of the mess of real life by quasi-magical fiat. He is a wish-fulfillment fantasy, and this gives him mythic potency.

While it would be mean-spirited indeed to let this evaluation spoil our enjoyment of Conan Doyle's stories (the quality of which is generally astonishing given the speed with which they were written), the fact is that Sherlock Holmes is not quite what his creator intended him to be. Nor, indeed, is the mythic Sherlock Holmes exactly the character whose exploits delighted the readers of *The Strand*. He is another fantastical beast, in actual fact, joining the ranks of Frankenstein's creature, Dracula, and Hyde—although (this was always Holmes's forte) he disguises it awfully well.

Holmes offers us a dream of order and explanation, of a comprehensible cause-and-effect world. In his day that ideal seemed both possible and desirable. But even Watson had his doubts, accusing his beloved companion of being cold and inhuman. Holmes allows us to contemplate what a person, what a world, governed entirely by reason would look like. We consider now the Holmes-machines we have built—our computers, rigorously logical, devoid of emotion, infallible—and to which we increasingly defer, and we wonder whether after all they are the kind of saviors we want.

———◆———

While H. G. Wells was drudging away in a drapery store in Southsea in 1882, less than a mile away a young, recently qualified doctor was attempting to set up his own business. He opened a surgical practice on Elm Grove and had a brass nameplate made up to look respectable. It announced: "Dr. Conan Doyle, Surgeon."

He didn't get many takers, even in this bustling residential area of Portsmouth. Perhaps this wasn't after all the best line of work for Arthur Conan Doyle, a twenty-three-year-old graduate in medicine from Edinburgh University. Maybe, he mused, I should be a writer.

Arthur, from a devout yet artistic Catholic family of Irish ancestry, had found his faith atrophying, first into Unitarianism and ultimately to agnosticism. After graduating in 1879 (he did not take his finals as a bachelor of medicine until 1881) he became an assistant to a doctor on the outskirts of Birmingham. There he already harbored ambitions of being an author, submitting short stories to magazines that were occasionally published for a small fee. Before selecting maritime Portsmouth as a promising location for his own practice, he served for a time as a ship surgeon on whalers and cargo ships.

There's not much indication that Conan Doyle saw writing as a deep calling; it seemed more a plausible way of supplementing a meager income. He was as utilitarian about selecting his genre as he was about deciding where to work. He noted that stories with a Gothic, occult, or supernatural theme seemed popular, and one of his first successes, "The Winning Shot," was of that ilk. Written in 1882 and published the following year in *Bow Bells* magazine, it earned him the grand sum of three guineas.

The medical business picked up a little in the mid-1880s, but still Conan Doyle strove to find a foothold in the literary world. In 1883 he wrote a rather Pooterish semi-autobiographical novel, *The Narrative of John Smith*, that got lost in the post on the way to the publisher. (He began to rewrite it from scratch but never finished it, perhaps for the better.) His next effort, *The Firm of Girdlestone* (1885), was rejected. (It was published in 1890 after Conan Doyle became a lucrative name.)

He decided he had to become more calculated in his efforts and noticed a public enthusiasm for crime fiction. There was, of course, Edgar Allan Poe's "The Murders in the Rue Morgue" (1841), with its ingenious young sleuth C. Auguste Dupin. "Detective" techniques were used by Walter Hartright in Wilkie Collins's *The Woman in White* (1859) and by Sergeant Cuff in Collins's *The Moonstone* (1868). There was Inspector Bucket from Dickens's *Bleak House* (1853); policemen and detectives

Sir Arthur Conan Doyle (1859–1930), photographed in 1914.

featured in works by Dumas and Balzac; and Samuel Warren's *The Experiences of a Barrister, and Confessions of an Attorney* (1853) contained the crime-solving attorney Samuel Ferret. The invention of detective fiction as a recognizable genre is often attributed to the French writer Émile Gaboriau, whose police officer Monsieur Lecoq had his exploits first serialized in the periodical *Le Siècle* in 1869.

Perhaps something in this vein would go down well?

There is a well-crafted and much repeated mythology about the genesis of Sherlock Holmes, which goes like this. At Edinburgh, Conan Doyle had been lectured by a surgeon named Joseph Bell, who displayed phenomenal powers of reasoning. Bell, Conan Doyle figured, could serve as the model for placing the sleuths of Poe and Gaboriau on a more scientific track. As he later explained:

> Gaboriau had rather attracted me by the neat dovetailing of his plots, and Poe's masterful detective, M. Dupin, had from boyhood been one of my heroes. But could I bring an addition of my own? I thought of my old teacher Joe Bell, of his eagle face, of his curious ways, of his eerie trick of spotting details. If he were a detective he would surely reduce this fascinating but unorganized business to something nearer to an exact science. I would try if I could get this effect. It was surely possible in real life, so why should I not make it plausible in fiction?

Bell was Conan Doyle's favorite lecturer: brisk, aquiline, and given to making rather theatrical diagnoses about his patients' circumstances: "Cobbler, I see," he might casually remark to one on their first meeting— or, "Would ye still be working in the linoleum factory?"

"From close observation and deduction, gentlemen," Bell allegedly told his students, "you can make a correct diagnosis of any and every case." Conan Doyle attributed to him also the very Holmesian observation that the type of labor performed by individuals can be deduced from their hands: "The scars of the miner differ from those of the quarryman. The carpenter's callosities are not those of the mason."

It's a good tale, and there is probably some truth in it. But it also serves the purpose of portraying Holmes as a more original creation than he really was. And Bell's deductive abilities seem, on Conan Doyle's account, scarcely less fantastical than those of Holmes.

Take, for example, the famous first meeting of Holmes and Watson (who narrates the stories) in *A Study in Scarlet* (1887). We are straightaway introduced to Holmes's trademark ability to reach startling conclusions from clues that are invisible to most mortals. Watson has just been demobbed as a military surgeon in the Second Afghan War, after a serious injury rendered him unfit for service. Alone and impecunious in London, he meets an old acquaintance named Stamford who tells him of "a fellow who is working at the chemical laboratory up at the hospital." This man (Holmes, of course) is looking for a fellow lodger to split the cost of lodgings he has found at Baker Street, which are more spacious than he needs or can afford. He is "a little too scientific for my tastes," Stamford admits, "—it approaches to cold-bloodedness." You could imagine him slipping a friend a bit of poison, he says, just to watch the effect.

The instant they are introduced, Holmes exclaims to Watson, "You have been in Afghanistan, I perceive."

"How on earth did you know that?" Watson replies. The explanation is deferred until a later conversation, after Watson has moved in with the detective:

> The train of reasoning ran: 'Here is a gentleman of a medical type, but with the air of a military man. Clearly an army doctor then. He has just come from the tropics, for his face is dark, and that is not the natural tint of his skin, for his wrists are fair. He has undergone hardship and sickness, as his haggard face says clearly. His left arm has been injured. He holds it in a stiff and unnatural manner. Where in the tropics could such an English army doctor have seen so much hardship and got his arm wounded? Clearly in Afghanistan.' The whole train of thought did not occupy a second.

Taken at face value, the reasoning makes excellent sense: the reader smiles at how inevitable the conclusion seems in retrospect. Conan Doyle claimed he once saw Joe Bell make a series of deductions of this sort on a patient who he figured, from visual inspection and a casual question or two, had been not long since discharged from a Highland regiment as a noncommissioned officer stationed in Barbados.

But a reader who shares Conan Doyle's literary enthusiasms might think they have heard this before. For Poe's "Murders in the Rue Morgue" opens with a similar set-piece for Dupin, who responds to the narrator's thoughts after they have been strolling in silence through

Paris, just as if he had spoken them out loud: "You're right. He is a very little fellow, that's true, and would do better for the *Théâtre des Variétés*."

The narrator is astonished at how Dupin knew he was indeed thinking just this of an actor named Chantilly. Dupin explains, and the explanation—running from a flower-seller they had just passed to cobblestones to the constellation of Orion—is far more absurdly improbable than anything Holmes comes up with. The format, however, is identical: tiny clues in appearance, circumstance, or action betray the facts with seemingly inevitable (but in fact rather improbable) logic. In "The Murders in the Rue Morgue" and the sequels "The Mystery of Marie Rôget" and "The Purloined Letter," the genre staples that the Sherlock Holmes stories crystallize are already present.

Conan Doyle understands that some readers of his tale might think the same, and he takes the matter in hand. "You remind me of Edgar Allan Poe's Dupin," Watson tells Holmes, only for the detective to assert his originality:

> In my opinion, Dupin was a very inferior fellow. That trick of his of breaking in on his friends' thoughts with an apropos remark after a quarter of an hour's silence is really very showy and superficial. He had some analytical genius, no doubt; but he was by no means such a phenomenon as Poe appeared to imagine.

It's a bold and rather witty preemptive strike (perhaps a little less so given that Poe had been dead for almost forty years and so couldn't defend his hero). For good measure, Conan Doyle heads off comparisons with Gaboriau's Lecoq: Holmes calls him "a miserable bungler," saying that Lecoq solved in six months what he could have cracked in twenty-four hours. In a delicious piece of metanarrative, Conan Doyle has Watson bristle at hearing his favorite fictional sleuths "treated in this cavalier style." This fellow "may be very clever," he says to himself (echoing, in all likelihood, the thoughts of the reader), "but he is certainly very conceited."

Conan Doyle's story of Joe Bell serves to further his goal of distinguishing Holmes from his predecessors by making him a man of science.

And so, when Watson first meets Holmes, the detective is engaged with some chemical experiment:

> Broad, low tables were scattered about, which bristled with retorts, test-tubes, and little Bunsen lamps with their blue flickering flames. [He] was bending over a distant table absorbed in his work. . . . "I've found it! I've found it," he shouted to my companion, running towards us with a test-tube in his hand. "I have found a re-agent which is precipitated by haemoglobin, and by nothing else." Had he discovered a gold mine, greater delight could not have shone upon his features.

Conan Doyle was on comfortable territory here, having been much interested in drugs at Edinburgh, where the eminent scientist Alexander Crum Brown lectured in chemistry. When he draws up a list of Holmes's traits, Watson rates his knowledge of literature, philosophy, astronomy, and politics as nil or feeble, but his knowledge of chemistry is "profound."

Making Holmes a kind of scientist was a timely decision. Forensic science was in its infancy, and public fascination with it had been fostered by some high-profile and sensational trials. Victorian Britain had been captivated by the investigation and trial of Mary Ann Cotton, who was convicted in 1873 of murdering her stepson Charles Edward Cotton by arsenic poisoning. It seems that she also killed three of her former four husbands this way (and possible several other people too) in order to collect their life insurance; poor Charles had been an obstacle to collecting the money on the last of these, his father Frederick. While the idea of a female serial killer sweetly putting arsenic-laced sugar into her husbands' tea captivated the public imagination, the trial was also a triumph of forensic chemistry. An analysis called the March test, devised in the 1830s, was used to detect arsenic in some of the tissues of Charles Cotton kept from the autopsy. No longer were criminal cases to be judged on circumstantial evidence; hard science could point the finger.

For that matter, so could fingers. Policemen and detectives had begun searching crime scenes for fingerprints, and in the early 1880s Francis Galton, Charles Darwin's cousin, showed that they could be used to identify individuals. As Galton himself admitted, the astronomer William Herschel had introduced fingerprinting three decades earlier as a civil servant in Bengal, to prevent impersonation and to establish authenticity of documents. But Galton established a system of classification for fingerprint patterns that is still used today.

Galton is best known now for his advocacy of eugenics, an idea that emerged from his studies of the Darwinian inheritance of traits. He tried to show that such traits could be deduced from fingerprints—a baseless idea, but one that echoed the view of criminologist Cesare Lombroso (see page 145) that "degenerate" and criminal characteristics could be detected in facial features.*

Holmes's experiment to prepare a precipitate from blood is clearly a forensic test of the same type. It didn't matter that he rarely needed to make use of techniques this clinical, solving his cases instead based on careful observation, a stray word, a scuff on a boot. Conan Doyle tells an audience already primed to hear it that thenceforth no criminal or miscreant can hope to escape the beacon glare of science.

———————————

Conan Doyle took some time to find the right name for his hero. At first he considered "Sherringford Holmes," the surname probably de-

---

* Conan Doyle's own predilection for simple-minded Galtonian thinking about breeding and inheritance is plain in his 1888 article "On the Geographical Distribution of British Intellect" for the journal *The Nineteenth Century*. Here he collects statistics for "intellectually distinguished men" born in different parts of the country and concludes that the Saxon is probably inherently superior to the Celt, and Londoners to those either north or south. With the characteristic prejudice of his times, he locates the "mental nadir" of the British Isles in western Ireland.

rived from the American jurist and physician Oliver Wendell Holmes, whose books *The Autocrat of the Breakfast-Table* and *The Professor at the Breakfast-Table* Conan Doyle admired. "Sherlock" might have come from Chief Inspector William Sherlock of the Metropolitan Police, a detective occasionally mentioned in the newspapers. Holmes's colleague was initially Ormond Sacker; Conan Doyle might have alighted instead on "Watson" from the name of another Southsea doctor, although its very ordinariness establishes the sidekick's Everyman status.

London was the obvious place to locate Holmes, but at first Conan Doyle didn't know the city at all. He worked out the locations and movements of his characters, he said, while sitting at his desk in Southsea with a post office map. The iconic address 221B Baker Street lies at the upper end, north of Marylebone Road, and is now the Sherlock Holmes Museum—notwithstanding the fact that Conan Doyle (presumably without any knowledge of the numbering system) placed the building at the street's southern end.

Like Wells and Stoker,* Conan Doyle owed his initial literary successes not so much to immediate evidence of talent as to dogged persistence with what was often rather mediocre work. He wrote *A Study in Scarlet* in 1886 and sent it to the *Cornhill Magazine*, which rejected it. So did two other publishers, but it was accepted by Ward, Lock and Company, a publishing house specializing in sensationalist fiction, for the paltry fee of £25 and no royalties. All the same, it pleased Conan Doyle that the story appeared in the company's popular *Beeton's Christmas Annual* of 1887.

In *A Study in Scarlet*, Holmes is already a fully formed character. "I have a trade of my own," he tells Watson. "I suppose I am the only one in the world. I'm a consulting detective, if you can understand what that is." Mostly he does his work without leaving his consulting room: "I listen

---

* All three were acquainted. Conan Doyle's play *A Straggler of '15* was staged in 1894 by Henry Irving's company (under the title *A Story of Waterloo*), stage-managed by Stoker.

to [the client's] story, they listen to my comments, and then I pocket my fee." Just occasionally, he admits, "a case turns up which is a little more complex. Then I have to bustle about and see things with my own eyes." All the mannerisms and abilities are laid out in this first story: the violin-playing (he is rather good), the skill at boxing and fencing, the pipe, the self-assurance bordering on insufferable arrogance.

Yet the story was never intended as the overture to a series, and Conan Doyle didn't plan a sequel to *A Study in Scarlet* until 1889, when he was commissioned at a literary dinner with American publisher Joseph Marshall Stoddart to write something for his magazine, *Lippincott's Monthly*. This became *The Sign of Four*, published in 1890.*

At the same time as working on these Holmes tales, Conan Doyle was writing a historical romance, *Micah Clarke*, which appeared to some acclaim in 1889. Considering his literary career now well and truly launched, he decided it was time to move to the center of that world. In 1891 he closed his Southsea surgery and opened a new practice in London. Apparently he never took on a single patient there, being able now to support himself from his writing alone.

If *A Study in Scarlet* laid out the parameters of the entire Holmes oeuvre, *The Sign of Four* began to give them real substance. It's here that we learn of Holmes's cocaine and morphine habit—introduced in the opening paragraph with what might seem to us now rather shocking candor:

> Sherlock Holmes took his bottle from the corner of the mantel-piece
> and his hypodermic syringe from its neat morocco case. With his
> long, white, nervous fingers he adjusted the delicate needle, and rolled
> back his left shirt-cuff. For some little time his eyes rested thought-
> fully upon the sinewy forearm and wrist all dotted and scarred with

---

* Oscar Wilde was also at that dinner; Conan Doyle knew him already, in part because of their shared Irish heritage. Stoddart commissioned him too, and Wilde went off to produce *The Picture of Dorian Gray*.

innumerable puncture-marks. Finally he thrust the sharp point home, pressed down the tiny piston, and sank back into the velvet-lined armchair with a long sigh of satisfaction.

When Watson expresses uptight disapproval, Holmes retorts that he finds the effects of cocaine "so transcendently stimulating and clarifying to the mind that its secondary action is a matter of small moment." But he takes the drug also to stave off an overwhelming sense of despair and angst when the world fails to provide him with challenges equal to his mental powers:

> I cannot live without brain-work. What else is there to live for? Stand at the window here. Was ever such a dreary, dismal, unprofitable world? See how the yellow fog swirls down the street and drifts across the dun-coloured houses. What could be more hopelessly prosaic and material? What is the use of having powers, doctor, when one has no field upon which to exert them? Crime is commonplace, existence is commonplace, and no qualities save those which are commonplace have any function upon earth.

We begin now to suspect that behind the façade of Holmes's logic and reason, an unquiet spirit struggles. There is real emotional tragedy—brushed aside in the romanticized Holmes of popular imagination—in the story's closing lines. "You have done all the work in this business," declares Watson:

> "I get a wife out of it, [police inspector] Jones gets the credit. Pray what remains for you?"
> "For me," said Sherlock Holmes, "there still remains the cocaine-bottle." And he stretched his long, white hand up for it.

It's at moments like that that you understand what elevated Conan Doyle above his many imitators.

But the dominant image of Holmes—the mythic Holmes—is the one Watson describes earlier in the story, when he remarks on the attractiveness of the detective's new client (and Watson's future wife) Miss Morstan. Holmes replies that he didn't notice, prompting Watson to burst out, "You really are an automaton—a calculating machine—There is something positively inhuman about you."

Holmes smiles languidly and responds:

> It is of the first importance not to allow your judgment to be biased by personal qualities. A client is to me a mere unit,—a factor in a problem. The emotional qualities are antagonistic to clear reasoning.

When Watson announces at the end of the tale that he has become betrothed to Miss Morstan, Holmes groans. She is charming and clever, he concedes, "but love is an emotional thing, and whatever is emotional is opposed to that true cold reason which I place above all things. I should never marry myself, lest I bias my judgment."

With *The Sign of Four*, Conan Doyle had truly arrived—if not perhaps at his intended destination of high literary acclaim. The newly launched magazine *The Strand* published "A Scandal in Bohemia" and "The Red-Headed League" in 1891, and commissioned six more tales. But already Conan Doyle was tired of the formula and felt the stories were a distraction from his true calling of historical novels. The next time *The Strand* came asking, Conan Doyle decided to stymie them by demanding a ridiculous fee: £1,000 for twelve stories. They accepted.

The Strand not only established Holmes in the public consciousness but supplied the iconography. The elegant, handsome Holmes depicted by the illustrator Sidney Paget is now precisely what comes to mind, despite Doyle's description of the man as having "a great hawks-bill of a nose and two small eyes set close together." The deerstalker didn't appear until the fourth short story, "The Boscombe Valley Mystery," where Holmes kits himself out for the countryside—Paget never shows him dressed this way in the city, where it would have been considered bizarre. The pipe was straight: only with Basil Rathbone's film portrayal did it acquire the curve of the "thinking man's pipe."

But Conan Doyle had soon had enough of Holmes. He had to die. "I am in the middle of the last Holmes story," Conan Doyle wrote in April

"TAKING UP A GLOWING CINDER WITH THE TONGS."

Holmes and Watson in "The Copper Beeches," as imagined by
Sidney Paget for The Strand *in 1892.*

1893, "after which the gentleman vanishes, never never to reappear. I
am weary of his name." During a stay in Lucerne, Switzerland, he told
a friend that "I have made up my mind to kill Sherlock Holmes; he is
becoming such a burden to me that it makes my life unbearable." Their
companion, a Methodist minister named W. J. Dawson, interjected
by proposing the nearby Reichenbach Falls as a good location for the
demise.

"The Final Problem" appeared in the *Strand* at the end of 1893. Here
Holmes meets his nemesis Professor Moriarty, the "Napoleon of crime":

He is the organizer of half that is evil and of nearly all that is unde-
tected in this great city. He is a genius, a philosopher, an abstract
thinker. He has a brain of the first order. He sits motionless, like a

spider in the centre of its web, but that web has a thousand radiations, and he knows well every quiver of each of them. . . . I tell you, Watson, in all seriousness, that if I could beat that man, if I could free society of him, I should feel that my own career had reached its summit, and I should be prepared to turn to some more placid line in life.

To this, Watson might reasonably have replied: But after all the adventures we have had fighting crime, why have we not come across this Moriarty before? The reason, of course, is that Conan Doyle has invented him on the spot to rid him of his tiresome detective.* He is, in truth, not a criminal genius at all but "a flimsy plot device," as literary scholar Benedick Turner points out: a "bumbling ruffian" who can think of no more ingenious way of disposing of Holmes than trying to wrestle him over a cliff.

The story is a study in unconscious ambivalence. Put it this way: if I chose to argue that Conan Doyle had planned all along to bring back Holmes eventually, "The Final Problem" gives me total license for that interpretation. After all, we never actually see Holmes plunge to his doom. And there is no body; the only evidence of his death is a note left for Watson on the mountain path. But I could go further and interpret the tale as an account of Holmes's scheme to fake his death and free himself from his former life.† For *we also never see Moriarty.* The only real evidence of his existence is Holmes's testimony to Watson—which,

---

* In a classic example of Conan Doyle's sloppiness of plotting, one of the "case book" stories he wrote later but set earlier than "The Final Problem" has Watson refer to Moriarty as "the famous scientific criminal." Some fans and "scholars" have displayed their characteristic determination to blur fact and fiction by lambasting *Watson*, not Conan Doyle, for his poor memory.

† That possibility was never going to escape the attentions of Sherlockian fans, who have advanced several variations on this theme: he planned to go away to cure his drug addiction, he was on a secret diplomatic mission, or he just needed a holiday. Equally, there are many theories about the "true" identity of Moriarty: J. Edgar Hoover, Fu Manchu, Count Dracula, and Satan are among the candidates. In fairness, the degree of earnestness with which these possibilities are mooted is variable.

THE DEATH OF SHERLOCK HOLMES.

Holmes and Moriarty fall to their doom at the Reichenbach Falls in Sidney Paget's illustration for "The Final Problem" (The Strand, *1893).*

in light of his previous silence on the matter, can easily be interpreted as invention. Yes, Conan Doyle surely did wish to be rid of Holmes—but I suspect he could not bear to kill the detective outright.

We can go further still: Moriarty is an anti-Holmes, the criminal genius the detective might have become had the scales of his constitution tipped the other way. And this much, surely, was part of Conan Doyle's design. In *The Sign of Four* Watson had already raised that specter: "I could not but think what a terrible criminal [Holmes] would have made had he turned his energy and sagacity against the law, instead of exerting them in its defense." Holmes, meanwhile, identifies with and even admires Moriarty: "My horror at his crimes was lost in my admiration at his skill." He anticipates Moriarty's moves by asking himself what he would do in the same situation. In line with Conan Doyle's crude Darwinian/Lombrosian notions about criminality, it is only innate biology that prevents Holmes from turning to the dark side: Moriarty "had hereditary tendencies of the most diabolical kind. A criminal strain ran in his blood." To press the point to its conclusion, Moriarty could be interpreted as yet another Hydean doppelgänger, and even Holmes seems to imply that his arch-enemy was summoned by his own imagination: "I must confess to a start when I saw the very man who had been so much in my thoughts standing there on my threshold."

The effect Holmes's demise had on his audience is legendary. Readers of *The Strand* were distraught and outraged. More than twenty thousand of them canceled their subscriptions; some took to wearing black armbands in mourning. Conan Doyle was unrepentant. "I have been much blamed for doing that gentleman to death," he wrote, "but I hold that it was not murder, but justifiable homicide in self-defence, since if I had not killed him, he would certainly have killed me."

His fans did not have to worry, for Holmes was now too famous to die. His first return was in a tale predating the struggle at the Reichenbach Falls, into which Holmes insinuated himself. For *The Hound of the Baskervilles* began in 1901 as a standard "creeper," inspired by the account of Conan Doyle's friend Bertram Fletcher Robinson of a giant dog said to have caused terror on Dartmoor in southwest England. He and Robinson carried out some location research, and Conan Doyle appropriated the name "Baskerville" from the coachman of his friend's family on their Dartmoor estate. Conan Doyle negotiated with *The Strand* for Robinson to be given a co-credit and fee for the serialized story. But in the course of preparing it, he discovered that only Holmes could solve the mystery of the diabolical hound.

Rumors of Holmes's death were in any case proving to be exaggerated. A play about the detective, performed by the American actor William Gillette, opened on the London stage at much the same time as the serialization of *The Hound of the Baskervilles* began, and several Holmes imitations had already been penned; one of the most celebrated was the anti-Holmes "gentleman thief" A. J. Raffles, created by Conan Doyle's brother-in-law Ernest William Hornung. But when the detective returned to *The Strand*, the magazine's sales increased by about thirty thousand.

Inevitably, Holmes didn't die in Switzerland after all. Offered huge fees by American publishers to revive him, Conan Doyle revealed that he had escaped Moriarty's clutches using a Japanese martial art called "baritsu" (a characteristically careless error in cultural borrowing—this was doubtless intended to be the martial art called bartitsu practiced in England at the time). In "The Adventure of the Empty House" (1903), Watson has an unfriendly encounter with an elderly bookkeeper who later reveals himself to be Holmes in disguise. It turns out that he had faked his death after all, deciding to lie low after killing Moriarty so that other of his criminal associates would not come seeking revenge.*

It seemed impossible to close the casebook. The final Holmes story was supposed to be "The Last Bow," published in *The Strand* in September 1917 after Conan Doyle felt obliged to do his bit for the war effort.

---

* It was a rather feeble excuse for Holmes' prolonged disappearance, given that one of Moriarty's henchmen, the sharpshooter Colonel Moran, had seen Holmes survive the fight and had even tried to drop rocks on him as he lay afterward on the mountain ledge.

Here Holmes has retired to the South Downs to study beekeeping, only to be recalled by the prime minister himself to track down a German spy. Conan Doyle was by this stage a knight of the realm and a literary eminence; his dinosaur novel *The Lost World* (1912) had been another big hit, establishing the intrepid explorer Professor George Edward Challenger. And Sir Arthur had become obsessed with spiritualism, touring the country giving public talks in support of the belief. More Holmes tales followed, taken from his earlier life and collected in *The Case-Book of Sherlock Holmes* in 1927, three years before his creator's death.

It never really mattered what the critics made of Sherlock Holmes. "Professional critics," says John Le Carré, "can't lay a glove on Conan

"HOLMES GAVE ME A SKETCH OF THE EVENTS."

*The classic Holmes: in cape and deerstalker for "The Adventure of Silver Blaze" (*The Strand*, December 1892).*

Doyle, and never could." The literary editor of *The Strand*, Greenhough Smith, put it as plainly as anyone needed, saying that the popular early story "A Scandal in Bohemia" demonstrated the "perfect art of telling a story." Perhaps we'd do better to heed the judgment of another fine storyteller who likewise befuddled literary critics for many years: fellow Edinburgh alumnus Robert Louis Stevenson, whose work Conan Doyle admired. Writing to him all the way from Samoa, Stevenson offered the decidedly ambiguous compliment that his was "the class of literature that I like when I have the toothache."

<hr />

The detective novel presents the world as a soluble place, in which all that happens has a clearly defined and accessible cause. No matter how baffling or weird events might look, their origins can be discerned under the clear light of reason. Theater director Peter Brooks argues that detective stories are thus exercises in *reconstructing stories*: they are puzzles in which the goal is to find out "what happened," to reduce all the options to a single narrative consistent with events. The related whodunit genre makes it the reader's job to do the work, once they are supplied with all the clues they should need.

In this game, Holmes is the archetypal Enlightenment man reconstructed for a late-Victorian sensibility, bringing order through rationality. His is the universe of Isaac Newton and Pierre-Simon Laplace, where every effect can be traced to a cause, in mind as much as in matter—it is all particles in action, governed by immutable and inevitable physical law. Holmes's famous dictum "Eliminate all other factors, and the one which remains must be the truth" does not permit of any real-world messiness in the way matters unfold. There is an explanation for everything, and we can understand the world completely if only our vision is unhindered by obstacles such as emotion. "From a drop of water," Holmes reads with approval in a magazine, "a logician could infer the possibility of an Atlantic or a Niagara without having seen or heard of one or the other."

Put another way, to every question there is a right answer. As literary

critic Sandra Kromm points out, there is something terrifyingly oppressive about that idea—for the first people to insist on it, she says, were
the Inquisition.

Yet Holmes's detective genius comes not just from his superbly deductive mind but also from his use of data and communications—from
being a native in the emerging Age of Information. He relies on timetables, shipping routes, telegrams, and newspapers, all the signifiers of
systematized contemporary life. He is a thoroughly modern man—more
so, really, than Conan Doyle, and so it is little wonder that the author
could not control him.

Yet the appeal of Holmes's powers of deduction is only superficially
predicated on this Newtonian universe of cause and effect. For the fact
is that Holmes is not really a rationalist at all, but a magician. Dupin's
reconstruction of his friend's train of thought in "The Murders in the
Rue Morgue" is obviously implausible, but so too are many of Holmes's
conclusions; Conan Doyle is simply rather better than Poe at masking
that implausibility. Here, unlike the whodunit, we can't be expected to
solve the mystery ourselves. Not only are the meager clues insufficient to
guide the reader to the correct deduction—crucial pieces of evidence
are often withheld until Holmes explains them—but they aren't even
specific enough for Holmes himself, who would require his reason to
be supplemented with something approaching clairvoyance. The enjoyment, then, comes not from the revelation of seeing what we missed, but
from the "reveal" of the magic trick. The puzzle is more ludic than logic.

Take that first meeting with Watson, when the detective divines the
doctor's past. There are plenty of alternative ways to account for Watson's appearance besides the hypothesis that he has just returned from
active service in Afghanistan. There is no reason, say, why Watson might
not have injured his arm in a nasty fall from a horse in a game of polo in
colonial Calcutta. The same implausibility applies to Holmes's deduction in *A Study in Scarlet* that the man glimpsed in the street from the
window of the Baker Street apartment is a retired sergeant of the Ma-
rines. The detective might sometimes amass evidence piece by piece as
a scientist would, but his feats delight readers because they suggest *sotto
voce* that magic and telepathy are possible.

And really, isn't this just what one might expect from Conan Doyle? To put it bluntly, Sherlock Holmes is a credulous person's vision of what a hard-nosed rational thinker is like. Conan Doyle is, after all, one of the most celebrated examples of the late-Victorian and Edwardian passion for the supernatural among the sober and respectable British public. His interest in spiritualism began around 1887, and despite declaring at first that he was "interested but very skeptical," he soon became a convert to the claims of mediums. He corresponded with the poet and classicist Frederic W. H. Myers, a founding member of the Society for Psychical Research—the network for investigators of the paranormal that included scientists William Crookes, Oliver Lodge, Lord Rayleigh, and J. J. Thomson along with John Ruskin, Alfred Tennyson, and the future prime minister Arthur Balfour.

We'd be wrong to suppose that, in pursuing these interests, Conan Doyle was merely gullible—for if he was, then so were these and other leading scientists of his day. Their willingness to entertain at least the possibility of spirit messages and invisible planes of existence invites and often receives condescension today. But such ideas were perfectly understandable in an age when so little was known about the human mind and when science was revealing all manner of hitherto unsuspected and invisible forces and influences—radio waves, X-rays, radioactivity— sometimes rendered visible by the new medium of photography. Crookes and Lodge suspected that the ether, the invisible and tenuous medium thought to carry the electromagnetic waves explained by James Clerk Maxwell in the 1860s, might be a bridge between our plane and others within which spiritlike intelligences existed. Maxwell's colleagues, the physicists Balfour Stewart and Peter Guthrie Tait, argued in *The Unseen Universe* (1879) that such an invisible realm might help to reconcile modern physics with articles of Christian faith such as the immortality of the soul. And after all, didn't radio waves now permit just the kind of remote, invisible communication of messages that spiritualists were claiming to channel? This association of radio telecommunication with the paranormal was explored in Rudyard Kipling's 1902 short story "Wireless."

Yet there's no avoiding the fact that Holmes's insistence on a rational and scientific view of the world sits uneasily with the more outré of his creator's inclinations, and the ambivalence is often apparent. In Conan Doyle's 1888 novel *The Mystery of Cloomber*—a piece of hokum about an army general living in seclusion in Scotland in fear of retribution for killing a Buddhist adept in India—his skepticism of science leaks out. As the narrator says:

> Science will tell you that there are no such powers as those claimed by
> the Eastern mystics. I, John Fothergill West, can confidently answer
> that science is wrong. For what is science? Science is the consensus
> of opinion of scientific men, and history has shown that it is slow to
> accept a truth. Science sneered at Newton for twenty years. Science
> proved mathematically that an iron ship could not swim, and science
> declared that a steamship could not cross the Atlantic.

Like many at the time, Conan Doyle could not easily relinquish a religious sensibility for the materialism that science increasingly seemed to demand. His 1895 novel *The Stark Munro Letters* was a semi-autobiographical account of a young doctor struggling to reconcile his scientific practice with his religion. Conan Doyle joined the Society for Psychical Research himself (as did H. G. Wells, who he met in 1898) and toyed with the idea of enlisting in the occult-revivalist Order of the Golden Dawn. He corresponded with Oliver Lodge about spiritualism, explored hypnotism (often still called mesmerism), and experimented with the popular craze of "spirit photography," which attempted to capture ghostly presences on camera. He set up a publishing company, the Psychic Press, which had its own bookshop, and in 1927 he opposed the motion "that spiritualism is repugnant to common sense" at the Cambridge Union. (Speaking for the motion was the biologist J. B. S. Haldane.) After Conan Doyle's death in 1930, his wife Jean Leckie insisted that she regularly consulted with the spirit of her husband on family affairs.

Even locating Conan Doyle within the currents of his time doesn't, however, excuse him from all charges of credulousness. The most notorious example of that trait arose at the height of his celebrity. In *The*

The photograph that fooled Conan Doyle: the Cottingley Fairies, as "caught on film" between 1917 and 1920 by the Yorkshire teenagers Elsie Wright and Frances Griffiths.

Coming of the Fairies (1922), he argued the case for why the "little folk . . . are destined to become just as solid and real as the Eskimos." Motivated by the faked photographs of the "Cottingley Fairies," which he revealed to the public in a 1921 article in *The Strand*, Conan Doyle suggested that these beings were perfectly consistent with modern science. He argued that they might simply be constituted of some material that was transparent to ordinary light but reflective to ultraviolet or infrared—"vibrations" to which mediums were perhaps also attuned. Despite being himself a keen photographer, Conan Doyle was unable to countenance the idea that the photos taken in the picturesque Yorkshire village between 1917 and 1920 by two teenage girls were in fact fakes made by cutting out pictures from a children's storybook.

This bathetic episode did Conan Doyle's reputation no favors. But *The Coming of the Fairies* also gives us a glimpse of undercurrents running through the Holmes tales. For all that the detective is a modern man, his creator yearned for a reenchantment of a world that he felt had become spiritually impoverished:

> The recognition of [the fairies'] existence will jolt the material twentieth-century mind out of its heavy ruts in the mud, and will

make it admit that there is a glamour and a mystery to life. Having discovered this, the world will not find it so difficult to accept the spiritual message supported by physical facts which has already been so convincingly put before it.

It is surely because the rationalistic Holmes was created by an author not in total command of his material but struggling to contain such conflicting impulses that he acquired mythopoeic potential. The duality seeps into Holmes himself, who in dressing-gown-clad, violin-playing, morphine-sedated mode is the archetypal late-Victorian Romantic mystic. For all his logic, he admits that intuition plays a part in his work too. And no one who proclaimed to be governed purely by science would be ignorant of the Copernican structure of the solar system, or be so indifferent to having it explained to him by the incredulous Watson. As Conan Doyle's biographer Andrew Lycett puts it:

> He tried hard to make his fictional 'calculating machine' the epitome of his scientific training. But the detective . . . refused to play the game. He was simply too wracked with contradictions and even emotions. The result is a character who reflects not just his changing times but also our own.

But did Conan Doyle really set out to make Holmes a living demonstration of reason's superiority to emotion? At times he seems instead to be trying to work out this equation for himself—to decide if his hero is truly so heroic, so estimable a man. The reader, after all, is encouraged to identify not with Holmes but with Watson: which of us would not, in that situation, be similarly inept, missing clues that the detective finds obvious, getting fooled by his disguises and exasperated at his smugness and intransigence? And Conan Doyle leaves us in no doubt that Holmes would be as dismissive and aloof with us as he is with poor Watson. Holmes doesn't always trust his great friend and sometimes excludes him from his confidence. When in a stroke of inspired metanarrative he has his hero read *A Study in Scarlet* (ostensibly written by Watson), Holmes judges it harshly:

Honestly, I cannot congratulate you upon it. Detection is, or ought to be, an exact science, and should be treated in the same cold and unemotional manner. You have attempted to tinge it with romanticism, which produces much the same effect as if you worked a love-story or an elopement into the fifth proposition of Euclid.

But of course *we* would rather have Watson's impassioned account than anything Holmes might produce, which would likely look more like a scientific paper than an entertaining novella.*

Conan Doyle seems, then, only to have half-believed that he was creating a series of stories celebrating the supremacy of science and reason as guardians of Western civilized society. Holmes has mythic potential precisely because his creator left us space to wonder, indeed to doubt, whether a society run on those lines would be so desirable after all.

———————— ◆ ————————

As the scion of reason, the mythical Holmes is often now presented as a character on the autistic spectrum—a man who does not just reject emotion but is incapable of comprehending it. The truth is that he is a poor caricature of such a person, for no one with autism speaks the way Holmes does. As far as that is concerned, he is simply a cipher for examining where a world governed by cold logic might take us.

The most famous fictional character motivated by emotionless logic was played by the actor who also portrayed Holmes in the 1975 movie *The Interior Motive* and in a 1976 stage adaptation of Conan Doyle's work: Mister Spock, lieutenant on the starship *Enterprise*, personified by Leonard Nimoy.

The logical Spock and William Shatner's impulsive Captain James T. Kirk weren't quite Holmes and Watson in space, but the debt to Conan

---

* We might perhaps anticipate something like the "little monograph on the ashes of 140 different varieties of pipe, cigar, and cigarette tobacco" that Holmes claims to have written in "The Boscombe Valley Mystery."

Doyle was clear—and explicit. The original *Star Trek* series created by Gene Roddenberry in 1966–1969 could be said to hinge on this interplay of logos and eros, the message plainly being that each is enhanced by the other. By the time *Star Trek VI: The Undiscovered Country* hove into view in 1991, screenwriter Nicholas Meyer decided it was time to come clean. "An ancestor of mine," says Spock, "maintained that if you eliminate the impossible, whatever remains, however unlikely, must be the truth."

Of course, what made Spock resonate with audiences was that he wasn't a purely logical Vulcan but was half human. Like Conan Doyle, he wanted his rational side to prevail but could not entirely suppress the impulses from his all-too-human heritage. Lieutenant Data, the android in *Star Trek: The Next Generation*, takes the Spock persona to the next level: he has a "positronic" computer brain that was initially devoid of all emotional register. Yet he too feels a need to experience this aspect of human behavior, and eventually he is given an "emotion chip" implant. In the episode "Lonely among Us," Data investigates a murder onboard

Holmes in space: Leonard Nimoy as Star Trek's Mister Spock (Paramount Pictures).

the *Enterprise* in the manner of Sherlock Holmes, and a later episode, "Elementary, Dear Data," goes full Conan Doyle. Data is enlisted for a holographic game in which he and chief engineer Geordi La Forge have to solve Holmes mysteries kitted out in full deerstalker and pipe. Data has read the entire Holmes oeuvre and completes the task easily, unaware that the fun is meant to be in the process. As a result, he is put back into the virtual-reality "Holodeck" to face a simulated adversary with the capacity to outwit him: Moriarty, who is so advanced a program that he can seize control of the computer and project himself outside of the Holodeck.

Here, pure logic and intelligence — personified now not by Data but by the computer system itself — is shown to pose a hazard. It is after all one of the dominant fears of the information age that we might build sentient machines whose reasoning is far superior to ours but which lack the empathy necessary to human existence. Holmes anticipates this modern techno-nightmare. Explaining in "The Mazarin Stone" (1921) why he eats so little, he tells Watson that he doesn't wish his active digestive system to deprive his brain of the blood it needs to function optimally. "I am a brain, Watson," he says. "The rest of me is a mere appendix." In "The Greek Interpreter" (1893), meanwhile, Watson admits that "sometimes I found myself regarding him as an isolated phenomenon, a brain without a heart, as deficient in human sympathy as he was preeminent in intelligence."

Holmes has at least once been "diagnosed" in faux-psychiatric analysis as autistic. His impatience, insensitivity, asexuality, and inconsiderate nature are all said to be characteristic of the condition, along with the body language and behavior that so baffle Watson — typifying the tendency to see autism simply as a catalogue of dysfunctions. We're encouraged too to regard the brilliance of Holmes's deductive mind as the positive side of the condition — what psychologist Stuart Murray has dubbed a "compensation cure." It's very much in this mode that Holmes is played by Benedict Cumberbatch in the TV series *Sherlock*: he is puzzled why a woman who had a stillborn child would still be thinking of her years later, and surprised at how upset Watson is discover that

Holmes has allowed him to believe for two years that he died in the fateful encounter with Moriarty.

We might choose to see all this as simply a reflection of how our neurotypical narrator, Watson, misunderstands and misreads the actions and motives of his Aspergic friend. But it's perhaps better to ascribe the misunderstanding to Conan Doyle—in whose time of course autistic people would have had to manage as best they could in a society that did not recognize their situation at all. If Holmes is indeed an attempt to portray this "neurotribe," he comes close to the popular caricature of the idiot savant. The truth, as Mark Haddon explores in his 2003 novel *The Curious Incident of the Dog in the Night*, is very different. The title of course comes straight from Conan Doyle; it alludes to the famous passage in "The Adventure of Silver Blaze" (1892) where Holmes makes a deduction about a crime from a dog that fails to bark when it might have been expected to. Haddon's autistic teenage hero Christopher Boone fancies himself a Holmes-style detective who sets out to solve the "mystery" of a murdered dog. But the reader sees how Christopher's difficulty in understanding the motives and intentions of others leads his reasoning astray and leaves him oblivious to the sad domestic drama unfolding around him. As Haddon shows, an autistic person could never do what Holmes does. On the contrary, the Holmes myth acts as a vehicle for the social stereotyping that neurotypical people often, in their discomfort and incomprehension, display toward those on the autistic spectrum.

In doing so it has helped secure that stereotype: allegedly Aspergic fictional detectives have flourished, probably to the detriment of public understanding of the condition. Spencer Reid from the CBS series *Criminal Minds* has an IQ of 187 and a photographic memory, but he is socially awkward and struggles to handle his emotions. Forensic scientist Temperance Brennan from Fox's *Bones* (loosely based on a character created by crime writer Kathy Reichs) has three doctorates but no social skills. Charlie Eppes (*Numb3rs*, from CBS) is a mathematical genius who lends his skills to the FBI for particularly difficult cases; he has odd mannerisms, obsessive tendencies, and gaping holes in his general knowledge. Clinician Gregory House (*House*, Fox) is a grouchy, drug-dependent misanthrope with only one friend (Wilson); he is explicitly inspired by Holmes, his name being a play on the homophone "Homes."

You get the idea: you can't be a crime-solving genius, we're told, without a strong dose of cognitive dysfunction. "I'm a high-functioning sociopath," Cumberbatch's Holmes tells an enemy before killing the man in cold blood.

It's a popular game among Sherlock aficionados to highlight the "myths" about Holmes—not least, he never said, "Elementary, my dear Watson." How they bristle and huff when their hero is misrepresented or miscast. But we know well enough by now that, once we accept Sherlock Holmes as a creature of myth, all this mutation, distortion, and reinvention is part and parcel.

And as proper myth requires, the Sherlock formula works anywhere, any time: in modern London, in the twenty-second century, in medieval Europe (William of Baskerville in Umberto Eco's 1980 debut novel *The Name of the Rose*), in outer space. The 1985 anthology *Sherlock Holmes through Space and Time* collected tribute stories from well-known science fiction writers including Philip José Farmer and Gene Wolfe, in which Holmes is a dog, an alien, a robot. Isaac Asimov based his space detective Wendell Urth on Holmes for a series of tales collected in *Asimov's Mysteries* (1968).

Holmes can be Chinese, female (Watson too: Lucy Liu's "Joan Watson" in the CBS TV series *Elementary* is both), a puppet (the Muppets' Gonzo). Perhaps he is gay—the cozy cohabitation of Holmes and Watson was never going to pass unnoticed. There's a spare bedroom available *if needed*, the landlady Mrs. Hudson tells the duo in the first episode of *Sherlock*, pointedly called "A Study in Pink." When they go for a meal, the waiter assumes they are on a date. "It's the gayest story in the history of television," said *Sherlock*'s Watson, Martin Freeman.

Holmes can fight any opponent: Jekyll and Hyde, Jack the Ripper,

Hitler, and (inevitably) Count Dracula.* Holmes the bohemian, muscle-bound action hero played by Robert Downey Jr. in Guy Ritchie's 2009 movie, paired with Jude Law's dashing Watson, might not be to every-one's taste, but the film reminds us that Holmes and Watson were the first "cop buddy" double-act. "These guys are sort of in love with each other," said Ritchie.

Yet no modern mythopoesis is more resolutely resisted by advocates of "authenticity" than this one. The cult of Sherlock hovers somewhere between the endearingly eccentric and the downright disturbing. Pil-grimages and tours to 221B Baker Street in London have become a sig-nificant tourist industry. Some enthusiasts take trips in period costume to the Reichenbach Falls in central Switzerland, where they converse in mock-Victorian lingo. The fan club called the Baker Street Irregulars, founded in 1934, maintains a quasi-Masonic exclusivity (including, until 1991, a prohibition on women members) and awards select members an official title ("investiture") corresponding to a title, phrase, or charac-ter taken from "the Canon," also known as the "Sacred Writings." The Irregulars maintain the affectation that Holmes and Watson were real people, represented by Conan Doyle as a literary agent. This pretense has a long tradition: Sherlock Holmes was the first fictional character to be awarded a "biography" written as if he were a real person, and historian of literature Michael Saler says there are now almost as many of them as of Conan Doyle himself—making Holmes arguably the first virtual-reality character. Even the UK's Royal Society of Chemistry has made him an honorary fellow.

Nicholas Meyer (who worked on two of the *Star Trek* movies) ex-

---

* There was nothing truly supernatural about "The Adventure of the Sussex Vampire," one of the tales Conan Doyle wrote in the 1920s for *The Strand* and collected in *The Case-Book of Sherlock Holmes*. What was mistaken for bloodsucking in a Sussex family turns out to have a perfectly rational (if far-fetched) explanation. All the same, Holmesian vampire tales have now almost become a subgenre, including Fred Saberhagen's *The Holmes-Dracula File* (1978) and Stephen Seitz's *Sherlock Holmes and the Plague of Dracula* (2006).

emplifies the devotees' insistence on authenticity. He has published four pastiche Holmes novels himself, the first being *The Seven-Per-Cent Solution*, in which Sigmund Freud helps the detective with his cocaine addiction. To Meyer, no actor—not even Basil Rathbone—quite succeeds in capturing the "real" Holmes on screen. Meyer turned down work on Steven Spielberg's *Young Sherlock Holmes* (1985), depicting a meeting between Holmes and Watson while they are at boarding school, because it "never happened"; in Meyer's view the screenwriter held the Holmes oeuvre in contempt.

It's true, of course, that there is nothing unique in this ultrapurist, cosplay fandom. People like to dress as Batman and Dracula, and they often have strong opinions about how books and stories dear to them should be staged and adapted. But Holmes is different, and this needs explaining.

There are clues to be found in *Sherlock Holmes: Victorian Sleuth to Modern Hero*, a 1996 book of talks presented at a meeting convened by the Baker Street Breakfast Club. Described as "a society of persons dedicated to the study of Sherlock Holmes," the club requires of its members "devotion to 'the best and wisest man whom I have ever known,'" as well as "a knowledge of the Canon and/or aspirations to become a Sherlockian scholar." The club might not be quite as secretive as the Baker Street Irregulars, but by goodness it has standards. What's more, it has lists: to enter the Sherlockian universe is a disorienting experience, entailing a cross between the trainspotter's delight in documenting minutiae and the conspiracy theorist's fantastical spinning of elaborate "interpretations" of events.

In her opening summary to the 1996 book, the head of the club, Sally Sugarman, claims that Holmes "is always himself. Both he and Watson are fixed points in a changing age that reassure us when we can find little else to believe in." Yet if this were really so, there would be no Sherlock myth at all—the works of Conan Doyle would be like those of Dickens or Austen, treasured but immutable.

Of course, there is simple nostalgia at play here—nostalgia for an imagined time outside the lived experience of fans, in particular for what writer Edward Docx calls a "vanished London." Holmes's world, says

Docx, "lives in the global imagination as the favourite fantasy of our capital city: the cobbles, the gaslights, the fog and horse-drawn cabs." It is the fantasy indulged in *Mary Poppins* and *Oliver!*, where the villains are cads and ruffians who can always be defeated by British decency, personified by bobbies with handlebar moustaches. This is not just some modern fantasy, Docx explains, but one that was being constructed even as Conan Doyle was writing, during a period when London was rapidly modernizing, the streets giving over to the motorcar, the gaslight being electrified, the slums cleared. By the end of the Second World War, that fantasy was a pile of rubble—and was in consequence needed more than ever.

Some of the reverence for the "pure" Holmes, meanwhile, reflects another kind of conservatism. Every crime he solves reinforces a status quo in which the British stiff upper lip defeats (sometimes damnably foreign) transgression, reason trumps passion, greed and sexual misdemeanor receive the wages of sin, and women stay domesticated. Rather few of the crimes Holmes tackles are violent; instead they threaten the bourgeois social order and the stable family.

The Holmes canon is easily manipulated for the purposes of romantic idealization. David Skene-Melvin, an anthologizer of crime fiction, offers the unlikely proposition that Holmes is a representation of the mythical Hero, who he alleges is often accompanied in "primal folkmyth" by a sidekick. Such idealization is evident, too, when Sugarman claims, "He represents for us our better selves, not only our rational selves but the selves who can contemplate the wonder of a rose or follow a problem to the end." Nothing in this Sherlockian world seems to have changed very much since Edgar Smith, leader of the Baker Street Irregulars, wrote in 1946 that Holmes "stands before us as a symbol . . . of all that we are not but ever would be. . . . We see him as the fine expression of our urge to trample evil and to set aright the wrongs with which the world is plagued."

"We need fear neither Moriarty nor Moran nor any other villains when we have his knowledge and skill to help us," Sugarman adds. "Like Arthur's sleeping knights, Holmes will be ready to help us when we need him." As Docx puts it:

Like all superheroes and detectives, Holmes plays directly to our
abiding human anxieties—reassuring us, righting wrongs. He overcomes violence. He restores order. He makes the foreign familiar. He
confronts our primal fears—that demon-hound baying out there in
the darkness—and he tames them.

This idealized Father-Hero is personified by Basil Rathbone in the most
celebrated of Holmes's early screen appearances. Rathbone's detective
is magisterially sanitized: his vowels clipped, his confidence and calm
impregnable, his character and dress impeccable (no cocaine injecting
here!), his private life more or less nonexistent. The tone is set by his
words to Watson (played for exaggerated contrast by Nigel Bruce as a
bumptious buffoon) at the end of *The Woman in Green* (1945):

> The stars keep watch in the heavens—and in our own little way, we
> too, old friend, are privileged to watch over our city.

That this personification of Holmes is so greatly at odds with the arrogant, repressed opium addict of the books alerts us to mythmaking
at work. To what extent Conan Doyle aimed to create a modern-day
Arthurian knight is debatable; what makes Holmes fit the bill so well is
not just that he is a pair of safe hands but that he reinforces a particular
worldview: one in which men are in charge, the British empire reigns,
criminals are thugs or crooked aristocrats, and you can be sure the adventure will end with everyone returned to their rightful social place and tea
being served by the fireside.

As Sandra Kromm puts it, "Male supremacist society is safe with
Sherlock as its guardian." What Holmes fears in women (and what he
disguises with aloofness) is the psychic disruption they threaten, and
the vulnerability. Watson copes with that problem in the time-honored
fashion Victorian men like Conan Doyle preferred: once he has married Miss Morstan, she is kept safely at home. English professor Stephen
Knight argues that Conan Doyle mined the anxieties of the predominantly male middle-class readers of *The Strand*, reflecting but never re-
solving them, so that they must be "assuaged again, and again—as each

Basil Rathbone and Nigel Bruce as the classic, cleaned-up screen Holmes and Watson.

monthly issue appears." For anyone "interested in seeing how dominant social groups use their literature to state and control fears," Knight says, "the Holmes stories are a fascinating source."

Control is the key. In Holmes, "the desire for control and mastery over others," says literary scholar James Maertens, "is a compensation for an ego cut off from embodied, loving connection." He personified the British Victorian male's need for public displays of mental superiority in order to dominate not only women and other races but *everyone*. Watson is never less than subservient, even in his blustering outbursts: he is the audience that the narcissistic Holmes requires. The battle with Moriarty is not so much a fight of good against evil—for the bad deeds Moriarty is supposed to have done, we have only Holmes' word—as a power struggle between alpha males. This quest for control extends even to the body and its passions, which Holmes treats with something like contempt. And this, after all, is the key to the genre. "The fantasy of detective fiction," says Maertens, "is that possessing and controlling the meaning of signs permits one to possess and control other people."

The Sherlock Holmes myth exists not because of but in spite of the advocacy of the Sherlockians, the Baker Street Irregulars, and the Baker Street Breakfast Club. Far from representing rationality as salvation, Sherlock Holmes created a space for interrogating a very modern dilemma: the limits of rationality in organizing human affairs. Even while he was presenting Holmes as the valiant and dependable knight who will always be there to rescue us, Conan Doyle sowed the concern that his hero's cold-blooded reason could as easily be tipped toward evil. "One day," says a police officer in *Sherlock*'s "A Study in Pink," "we'll be standing around a body and it will be Sherlock Holmes that put it there." Later in the series it is revealed that Holmes does indeed have criminality "in the blood," in the form of his brilliant, psychopathic younger sister Eurus, who has brought the entire psychiatric institution in which she is incarcerated under her will. (The place is called Sherrinford.)

Can our lone protector be trusted? That question became increasingly urgent as ever greater numbers of citizens dwelt in oppressive, anonymous urban environments that were prey to organized crime—or at any rate, to fears of it—in the early twentieth century. The urbane gentleman-detective seemed now to be a figure from another time. A new kind of hero was needed, in whom the dark was as strong as the light.

Chapter 8

I AM THE LAW

BATMAN (1939–)

A lone individual, operating incognito under
an alias and supported by great wealth, wages a bitter
and often violent vigilante war on crime.

Who is the Batman? That's what everyone in Gotham City is asking, and it's a good question. We know, as they do not, that the caped crusader, the dark knight, the lone vigilante, is the crime-fighting alter ego of millionaire playboy Bruce Wayne. But that's not really the question, is it?

Who is the Batman? Is he a bringer of law and decency, or is he a savage avenger outside the law's capacity to govern and command? Is he a brilliant and implacable sleuth or a crazed obsessive determined to eradicate crime whatever the cost to himself and others? Is he really to be trusted? *Is he on our side?*

The story of the Batman is (once again) routinely identified as mythic without serious reflection on what that means. Dennis "Denny" O'Neil, editor of the Batman titles at DC Comics from 1986 to 2000, puts it succinctly:

All these years, I've considered myself to be just writing stories. I know now that that's wrong. That Batman and Robin are a part of our folklore.... They are much deeper in our collective psyches than I had thought.... They have some of the effect on people that mythology used to.

And "if you get into that," O'Neil adds, "you can't avoid the question of religion."

Batman is unique among the myths I've collected here—although I suspect he is representative of how modern myths will now evolve—in that he does not spring from a literary text. Neither is the canon from which the myth springs the work of a single author. The myth is entirely open-ended—not just in the sense that its canon continues to grow without natural limit, but because the character himself (and not just others based on him) is constantly diversifying. He has become a pirate, a vampire, an old man, a spacefarer. "Batman came out of the darkness, out of the collective unconscious where visions of avenging angels dwell," says historian of popular culture Les Daniels, with an excusably melodramatic flourish.

This might seem to imply that Batman is too amorphous a creation to be truly mythic. But there is a core narrative to which all Batman tales must stay rooted: a central mythos about his origin and his quest. It is codified, to a degree, in O'Neil's unofficial "Bat Bible," where he lists the basic tenets that cannot be violated in any Batman story published by DC. The company seeks to maintain these traits wherever it retains some creative control over the hero's manifestations, in particular in the films made by agreement with Warner Brothers. There are things Batman can and cannot do or become. At least, there *should* be.

The world of comic books and superheroes is notoriously fractious. While fans of Sherlock Holmes can be obsessive and possessive, fandom in comic-land has a degree of genuine creative input into the way its stories evolve. Fans often feel they have a role in the decision-making process, and sometimes the editors of comics truly grant them that, whether by responding to complaints and pressure or by handing over choices of plotline to popular vote.

Arguments exist too among the writers and artists. Frank Miller, whose concept of the Dark Knight revitalized Batman in the 1980s, has voiced strong criticisms of the ethos and vision of both DC and Marvel, considering them at times to be an Establishment that treats its heroes too conservatively ("You really have to keep in mind that [they] are scared chickens"). Batman is the product of a diverse pool of writers and illustrators, who do not and need not agree on what he represents or what he has done.

This might seem confusing. When we're talking about the Batman myth, whose version should we then address? But it is this very plurality of visions that allows the myth to flourish. Such disputes are the ingredients of the same sort of stew that, for the earlier myths in this book, was cooked from a single author's inner conflicts and unacknowledged impulses.

Batman persists for the same reasons Frankenstein persists: because we all know the story but can't agree on what it is about. But as with all these myths, we are not totally in the dark. Batman is a product of mid-twentieth-century urban life; there is nothing remotely Victorian or Edwardian about him. He condensed from the fear that, in the city's dim alleys and underpasses and backroom joints, teeming with hoodlums and gangland bosses, the rule of law will not be enough to protect us. In Bruce Wayne's youth it was not enough to protect *him*. By avenging his childhood trauma, he may become a savior to us all.

Yet we are not fully convinced that we can trust him. As a vigilante, he blurs the boundaries of hero and villain. Yes, he fights for law and order, but he is not constrained by them. His own bizarreness, behaviorally and sartorially, has confined him to the borderlands of society, where his anonymity offers freedom from social and legal norms. Only a man driven by revenge is ruthless and determined enough to battle the evil forces in modern society—but can that motivation be sustained without curdling the soul? Maybe a young companion will foster social bonds and thus temper his steel? We hope so, but we can't be sure.

All of this makes Batman* a hero for our times, with not gold but iron in his soul. He is the silhouette on the skyline, simultaneously the city's guardian angel and a dark demon.

＊

For a comic character, Batman is no bundle of laughs. It's an obvious but oddly neglected fact that "comics" began as funny stories told in graphic form. The comic strip was precisely that: a short series of frames published in newspapers as light, perhaps satirical entertainment. Newspaper comics began around the end of the nineteenth century, and by the first decade of the twentieth were a regular feature of the dailies and their color supplements, often called "the funnies."

And that became the title of the periodical-style collections of comic strips produced by George T. Delacorte's Dell Publishing in the United States, starting in 1929. *The Funnies* ran for thirty-six issues, sold as a stand-alone with its own cover price on newsstands. It was the first comic book, imitated in the early 1930s by titles such as *Funnies on Parade* and *Famous Funnies*. There was clearly a market for this kind of thing.

In 1935, National Allied Publications in New York issued the comic periodical *New Fun*, soon renamed *More Fun*. Since the late 1920s, newspapers had been running comic strips with dramatic rather than humorous content, in particular action-packed stories from the popular Western and detective genres of pulp fiction. In late 1937 National Allied Publications brought these too into the comic-book format. *Detective Comics* was the idea of the company's new owner, Harry Donenfeld, previously a publisher of lurid pulp such as *Spicy Adventure* and *Spicy Detective*, the racy covers and content of which nearly landed him in

---

* Let me clear up the name. Purists might have been pleased to see me using the definite article at the outset, and now I risk offending them by dropping it. He was certainly "*The* Batman" when he began, but popular usage has eroded that distinction, and it's rare today to find this nicety observed. Henceforth Bruce Wayne is, when in full battle garb, Batman. If it's good enough for O'Neil, it should be good enough for us all.

jail for obscenity. Donenfeld rebranded the whole company under the initials of his new title: DC Comics.

In 1938, DC launched a new series called *Action Comics*. Its first issue introduced a character invented by writer Jerry Siegel and artist Joel Shuster: a hero with superhuman powers of strength, flight, and X-ray vision, who arrived on Earth from another planet and spends his days in disguise as a mild-mannered news reporter named Clark Kent.

Superman appears on the cover of *Action Comics* #1 in all his outlandish splendor: tight blue costume, dinky red boots, and ridiculous groin-hugging red shorts, apparently inspired by those worn by the strongman archetypes of wrestlers and circus weightlifters. The red cape was a nod toward swashbuckling heroes such as Zorro and the Scarlet Pimpernel; across his mighty pectorals was emblazoned the shield-shaped "Superman" monogram. It's a decidedly weird hodgepodge, nodding also to the skintight clothing often worn by science-fiction heroes such as Flash Gordon and Buck Rogers, who already had their own popular comic strips and were both portrayed in 1930s film serials by Buster Crabbe.

No one seemed to mind Superman's questionable garb and appearance. Indeed, it established the template for all future superheroes: garishly colored, muscle-bound, and vaguely kinky. *Action Comics* was a hit from the get-go, and the DC bosses wanted more.

They repeated the winning formula in *Detective Comics* #27 in 1939 with a character devised by artist Bob Kane and writer Bill Finger. In contrast to the seemingly endless stream of subsequent DC and Marvel superheroes with quasi-magical powers (generally justified with a smattering of crude pseudoscience involving radioactivity or failed experiments), Batman has nothing but brawn, wit, and wealth to assist him in his life of crime-fighting. He was initially imagined as a true detective of sorts, albeit with the kind of enhanced abilities familiar from other mysterious sleuths of pulp fiction, such as the Shadow and Doc Savage: skill in martial arts and physical endurance, encyclopedic knowledge, and a hint of occult powers achieved by esoteric instruction.

The Shadow is Batman's most obvious progenitor. He began life as the voice narrating a radio series called *Detective Story Hour* in 1930; his popularity led the show's sponsors to commission pulp writer Walter B.

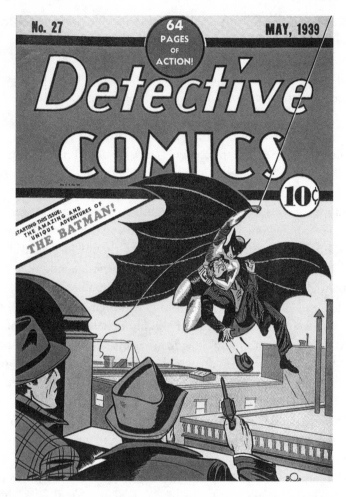

Batman debuts in Detective Comics #27 (1939).

Gibson to flesh him out into a fictional character. Gibson created an entire mythos for the mysterious detective: he was the alter ego of wealthy playboy Lamont Cranston, who had learned his secret powers of concealment in the East. Gibson churned this stuff out prodigiously: his peak productivity of more than a million and a half words per year dwarfs that of Arthur Conan Doyle or H. G. Wells. The Shadow became hugely popular, appearing not just in pulp novels but in comic strips and movies. The aura of magic that surrounded him morphed into full-blown supernatural superpowers for Batman's contemporary vigilante the Spirit, the comic-book masked detective created by Will Eisner, who debuted in 1940.

Kane was born Robert Kahn in 1916, the son of an engraver for the New York *Daily News*. Although he is sometimes accused of claiming most of the credit for the invention of Batman, he and Finger were a team from the start. They were old school friends from the Bronx, both still in their early twenties when they conceived of their most famous creation. Kane had been working as an illustrator for comics since 1936, and had collaborated with Finger on previous DC stories such as "Rusty and His Pals" and the square-jawed soldier of fortune Clip Carson. Division of duties between an artist and a writer was, and still is, the norm for comics, and in recent years produced the "star pairings" that created the most iconic incarnations of Batman. The job of illustrating was conventionally subdivided further, with one artist responsible for the design and line drawings and another—the inker—for laying on the colors. That rather odd arrangement began as a means to mass-produce the needed images more quickly and efficiently, but it became a part of the tradition.

Both the Shadow and the Spirit dressed like ordinary detectives: broad-brimmed fedora, dark suit, long raincoat. Batman chose another path: to fight criminals, he must first put on his superhero's fancy dress. Where did Batman get that rather grandiose sense of theatricality? The cape, like Superman's, was inspired by Zorro and the Scarlet Pimpernel, as was the identity-concealing mask, which allowed those earlier vigilantes to mix in everyday society without their actions getting them into trouble. Zorro was created by American pulp writer Johnston McCulley in 1919, but Kane knew him mostly through the 1920 movie *The Mark of Zorro* starring Douglas Fairbanks, in which Zorro's alter ego is a wealthy fop. Kane also claimed that Batman's winglike cape was inspired by Leonardo da Vinci's sketches for a flying machine.

The bat imagery lends Batman an air of the supernatural, even the demonic. Kane claimed inspiration from the 1930 movie thriller *The Bat Whispers*, in which a detective carries out murders while disguised in a bat costume. That movie was the second adaption of a 1908 mystery novel by Mary Roberts Rinehart called *The Circular Staircase*, which Rinehart and Avery Hopwood turned into a play called *The Bat* in 1920; it ran successfully on Broadway. The first film adaptation was a 1926 silent movie of the same name, which was remade in 1959 with cinema's then-current Dracula, Vincent Price, in the title role. *The Bat* was also an inspiration for the 1931 *Dracula* with Bela Lugosi—which Kane says also fed into his vision for Batman.

There's more; Batman's predecessors offer an embarrassment of riches. There was at that time not one but two crime-fighters known as the Black Bat: a pulp hero of 1933 whose nickname was never explained, and a character in mask and cape who appeared in the comic books of Thrilling Publications (also known as Better Publications) more or less at the same time as Batman. This second Black Bat was a former district attorney, Anthony Quinn, who took the law into his own hands and left

The ingredients of Batman include the 1926 film The Bat.

his dead victims emblazoned with the sticker of a bat to strike fear into other lawbreakers. His mask served to hide a disfigurement caused by acid—prefiguring the storyline of Batman villain Harvey "Two-Face" Dent. Batman and the Black Bat were in fact so similar that DC and Thrilling Comics threatened to sue one another for plagiarism, until DC editor Whitney Ellsworth, who had once worked for the rival firm, intervened and smoothed the waters.

In short, there was at first little that was original about Batman. Why, then, did he persist when his rivals did not? What made him so special?

Perhaps that's the wrong way to frame the question. As we have seen with other modern myths, particularly Frankenstein, Jekyll and Hyde, and Dracula, the myth is triggered not by any radical new innovation but by the right combination of preexisting images, archetypes, narratives, and context. To understand it, we need to examine the ingredients.

The first Batman stories offered no explanation for why he had chosen his odd outfit, nor why he was fighting crime in (at that stage) New York City in the first place. The answer arrived in *Detective Comics* #33, published in late 1938, and it was a disturbing tale.

Bruce Wayne is the son of Thomas and Martha Wayne, wealthy citizens of what was shortly to become Gotham City. While visiting the opera one night, the family is held up at gunpoint by thugs, and in the ensuing scuffle the parents are shot dead. The next frame shows "a curious and strange scene," as Bruce kneels by his bed in candlelight and, in lieu of a prayer, says, "I swear by the spirits of my parents to avenge their deaths by spending the rest of my life warring on all criminals." Then we see the adult Wayne making himself a "master scientist" and training his body "to physical perfection until he is able to perform amazing athletic feats."

As he sits brooding one evening in the mansion he has inherited, he decides he needs a disguise. "Criminals are a superstitious cowardly lot," he muses. "So my disguise must be able to strike terror into their hearts.

Batman's origin: the slaying of Bruce Wayne's parents.

I must be a creature of the night, black, terrible . . . a . . . a . . ." At that moment, "as if in answer, a huge bat flies in the open window":

> *And thus is born this weird figure of the dark . . . this avenger of evil, 'The Batman.'*

We find out later that the murderers were no random thugs, but the henchman of criminal boss Joe Chill. In due course (it took until 1948), Batman discovers the truth and tracks down Mr. Chill, who is shot by his own gang for having drawn down the wrath of the Caped Crusader.

Of course, we should hesitate to read too much into a pulp backstory hardly designed for such psychological probing.* But it's the very artlessness of this origin tale that makes it repay the effort—for we should know well enough by now that myth is mostly born when the unconscious is doing much of the work. The murder of Bruce Wayne's parents has become an essential part of the Batman myth: whatever else gets mutated in the story, this core remains. Why is it so important?

Kane and Finger must have sensed there was something peculiar about the plot they were unfolding. After all, a boy pledging revenge might seem sad and unsettling, but why "curious and strange"? What *is* strange is that the young Bruce makes this commitment in the attitude of prayer, as if seeking divine consent. Stranger still, the boy does not, as one might conventionally expect, seek to join the police or justice service; he decides that his promise will be fulfilled in secret, outside the law. The implication is that the law is not enough: crime will be prevented only by going beyond it. Even Sherlock Holmes has official sanction and works openly in collaboration with the police. But Batman does his crime-fighting incognito and in hiding.

Straightaway we are in a moral gray zone, not entirely sure that Batman's activities are licit. We see he is a loner, turning his back on anyone

---

* Bob Kane attested, however, that an attack by a gang of thugs he himself suffered in the Bronx when he was fifteen years old played a part in the genesis of the tale.

who might help (or encourage restraint), pursing a personal vendetta rather than acting for the common good. Already, he does not seem fully sane; he should, Kane later explained to moviemakers, "be played as a psychologically disturbed eccentric."

His disguise confirms that we are right to be suspicious. He could, like the Shadow, simply hide his face with a scarf. But he chooses to create an entire costume, head to toe—and one that throws into doubt his very humanity. He will become a "creature," a bestial alter ego. This disguise is not simply about concealment, but first and foremost about terror. It's not enough to stop the criminals; he wants to *terrify* them. Kane seems to sense the oddness of it all: Batman isn't just anonymous, nor even mysterious, but a "weird figure of the dark," the kind of phrase we might use to describe a devil or an apparition. In a single page, Batman's origin story shows Bruce Wayne transforming from a distraught child into a lonely, obsessive outcast. It seems the blight of crime in the dangerous metropolis can no longer be fought with Sherlock Holmes's cunning and rigor but requires a touch of madness. In becoming such a sacrosanct part of the myth, Batman's origin makes this aspect of his character explicit and indelible.

It's a testament to the significance of the origin story that the Batman writers keep returning to it, elaborating here and there as if to scratch an itch, or to hone it into the shape that the developing mythos requires. In "The First Batman" (DC #235, 1956) it is revealed that Wayne's parents' killing was no random hit-and-run by Chill's thugs; he was contracted to do the job. Sometime before, Thomas Wayne, a renowned doctor, had been kidnapped while at a society fancy-dress ball dressed as a bat. He was brought to gangland boss Lew Moxon, who demanded that Wayne remove a bullet from his shoulder. Wayne escaped and called the police, and Moxon vowed revenge; he hired Chill, who used the holdup as a cover ploy so that the killing would look random. Bruce now realizes that he was spared on purpose, so that his testimony would confirm there was nothing more to it than a routine robbery, averting any deeper investigation.

After discovering this plot, Batman traces Moxon to "Coastal City." But Moxon has had a head injury and can't remember anything about

Thomas Wayne—until Bruce puts on his father's old bat costume, whereupon a terrified Moxon flees and is struck dead by a car.

Frank Miller's "Batman: Year One" series of 1987 picks up the tale of Bruce's Shadow-style esoteric training in preparation for his life of crime-fighting. Aged thirteen, he leaves Gotham City to travel afar and learn the martial and magical arts, returning when he is twenty-five to find the city steeped in crime and corruption and desperately in need of a hero, not just on the streets but in the corridors of power. After Nixon, foul play could go right to the top.

These and other permutations and additions to the story of how Bruce Wayne became Batman, while hardly offering a unified or coherent psychological foundation, disclose an anxiety about his motives. They feel like the probing of a therapist: I think there is more you're not telling me! At the same time, they creep ever closer to a sense of pre-destiny, making Batman not just a man mildly deranged by a childhood trauma but a hero defined by fate or supernatural forces. Look, they say: Batman might seem weird, but he is no mere man; he is a force of nature.

***

Superheroes are modeled on myth and legend, with their fantastical powers, extravagant apparel, and complicated social networks. Some, such as Marvel's Thor and Loki, merely recycle from the ancient stories. They are an infinitely adaptable community, easily tailored to meet the demands of the moment—to acknowledge the presence of feminism, say, and to redress their initial racial homogeneity. But I don't believe that any except Batman truly earns mythical status.

Why not Superman? He will never be a myth for one simple reason: he is a crushing bore. There is no moral shading in the Man of Steel: he is the blandest force for good you can imagine. The first DC superhero might be seen as an embodiment of the *Ubermensch*—the ominous icon that, when Superman first emerged, was being glorified across the Atlantic in the Third Reich. As Ursula Le Guin puts it, "His father was Nietzsche and his mother was a funnybook." Of course, he is also the all-American male, brawny and noble and upholding standards of decency.

Yet under that steely skin, we sense, lurks a deeply conservative reactionary. No feelings can penetrate it. He is dependable and utterly without
imagination, brave because he can't be touched, a creature at ease in a
male world (but flustered by Lois Lane). He is, says political scientist
Alan Baily, an "assimilated alien," while in contrast Batman is an "alienated citizen."

Frank Miller gets Superman's true measure in *The Dark Knight Returns* (1986), when he, now a government lackey, confronts Batman on
the orders of the US president. Batman has come out of retirement at
the age of fifty-five to rid Gotham City of a gang called the Mutants,
but he ends up leading some of them in a vigilante gang to keep the city
functioning after a blackout. Superman is dispatched to arrest the errant
old knight. "You, with your wild obsession," he scoffs.

Superman is a law-keeper, and his head is held high. But Batman has
his head bowed in shadow, and he is an avenger. It is Batman who really
carries the spirit of Nietzsche into the urban jungle of modern America.
The crooks killed his parents, but they did not kill him—and that made
him stronger.

In the early stories, Batman was literally murderous: he did not hesitate to take the lives of his victims. When the criminal boss he appre-
hends in his first adventure falls into a vat of acid, Batman shrugs: "A
fitting end for his kind." In the first issue of Batman's eponymous comic,
in 1940, he dispatches a victim with only halfhearted regret, saying,
"Much as I hate to take human life, I'm afraid *this time* it's necessary."
One story shows him with a smoking gun in hand, and it's a testament
to the way his character has been managed subsequently that the sight
looks strange and even shocking. For Kane soon realized that not only
should Batman *not* be a ruthless killer but *he should not use a gun*. DC
editor Whit Ellsworth told Finger sternly, "Never let us have Batman
carry a gun again." These became two of the rules rigidly imposed on
subsequent story writers.

Even with those restraints, Batman looked just a bit *too* unstable as
the lone vigilante. He needed civilizing; he needed friends. The police
commissioner James Gordon was his first ally, ready to tolerate Batman's
lone-wolf activities as a supplement to his own law force. But in *DC #38*

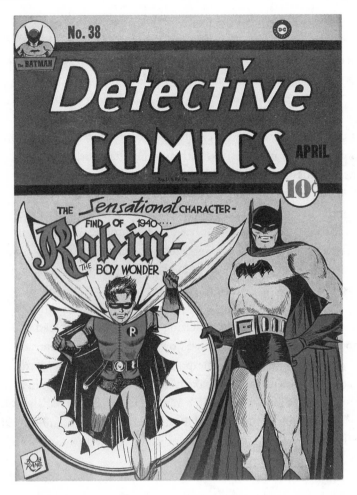

Enter the Boy Wonder, in DC #38 (1940).

in April 1940, Batman acquired his most significant companion: Dick Grayson, a teenager from a family of trapeze artists, whose parents were also murdered by hoodlums. Bruce Wayne adopts Dick and initiates him into the life of crime-fighting, where he takes on the alter ego of Robin the Boy Wonder.

In 1943 Batman gained another faithful companion: his butler Alfred, who attends at Wayne Manor and is the only other person privy to

Wayne's secret life. He also acquired the battery of gadgets that made him so appealing to children: the utility belt, the Batmobile, Batgyro (a helicopter), Batmerang—and the wonderful Batcave, seducing youngsters with the universal allure of a secret den.

One of the obvious functions of the adolescent, ebullient Robin was to give the teenage boys who were DC's main readership someone to identify with. He was also useful for plot exposition: Batman no longer had to account for his actions with an inner dialogue but could explain himself to his apprentice. With Robin around, Batman could lighten up a little: the two would engage in friendly banter, and Batman might even crack a smile now and again. This was partly a buddy narrative, partly a father-son relationship. It's not hard to spot problems on the horizon.

The appearance of Robin suggests that Kane and Finger were rather alarmed by the monster they'd created, and wanted to socialize him. The DC editors, meanwhile, were concerned that parents would stop their children reading Batman if he was too violent: Kane says they thought "that mothers would object to letting their kids see and read about . . . shootings." So Batman was toned down on all fronts, portrayed almost as an honorary member of the police force; even the pointed ears of his costume were shrunk to make him look less satanic. Robin injected the sense of a caper rather than a vendetta: "Say, I can hardly wait till we go on our next case. I bet it'll be a corker!" In August 1940 Batman proclaims a kind of manifesto evidently designed to turn Robin and him into decent, patriotic role models:

> I think Robin and I make it pretty clear that WE HATE CRIME AND
> CRIMINALS! . . . Why? Because we're proud of being AMERICANS—
> and we know there's no place in this great country of ours for law-
> breakers! . . . We'd like to feel that our efforts may help every youngster
> to grow up into an honest, useful citizen. . . . If you do this, if you are
> definitely on the side of Law and Order, then Robin and I salute you
> and are glad to number you among our friends!

This toe-curling piece of damage limitation is a testament to the dangerous power and appeal of the figure that Kane and Finger had unleashed.

It also shows that a struggle was already developing over authorial control of Batman.

But could an upright, law-abiding Batman really do the job that needed doing? Gotham City—so named from 1941 to give it a universal quality that New York (for which of course Gotham was an old nickname) could not supply—was a madhouse, plagued with an endless succession of colorful crooks: Doctor Death, Two-Face, Clayface. They seem so much a product of the comic-book world that it comes as a shock to realize American culture was at that time familiar with real-life criminals bearing similarly flamboyant nicknames: "Pretty Boy" Floyd, "Machine Gun" Kelly, "Dutch" Schultz, Bonnie and Clyde. The world Batman inhabits is only an exaggerated form of the crime-ridden, gang-dominated cities created by the eras of Prohibition and the Great Depression.

While Batman was inevitably drafted into the war effort in the 1940s, media scholar Will Brooker points out that he was rather resistant to those demands: he remained firmly rooted in Gotham, where Axis agents had to come to *him*. Batman's first cinematic venture was the 1943 Columbia serial* *The Batman*, with the portly Lewis Wilson in the title role,† in which he "nips the most sinister of Nipponese plots" in a vehicle both kitsch and excruciatingly racist. But the comics themselves continued in much the same vein as ever during wartime, with their parade of home-grown deranged criminals.

The first issue of *Batman* in the spring of 1940 introduced the most celebrated of these foes: the Joker, his bone-white clown-face permanently deformed into a ghoulish, maniacal grin, in twisted imitation of the psychotic Mr. Punch. He is everything Batman despises, an agent of chaos whose battle with the Caped Crusader seems more a struggle

---

* These episodes, each lasting thirty minutes, were screened weekly in cinemas before television was widespread.

† Wilson's belly constantly threatens to overwhelm his understocked utility belt; Bob Kane grumbled that the actor "should have been forced to go on a diet before taking the role."

of will and ego than a fight against crime. He is an anti-Batman, right down to his utility belt filled with jokes. As Tim Burton attested while making his 1989 movie with Jack Nicholson playing the Joker, "It's a fight between two disfigured people."

No other villain has quite matched the Joker either in popularity or in potential for probing the depths of Batman's damaged psyche. All the same, the parade of weird misfits—the Penguin, the Riddler, the Scarecrow—inhabits a very different universe from the cads and corrupt generals defeated by Sherlock Holmes. They burst forth like figures of nightmare, the deformed ancestors of Freddy Krueger, Pennywise the clown, and *Halloween*'s Michael Myers. As Les Daniels writes:

> A peculiarly American form of expressionism developed, in which characters lived surrounded by countless emblems of their obsessions, treated crime as a series of publicity stunts, and dressed up in crazy costumes as they struggled to dominate the night.

Batman was a response to this world, thronging with crazed psychopaths from the urban badlands and perverted scientists whose promise of a bright new future was starting to tarnish. The Batman stories, writes Daniels, "had become a seething nuclear stockpile of a society's dark dreams and desires." The world is a dangerous place; evil lurks in the hearts of men. Part wish fulfillment, part *Death Wish* revenge fantasy, Batman personifies not a solution to that predicament so much as an anxious question: might we not sometimes need a protector willing to ignore the law because he himself has come from a place of death and darkness?

Comic and pulp fiction always troubled guardians of public morality. The scantily clad female victims on the covers of 1930s detective and science-fiction pulps seemed, however, to worry these watchdogs not so much for their insinuations of molestation and objectification of woman as for the perverting effect they were imagined to have on the boys who constituted the main readership.

Little of that was evident in *Batman*, although some critics condemned its violence and grotesque villains. But in his 1954 book *Seduction of the Innocent*, American psychologist Fredric Wertham raised another fear: Batman and Robin might be gay.

Wertham's book was a blistering attack on comic books in general, primarily for the corrosive moral effect of their violence on their young readers. This could easily sound like so much reactionary bluster, but in fact Wertham is sometimes perceptive about recurring comic tropes that would make us blanch today. Rape scenes weren't uncommon, nor were sadomasochism, torture, and borderline pedophilia. Wertham complained about a scene in which a man attacks a schoolgirl ("All I want is a little kiss! C'mon!") and chokes her to death; his book included images from a Western comic of a young woman being spanked ("This'll maybe help you to grow up faster'n anythin' I know about") that could have (and perhaps did) come straight out of an S&M mag. The comics were filled with adverts for toy guns.

Wertham's remark that comics were "a precipitate of fragmentary scenes, violent, destructive and smart-alecky cynical" might be unfair as a generalization, but was often not wide of the mark. These were the products of a culture that had yet to find anything problematic in serving up images of young women placed in physical and sexual jeopardy, often indulging oppressive male fantasies, as entertainment.

But *Seduction of the Innocent* is remembered now mostly for the short section in which Wertham finds evidence of homosexuality in the Dynamic Duo's cohabitation. He begins by suggesting that all comic-book superheroes are presented with homoerotic overtones: "The muscular male supertype, whose primary sex characteristics are usually well emphasized, is in the setting of certain stories the object of homoerotic sexual curiosity and stimulation." At Wayne Manor, he says, the situation has become far more grave:

> Only someone ignorant of the fundamentals of psychiatry and of
> the psychopathology of sex can fail to realize a subtle atmosphere of

homoeroticism which pervades the adventures of the mature "Batman" and his young friend "Robin." . . . Sometimes Batman ends up in bed injured and young Robin is shown sitting next to him. At home they lead an idyllic life. . . . Bruce Wayne is described as a "socialite" and the official relationship is that Dick is Bruce's ward. They live in sumptuous quarters, with beautiful flowers in large vases, and have a butler, Alfred. Batman is sometimes shown in a dressing gown. . . . It is like a wish dream of two homosexuals living together. Sometimes they are shown on a couch, Bruce reclining and Dick sitting next to him, jacket off, collar open, and his hand on his friend's arm.

Taking the then-conventional view that homosexuality was a phase that many boys passed through on their way to a healthy attraction to the opposite sex, Wertham worried that the Batman comics, reaching them at that vulnerable age, might lead them down the path to sexual deviance. In support of this warning, Wertham reported the testimony of some of his patients at a clinic ominously called the Readjustment Center. "I found my liking, my sexual desires, in comic books," said one:

> I think I put myself in the position of Robin. I did want to have relations with Batman. . . . I remember the first time I came across the page mentioning the "secret batcave." The thought of Batman and Robin living together and possibly having sex relations came to my mind.

DC writer E. Nelson Bridwell called Wertham's accusation "one of the most irresponsible slurs ever cast upon our heroes." And this tells you all you need to know about the tenor of the debate: on the whole, both critics and defenders of Batman were united in their view that being gay was a terrible thing, and the real question was "are they or aren't they?"

Homophobia clouds the discussion even now. Suggestions that Batman and Robin are in a gay relationship can still infuriate some fans, as though the idea besmirches their hero. But at the same time, indignant accusations that Wertham's suggestion was driven purely by the ugly prejudices and paranoid imaginings of his time collide with the testimony of some gay men from the 1950s that they very much identified with the fantasy he depicted.

While a gay relationship between Holmes and Watson seems a perfectly reasonable and interesting way to expand that myth today, in Edwardian times there would have seemed nothing untoward about two men bonded by companionship and camaraderie. (That is not, of course, to deny that these relationships might nevertheless include a homoerotic element.) Batman and Robin were quite another matter. Living secret lives with no social ties, they dressed in costumes that were indisputably queer in at least one sense of the word. Images of Bruce and Dick reclining on the sofas of Wayne Manor, or being trussed intimately together when their enemies hold them captive, seemed positively to invite gay readings, and some gay young men certainly interpreted them that way.

Wertham has been demonized as a rabid McCarthyite witchfinder determined to eradicate homosexuality from American life. In fact he

Sometimes it is hard to imagine that the creators of Batman and Robin did not have gay themes in mind.

was a more complicated figure, liberal by most standards of his time. *Seduction of the Innocent*, for all the judgmental moralism of the title, seeks more to understand homosexual feelings and the anguish they can cause young men than to pronounce them wicked. Wertham also criticizes the misogyny in comic books that "instill the idea that girls are only good for being banged around or used as decoys." He complains of the way all the advertisements for women and girls insist that "it is the woman with graceful, appealing figure lines who enjoys social and romantic triumph." And he excoriates the pervasive racism he found: "In many comic books dark-skinned people are depicted in rape-like situations with white girls," Asians are presented with "an unnatural yellow color," and, as in the "exotic travel" literature of the day, only Black women are ever shown with their breasts exposed.

Yet there's no denying that Wertham's book is steeped in the prejudice of its times. Media scholar Andy Medhurst has been a particularly vocal critic, accusing the psychologist of "shrill witch-hunting" and calling his book "flamboyant melodrama masquerading as social psychology." It's complicated, though. Yes, the idea that flowers and dressing-gowns are irredeemably gay sounds like the crudest of stereotyping—but Medhurst himself admits that certain dress codes served as "semiotic systems adopted by gay people themselves, as a way of signaling the otherwise invisible fact of sexual preference." While on the one hand he derides the idea that comic books are somehow turning impressionable youngsters gay, Medhurst says of the testimony quoted earlier by one of Wertham's "patients" that "reading it as a gay man today I find it rather moving and also highly recognizable."

Which is a way of saying that Wertham was on to something. Intentional or (almost certainly) not, Bruce and Dick's relationship was ripe for projection as a gay fantasy. Wertham was not wrong to suggest as much, but he was blinkered by the mores of his time in pronouncing it a problem.

Denny O'Neil's "Bat Bible" insists that Batman will never be explicitly gay—but the rule seems motivated more by concerns about a media and fan backlash than by any intrinsic disapproval of the hero's being depicted that way. I suspect O'Neil was canny enough to realize that it is precisely by leaving possibilities open that such myths acquire their potency. Homosexuality—and society's attitude to it—is at the core of the Batman myth. It continues to resonate in the Dark Knight era, albeit not always in ways we might welcome. The Joker now taunts Batman with a sexual undercurrent: in *Arkham Asylum*, a Batman graphic novel written by Grant Morrison, he calls Batman "honey pie" and asks "How is the Boy Wonder? Started shaving yet?" only for Batman to call him a "filthy degenerate." When Batman beats the Joker during the interrogation sequence of Christopher Nolan's *The Dark Knight*, it looks worryingly like an abusive relationship, the dialogue disturbingly intimate and full of double entendre. "You wanted me. Here I am," Batman rasps. "You complete me," the Joker tells him. "To them you're just a freak—like me." Batman's furiously violent response makes us question our loyalties; it's as though the Joker has touched a nerve.

"The homophobic nightmare is very much a part of the Batman/ Joker mythos," says Frank Miller. "It's always been there, I just spelled it out a little more plainly." He adds that Batman "would be *much* healthier if he were gay."

<hr/>

The title of Wertham's book was unintentionally revealing. It is one of the grains of grit in the Batman myth that, while it is ostensibly for kids, it can never be reconciled with idealized views of childhood innocence. You can tame *Crusoe* and *Gulliver* so that they will sit respectably on the nursery shelf (you can even, it seems, persuade yourself that this is where they always belonged). But *Batman* is a modern Punch and Judy, garish and comic on the surface but murderous and violent at its heart.

Modern myth always seems to get refracted through the prism of childhood. If it doesn't (appear to) start there, as Batman did, then we try to file it away there anyway: the "young adult" section of my local bookshop has on its shelves *Frankenstein*, *Dracula*, *Jekyll and Hyde*, and *Robinson Crusoe*. I know that YA fiction can be gritty fare these days, but all the same this feels like a containment exercise: read this stuff before you grow up. They're only monster tales.

Perhaps this (dis)placement is not surprising, for the children's section is often the best place to go looking, too, for the ancient myths. (For adults they must these days be "reinterpreted" by well-known writers, although I've no doubt those writers themselves know why myths are important enough to be worth the effort.) The indignation instilled among conservative critics by *Detective Comics*, by Adam West's camp capers as the 1960s Batman, and by the dour savagery of the Dark Knight, relies in the end on another myth: that childhood is innocent, or used to be before such things came along. Some of Wertham's ire at comics was directed at the "childish" images themselves — "the corruption of the art of drawing," he said, speaking like a man who has never seen a Goya or an Otto Dix. It is not, of course, that there aren't things from which children need to be protected — but modern myths are not among them, and neither will these stories be so easily neutered by turning them into children's tales. It is a common trick to pretend that we are trying to protect children from what we fear ourselves.

Wertham's denunciation of comic books came as a bombshell to the industry. The matter was debated in the US Senate, sales of comics fell as parents began to forbid them, and the Comics Code Authority was established in 1954 by the Comics Magazine Association of America in an attempt to ward off government regulation. Its focus was on gore, violence, and sex, which were restricted by guidelines that came to be known as the Comics Code. Adherence was voluntary but advised, not least since some stores would not stock comics that did not bear the CCA's seal of approval. Implied homosexuality was forbidden under a clause that prohibited depictions of "sex perversion," "sexual abnormalities," and "illicit sex relations."

Batman and Robin had now to prove themselves to be beyond suspicion. The solution was to make them part of a nuclear "Bat-family." It was gruesome stuff. First to appear was the dog Ace, a "Bat-Hound" complete with mask. July 1956 saw the debut of Batwoman in her sexy catsuit (perhaps we should say batsuit to avoid confusion with the villainess Catwoman), a former circus acrobat named Kathy Kane who carried feminized weaponry such as lipstick filled with tear gas. Poor Robin is tormented by visions of Batman getting married, first in "Bride of Batman" (autumn 1953)* and then in a nightmare in which Batman marries Batwoman (March 1959). "What'll become of me now?" he frets as the couple leaves the church.

If only he knew that there was worse in store. By 1961 he is being harassed by Batgirl—Kathy Kane's niece Betty, who makes her own costume, imposes herself on the family, and takes to giving a red-faced Robin kisses: "Oh Robin—then it's all right for me to kiss you now." ("If a comic book could actually turn people gay as Dr. Wertham had suggested," writes Les Daniel, "this one might have had the power to do it.") Piling on the indignities, the DC team invented a "cute" baby-substitute called the Bat-Mite, a Disney-esque being from another dimension. "It's an elf," Robin exclaims. "An elf dressed in a crazy-looking Batman costume!" And sadly, it was.

In the stifling conformism of McCarthy-era America, the Batman narratives were increasingly infantilized—literally so in 1962, when Batman was changed into a "Bat-Baby." This was just one of a series of gimmicky sci-fi transformations: the Giant Batman, the Phantom Batman, the Zebra Batman. His adventures had become increasingly fantastical and science-fictional even by the end of the 1940s: he and Robin would time-travel to ancient Rome or the Greek Olympics, or go swashbuckling with the Three Musketeers. His crusade to rid Gotham of crime seemed to have lost its urgency.

But even if the strictures of the post-Wertham era saddled Batman with an excruciating domesticity, it's far from clear how seriously the DC writers took their new obligations. One can hardly read some of those tales from the 1950s now without suspecting wry satire on the idea

---

* But what, really, is going on here in the collective subconscious at DC? A deeply uncomfortable Batman sweats and loosens his collar as his "bride" smirks at him covetously, while a Gotham bystander thinks, "I wonder why he's doin' it? I never thought he'd fall for a girl!"

William Dozier's 1960s *Batman* television series (we're coming to it) plays up the hero's uneasy response to heterosexual overtures and Robin's anxiety about losing his companion to a woman. When Adam West's Batman encounters slinky, Lycra-clad Catwoman (a wasp-waisted Julie Newmar), the tongue-tied hero is utterly (and hilariously) out of his depth. And as the couple stroll past arm in arm, a disappointed-looking Robin exclaims, "Holy mush!"

The sartorial dilemmas of the "Rainbow Batman" (1957).

of a gay Batman. Take *DC* #241 in 1957: "The Rainbow Batman." The cover shows the brawny hero clad in a pink outfit and fastening a fuschia cloak, while behind him in the Batcave hang other outfits in every bright color you might care for. "Batman," exclaims Robin, "last night you wore the *green costume*—and tonight you're wearing the *red*! Why?"

"I must, Robin," he replies. "I must wear a different-colored Batman costume each night!" There could hardly be a more stereotypical portrayal of the gay persona, especially in an age when any interest from a man in "color schemes" would render him an object of suspicion.

Consider too how Batwoman reacts after an encounter with Clayface in *Batman* #159 (1963). "Oh Batman," she swoons, as if in a Roy Lichtenstein painting, "I'm afraid you'll just have to hold me. I'm still shaky after fighting Clayface—and you're so stron-nn-g!" Can this be intended as anything but parody?

Will Brooker argues that the Bat-family of the 1950s and 1960s, far from serving to deflect accusations of homosexuality, resembles a portrait of gay men being forced by social expectations to imitate heterosexual norms. Batman and Robin scarcely seem to welcome the overtures of the females, but instead look panicked and trapped—"Ulp!"— whenever a romantic response is required of them. That's not to imply that the DC editors intended anything so explicit; it seems far more likely that this is yet another example of the psychological leakage modern myths incur. Kane attests that he and his colleagues might indeed have decided to play a little with the can of worms Wertham had opened.

<hr>

While, then, it's a common complaint that William Dozier's iconic *Batman* TV series of the 1960s turned the comic-book character into a parody and distorted a serious tale for cheap laughs, the DC comics of the 1950s and early 1960s had already become high camp before Adam West squeezed into his tights.

Defining the contours of camp in her seminal 1964 essay "Notes on Camp," Susan Sontag fretted that to try to write seriously about it was to risk producing a "very inferior piece of Camp" in itself. But that's Sontag for you: fearing that only seriousness will do the job, that the alternative is mere frivolity. That attitude is not going to work for Batman, nor for talking about superheroes generally—nor, I suspect, for engaging with most modern myths.

If we can't acknowledge that Batman is fundamentally silly, we have painted ourselves into a corner. There is no getting away from the fact that Batman wears a ridiculous costume, that he is dressing up pantomime-style, and that this is funny. It is what makes him such a popular fancy-dress choice for small children; it is why spoofs like *Mad* magazine's "Bats-man" and skits like Del Boy and Rodney's superhero turn in the British TV sitcom *Only Fools and Horses* are inevitable. And it is why Dozier's knowingly camp series was not an aberration but a natural progression. The outraged fan pursing lips at the fetishized, leather-clad buttocks and bulging codpieces of the Dynamic Duo in Joel Schumacher's much reviled 1995 film *Batman Forever** is falling prey to the same

<hr>

* This and Schumacher's subsequent *Batman and Robin* (1997) bring the camp sensibility of the Adam West version into a climate no longer so terrified of homoeroticism. With nipples modeled onto his black leather muscle-suit, Chris O'Donnell was a Robin very much over the age of consent and ready for action; he became the cover pinup for the gay magazine *Attitude* in 1995.

self-importance as Sontag—although where the fan disapproves, Sontag somberly analyzes.

So let's just put it out there that Batman is absurd, and then we can start to take him seriously again. He is, after all, no more absurd and no less serious than life itself. There is no more dressing up in Batman than there is in the Catholic church, the British monarchy, or pretty much any state ceremony you care to mention. His response to his trauma and his demands for vengeance are no crazier than ours would be (given the chance), and his lavish excesses underneath Wayne Manor are tame compared to how some billionaires spend their money. Batman's absurdity doesn't distance him from life but makes him a good vehicle for commenting on it. And if we can't be playful in analyzing the way he plays that role, we might as well pack up now.

All this is why we need not—as some fans and commentators would prefer—dismiss West's Batman as an aberration, a lapse of taste between the original grim vigilante and the grittiness of the Dark Knight. There is at least one good reason both why the TV series of the 1960s was made at all and why it was such a hit—which is that it allowed us finally to acknowledge the one aspect of Batman that had remained latent, had become almost taboo, since his inception: he is a funny man in a bat suit.

"The pure examples of Camp are unintentional," said Sontag. "They are deadly serious." It's a matter of debate whether this disqualifies Dozier's series, which ran for two years from 1966, as "pure," for there was an element of knowingness in it all along—a determination to capitalize on the pop-art culture that was eliminating distinctions of high and low art. On the other hand, West and his sidekick Burt Ward brought a touching ingenuousness to their roles; they believed they were capturing the essence of Batman. "I reveled in the power and isolation and beauty of the character," West said early in the filming.*

Dozier established his reputation with serious dramas, and he

---

* You have to suspect, though, that West sometimes had his tongue in his cheek. "If you'll notice, my Batman was always moving," he attested. "The tights were itchy."

Adam West's Batman and Burt Ward's Robin, ready for action.

knew nothing about comics. When he was offered the series, at first he thought it a ridiculous project. But then he realized that it could work on several levels at once, providing thrills and adventure for kids while winking over their heads to the adult audience in the manner introduced by satirical shows like *The Flintstones* and *The Munsters*—which is now de rigueur in children's movies. This wasn't just about winning some postmodern critical respectability; broadening the audience was a sound economic move.

Dozier's *Batman* is quintessentially postmodern, full of self-referential irony but played straight enough to leave you guessing. Like many who saw it as a child, I'm astonished now at how my younger self could have missed the in-jokes and could ever have imagined that Adam West was a daring crime-fighter. Yet perhaps the children watching the show picked up on West's earnestness and trusted it—he really did think that his bizarre speech patterns, with their peculiar pauses and sudden accelerations, were the hallmark of fine acting. To us kids, the comic-strip sound captions—POW! ZWAK!—during fight scenes were just part of the show, not sly nods to the pop art of Roy Lichtenstein.

On one level, the *Batman* series was pure escapism for American audiences in the grim days of Vietnam. It made people "feel secure and happy in times of chaos and war," said West. But some regarded the satire as subversive. The *New York Times* complained that, in seeming to mock heroic values, the show was un-American. Stern newspaper editorials revived the old accusations that comic-book fare would pollute children's minds with images of crude violence.

But Bob Kane, at least, didn't seem to mind. While acknowledging that it was a far cry from the "original conception of Batman as a lone, mysterious vigilante," he called it "a marvelous spoof, and great for what it was." As long as DC Comics was able to insist that its "Batman rules"—no guns, no killings—were respected, the company seemed happy. And after all, the TV series was the saving of Batman. The comics had been suffering a sales decline for years by the mid-1960s, but with Batman back on the screen on glorious Technicolor, he became a phenomenon again.

Dozier's *Batman* wasn't coy about its intentions; they were out of the closet from the beginning. The director courted pop culture; the launch party for the series was held at the trendy New York disco Harlow's, with Andy Warhol in attendance. This conscious mobilization of camp sensibility continued right up to the final episode in 1968, which featured aging gay icon Zsa Zsa Gabor as a wicked beautician whose hair dryers can read minds. "Pop art and the cult of Camp have turned Superman and Batman into members of the intellectual community," said *Life* magazine in 1966, "and what the kids used to devour in comic books has become a staple in avant-garde art."

It wasn't possible, then, for critics simply to dismiss West's Batman as trashy, because now trash was culture too. Comics were junk and television was junk and so, according to the *Saturday Evening Post*, the Batman series was "junk squared." And to the dismay of critics "for whom good and bad are art's only poles," the *Post* admitted that the series "can be surprisingly likeable."

This blurring of "high" and "low," of the satirical with the deadpan, had not yet tipped toward self-conscious irony. Dozier's Batman never had the flavor of a calculated spoof or satire—this was not *Austin Powers*, nor did it suffer from the knowingness of Schumacher's Batman movies of the 1990s, in which famous actors take turns sending themselves up and everyone is in on the joke. The TV show shows us, rather sweetly and dopily, where one branch of this modern myth can lead. As Dozier hoped, the show appealed to adults as much as to children, although for very different reasons: the kids might angrily admonish their parents for inexplicably *laughing* during the exciting fight scenes.

It *is* funny—but Batman's relationship with comedy has always been complicated. On the face of it, there is little amusing in this story of a boy turned into an obsessive, weird vigilante by his parents' murder. But the villains were always clowns as much as they were demons: the Penguin, the Riddler, and most of all the Joker. They are Punch-like in their efforts to entertain with crazed violence, seeming less concerned with killing Batman than with mocking him for being so po-faced about his crime-fighting. They are happily absurd, while he seems to have no idea how silly he looks. For Batman, humor is frivolous; for his enemies it is deadly serious. Alan Moore nods toward this determination to find hilarity in death and chaos in his seminal 1988 story *The Killing Joke*, while Todd Phillips's 2019 movie *Joker* takes Moore's backstory of the Joker's origins as a failed stand-up comedian, with its connection between laughter and terror, to the darkest place yet. This Batman-(almost) *sans*-Batman tale casts Robert de Niro as an odious talent-show host: precisely the kind of role, reviewers noted, that would have been coveted by de Niro's delusional Rupert Pupkin in Martin Scorsese's grim fable *The King of Comedy*, to which *Joker* pays homage. Yet for all the unnerving brilliance of Joaquin Phoenix's Oscar-winning performance, too literal an "origin" story for the Joker seems only to diminish his power as the amoral, inhuman agent of chaos that Heath Ledger created (in another Oscar-winning performance) for Nolan's *The Dark Knight*. Ledger's Joker spun multiple, contradictory origin stories, until the whole question became irrelevant; he was a nightmare made flesh, and that is all we could (and needed to) know.

This engagement with the comedic isn't found much in the other modern myths I have examined, but it is a vital part of the tensions explored in Batman. The myth acknowledges the old saw that there is something both sad and dangerous about funny men—the circus performers who have us in stitches before necking cheap scotch and abusing their assistants when the big top has gone dark. It is almost as scary to see how grimly devoid of humor Batman is himself, and how seriously he takes the weird pantomime he has created. We can't help sympathizing with the Joker as he taunts Batman in *Arkham Asylum*: "Look who's laughing now!"

"Maybe every ten years Batman has to go through an evolution to keep up with the times," Bob Kane once suggested. West's goofily optimistic hero could not live beyond the rosy counterculture of the sixties. Things are different now, and so is Batman.

The TV series gave the Caped Crusader a temporary boost, but it also imposed a kind of tyranny on the comics. Editor Julius Schwartz was brought to DC in 1964 to salvage the character from the travesty he had become, and he jettisoned the puerile Bat-family along with all the alien-themed and fantastical flotsam that had accompanied it. Batman and Robin got back to basics, taking on new villains such as Poison Ivy (1966). But the comics were now partly in thrall to the popular series: the tail was wagging the dog. Alfred the butler was killed off in 1964, only to be brought back by an unlikely *deus ex machina* two years later because the TV series demanded his presence. Batman's bat-logo got its distinctive yellow oval from the series (although it seemed less than impressive on Adam West's chest). When the television executives demanded a new "Batgirl"—not the coy Betty but Commissioner

Gordon's daughter Barbara, played in a full-on S&M catsuit by Yvonne Craig—Schwartz had to fall in line. He came under pressure to emulate the campy tone of the TV show in the comics.

Those were tough times for DC. The graphics might look great—artist Neal Adams introduced a menacing style in the late 1960s, soon to be assisted by the hard-boiled narrative skills of Denny O'Neil—but interest was waning. Bob Kane himself quit the company in 1967, and by the 1970s Batman's future was far from assured.

He survived by embracing his shadows. Thanks to illustrators such as Jim Lee and Brian Bolland, and writers like Frank Miller and Alan Moore, Batman today is more popular and ambiguous than ever: implacably violent, probably deranged, and, some feel, on the brink of fascism. The Gotham he fights to protect is riddled not just with crime but with corruption, and Wayne's vigilante operates beyond the jurisdiction of authorities, his only ally the incorruptible Commissioner Gordon. In Miller's *The Dark Knight Returns* he rides horseback against evil, a caped Arthurian knight uttering the defiant, Nietzschian declaration: "Tonight, *I* am the law!"

Miller and Moore did not have to invent the dark side of Batman—it was always there. Bruce Wayne alleges that his costume serves only to terrify lawbreakers, but it is demonic enough to make honest citizens uneasy too. The entire *raison d'être* of the police force is that they are accountable servants of the state, created precisely so that we will not be reliant on vigilante justice. Batman tells us that this might not be enough. There are crimes the police can't crack, criminals they can't catch. They are powerless before the feral anarchy of the Joker or the venality of Harvey "Two-Face" Dent.

Miller, like his Dark Knight, is skeptical of the ability of governments, committees, and institutions to keep us safe. "I think we're at our best when we're autonomous," he says. But when our protectors are themselves beyond the law, can we trust them? "What gives you the right?" cries one of the failed Batman wannabes who has ineptly tried to take the law into his own hands in Nolan's *The Dark Knight*. "What's the difference between you and me?" Christian Bale's Batman replies, "I'm not wearing hockey pads." Or more properly: I can be a vigilante because I have the wealth and resources to support it. There is nothing egalitarian about Batman.

He is the hero we deserve—which is to say, perhaps no hero at all, but just what Miller calls "an extremely violent character" with the right connections. "Heroes have to work in the society around them," Miller says, "and Batman works best in a society that's gone to hell. That's the only way he's ever worked. He was created when the world was going to hell, and *Dark Knight* came out when the world went to hell." A post-heroic hero, indeed.

And perhaps this is nothing more than a modern update of the Gothic tradition. That continuity is indeed what cultural historian Andreas Reichstein perceives in Batman, whom he regards as the next stage of a narrative running from Jekyll and Hyde to Dorian Gray and Doctor Moreau; he is, in fact, "an American Mr. Hyde," the brutal id of Bruce Wayne. "He is the American answer to the Victorian fear of losing control," Reichstein says. Or looked at another way, "the Batman mythos has been about the pressing of Gothic Fear into the service of heroic justice," according to cultural critic Mark Fisher.

We're not sure we can endorse what Batman does, as he pushes the Joker into a vat of toxic waste or beats him around in a police cell. But we're glad to have somewhere to place the blame—for after all, it's not as if the authorities have endorsed him, is it? "They'll hate you for it, but that's the point of Batman," Alfred tells Wayne in Nolan's *The Dark Knight*. "He can be the outcast." The devils that work in him are not so far from those that drive embittered, alienated individuals in our own world to wage war on "the state" and write crazed manifestos of power and blood. But he manages—just—to "make his devils work for the common good," as Miller says. To set him up as a role model for the nation's youth is, in Miller's view, absurd—for that idea "castrates every hero or villain who falls under the saw of the censors."

Miller has been accused of flirting with quasi-fascist ideas in his Dark Knight stories—but that is the point. Batman has to be a loner, he says, for if he were in a position of authority "he would be a tyrant." In other words, his behavior "has a lot to tell us about the complications sur-rounding authority and its expression in a modern democracy." Nolan's

The Dark Knight Rises (2012) was interpreted by some as a critique of the Occupy movement that protested against the corruption and abuses of the financial system, as the villain Bane leads a mob to attack the financiers of Wall Street and promises to return Gotham City to "the people." Has Batman then become a right-wing defender of capitalism? Well, you didn't ever imagine he was a liberal, did you? Have you *seen* that mansion?

To find out what would happen were Batman to explicitly relinquish the life of the loner and pledge his allegiance to the state, look no further than Judge Dredd, the brutal enforcer of Mega-City One in the British comic *2000 AD*. One of Batman's most obvious progeny, Dredd too seems to be granted license to take matters into his own hands as he fights exotic maniacs with his face half-hidden behind a masklike visor. "I am the law!" is one of Dredd's trademark declarations—but now that chilling phrase has the full power and authority of the state behind it. No, it is a mercy that, as Denny O'Neil has attested, Batman is *not* Judge Dredd.*

<hr />

This darkness has left no room for Robin, who after all was introduced to lighten up his mentor. Robin's role was never a happy one in the grim new world. He is gone from Frank Miller's *The Dark Knight Returns*, replaced by a thirteen-year-old girl named Carrie Kelley, for whom Batman becomes a father figure after she is neglected by her parents. And when Dick Grayson does show up in *The Dark Knight Strikes Again* (2001–2002), things have gone badly wrong for him: he went insane after Batman rejected him because he "couldn't cut the mustard," and has become a killer modeled on the Joker. Grayson attacks and nearly kills Carrie in what could easily be read as piqued fury at his rejection and

---

* The kinship is affirmed, however, when the two figures battle together in the 1991 collaboration *Batman and Judge Dredd: Judgment on Gotham*.

replacement by his former lover. Batman finally kills him by casting him into a lava pit beneath the Batcave.

Grayson had in fact already been usurped back in 1983 (*Batman* #357) by a new assistant called Jason Todd, who was so irritating that fans rejected *him*, leading DC to conduct a phone poll asking whether he should be killed off. That ploy—condemned by Miller as "the most cynical thing that a publisher has ever done. . . . An actual toll-free number where fans can call in to put the axe to the little boy's head"—sealed Todd's fate, and he died in the 1988 story "A Death in the Family." The death scene is homoerotically charged, as Batman hugs Todd/Robin's body to his breast ("I've lost him")—a trope repeated even more suggestively in the 1997 *Batman: Legends of the Dark Knight* ("We're *together* at least. One *last* time together . . . as it *should* be.")

But events get complicated in the DC Universe. After giving up his role as Robin, Grayson mutated into the character Nightwing—or rather, he took that name, used originally as an alter ego by Superman during an adventure on his native planet Krypton. . . . Well, you don't really need to know the details. Grayson's Nightwing even, in oedipal fashion, takes up Batman's cape himself for a time after Bruce Wayne's apparent death (deaths in the DC Universe must always be taken with a pinch of salt), fighting for justice in Grant Morrison's *Batman and Robin* series (2009–2011) alongside Bruce's brooding son Damian Wayne. The plot thickens, seemingly without limit—although one can't quite shake off the suspicion that the writers are struggling to find a job for a rather embarrassing adolescent they no longer need.

After more than eight decades of writing by many authors tasked with conflicting objectives and targeting diverse audiences, Batman's story is less a tangled tale and more a web of bewildering complexity that even the most dedicated fans struggle to stay abreast of. The DC Universe is a busy social network where Batman battles Superman or teams up with Wonder Woman. Every new tale must enter into a dialogue with the past—most notably, perhaps, in Morrison's story *Batman R.I.P.* (2008), which attempts to make sense of some of the more bizarre narratives of the 1950s and 1960s sci-fi era* by having Bruce Wayne undergo a breakdown and become a homeless, filthy wretch who might or might not be hallucinating the "Bat-Mite." Did he just imagine all that crazy stuff? Did we?

Everything is now possible; nothing is out of bounds. All aspects of Batman are open to exploration: good and bad, past, present, and future. The pretense of maintaining a single, consistent narrative arc has long since been abandoned. And this, after all, is exactly what must happen when the generative story is still being told alongside the myth: when the canon was never closed. There is a final page for Crusoe and Frankenstein, and even (eventually) for Sherlock Holmes. But for Batman there never was: the myth hasn't grown out of the story so much as overtaken it.

DC Comics has just launched an African American Batman: Tim Fox, the son of Bruce Wayne's business manager Lucius Fox, who now owns Wayne's fortune and resources. There will of course be the predictable bigoted outcry, but it's a measure of Batman's now-mythical status that this storyline makes perfect sense. It won't be (or should not be) an exercise in "political correctness," but rather an acknowledgement that the Batman myth is bursting with potential to explore the tensions about race in modern America. Imagine it: a Black vigilante in a hooded mask working in uneasy relationship with a militarized police force as he takes vengeance on white hoodlums, perhaps even psychotic old Whiteface himself, the Joker. For what, after all, does Batman say in Morrison's far-future tale *DC: One Million* (1998)? "One Batman? You believe there can only be one Batman? Batman is not a *man*. I'm an *ideal*."

It's coincidentally apt, then, to recall how open-ended was the account of the Batman mythos in the 1996 limited-edition series *Batman Black and White* (the title was no racial allusion but referred to the experimental absence of color from the illustrations). Here he sounds

---

* Those stories from Batman's so-called Silver Age were collected in the *Black Casebook*, issued by DC in 2009 as "The stories that inspired *Batman R.I.P.*" with an introduction by Morrison.

closer to Michael Moorcock's protean Eternal Champion than to the slightly obsessive, vengeful rich guy in a funny suit:

> Once upon a time, the city was a gothic realm of eternal night. . . . And then, from the darkest pit of hell, a champion arose. No one knew how, or from where. He was as black as the villainy he sought, and he stood in harm's way, between the good and the evil. He did things no mortal man could do. Some say he could fly. Some say he could breathe underwater. Some say he had a secret fortress and slept there to recover from his never-ending battles. Some say he never slept at all.

Needless to say, the folks who will holler the hardest about Batman being Black will probably not have batted an eyelid at the many transformations we have witnessed already. Consider also what has become of Batman's dark side, which was never far beneath the surface. Now that public morality no longer insists he be an all-American role model, the writers are free to show us more of Batman's demons. In particular, we can return to the inspirational roots that Bob Kane found in vampire tales. In *Batman* #351 (1982), writer Gerry Conway had Batman turn temporarily into an actual vampire—a transformation easily read as a liberation of the murderous id that always threatened to emerge in his quest to avenge his parents. After reminding us of his orphaned youth, the story tells us of the mutated superhero:

> His fangs drip blood, his eyes glazed in the grip of a living nightmare. . . . Somewhere, deep within, the shards of his almost-broken soul cry out in pain . . . and despair of all hope.

This vampiric Batman was always latent. Frank Miller added an occult element to the origin story when he showed a young Bruce falling into a cave beneath Wayne Manor filled with bats—and also with some unnamed, unseen creature "gliding with ancient grace . . . breath hot with the taste of fallen foes . . . the stench of dead things, damned things . . . glaring, hating . . . claiming me as his own." There's little doubt that it was a vampire down there in the dark with the frightened boy.

Batman's vampire alter ego was given free rein in DC's spin-off Elseworld series of alternative superhero tales, which included the "Batman and Dracula" trilogy written by Doug Moench and drawn by Kelley Jones: *Red Rain* (1991), *Bloodstorm* (1994), and *Crimson Mist* (1998). Here Batman becomes a vampire by choice in order to fight Dracula, who is preying on Gotham City.

Far from being whimsical "alternative takes" on the Batman myth, these Elseworld tales act as a vehicle for delving explicitly into what is otherwise only hinted at in the core canon. Here Batman is permitted to lose the self-control to which he clings tenuously in the original Gotham, and his bloodlust emerges raging. Pursuing the homoerotic vein of his conflicts with the Joker, he is seen in *Bloodstorm* draining his enduring foe of his blood. He calls on the faithful Alfred to curb this now uncontrollable passion with a stake through his heart. But he emerges again to fight a new crime wave, before destroying himself by walking into the sunlight: it's only by yielding to the light that Batman can ultimately be destroyed.

The further the universe of Batman leaks into that of his own mythic cousins, the more it opens up to the darkness of Gothic horror. In *Batman and Judge Dredd: Judgment on Gotham*, the sex and horror motifs curdle into atavistic savagery. Never mind any Comic Code now; hands get stuck through skulls, heads are bitten off, sexy lovers are slaughtered by a quasi-alien ghoul whose claws, like Freddy Krueger's finger-blades, penetrate their faces. We are in a postheroic age indeed, where there is no trustworthy authority, our saviors are on the verge of breakdown, and morality must compromise with might. "What kind of lunatic world have I stumbled into?" Batman wonders.

The Elseworld series offers a literal illustration of what modern myths have now become. They exemplify the merging metanarratives through which these stories burst their bounds and demand multiple readings, cross-references, internal dialogue. They demonstrate too why comic books and graphic novels are a perfect breeding ground for modern myths. Because so much of their work is done visually, these stories may more quickly establish the basic *grain* of the narrative, along with the iconography that is always so vital to a modern myth's psychic work—as we have seen with Whale's *Frankenstein* and Sidney Paget's representations of Sherlock Holmes, in the movie transformation scenes of Jekyll and Hyde and the looming tripods of the Martians. Images speak directly to the unconscious, and they do not have to explain themselves. Mary Shelley wrote cinematically, almost a century too soon.

Unlike novels, comic books are acutely sensitive to audience feedback, evolving to the shape of public demand and amplifying what resonates without too much consideration of the implications. The relative crudeness of their narratives, psychology, and presentation in their early days recapitulated some of those characteristics in the yarns rather cavalierly patched together by Defoe, Stoker, and Conan Doyle. That's not to deny other kinds of sophistication in all of those works, which gain impact and depth from their intuitive immediacy. Crude they may be, but they are far from incompetent. Conan Doyle knew just what his readers wanted, and so did the editors of DC Comics (and when they did not, they only had to ask their readers). Because these works are not striving to be "high art," they have the luxury of bluntness. They are without guile, as myths need to be. And they do not need to reshape some generative text, for the text is itself continually co-created, adapted, and modified. It is this relinquishing of narrative control to society that is likely to shape the modern myths of the future, which is why they seem likely to arise in media other than traditional literature. Let's see what form they might take.

꙲

Chapter 9

MYTHS IN THE MAKING, MYTHS TO COME

O ne way to spot modern myths today is through their potential for cross-fertilization. Batman stalwarts Alan Moore and Kevin O'Neill have united Dracula, Jekyll and Hyde, Sherlock Holmes, and the War of the Worlds, along with Wells's Invisible Man, Jules Verne's Captain Nemo, and others, in their comic series *The League of Extraordinary Gentlemen*, a queasy, delirious dream of the British empire's apogee. The British-American television series *Penny Dreadful* is similarly populated by characters from *Dracula*, *Frankenstein*, and *Dr. Jekyll and Mr. Hyde*, along with Oscar Wilde's Dorian Gray. Kim Newman's *Anno Dracula* (1992; see page 220) combines a comparable mythical cast with real-life contemporaries such as George Bernard Shaw and Lewis Carroll. Newman's sequels, ranging through the twentieth century, embrace iconic characters real and fictional including Count Orlok, Dr. Caligari, Philip Marlowe, and Orson Welles, ending with a dizzyingly postmodern mash-up that alludes to Andy Warhol, a fictional Francis Ford Coppola film of *Dracula* (which, starring Marlon Brando and Martin Sheen, is evidently a twisted variant of *Apocalypse Now*), and screen vampire tales such as *The Lost Boys* and *Buffy the Vampire Slayer*. It is striking how easily and naturally, in these amalgams, the myths sit alongside one another.

This seems to be the final, validating stage of a modern myth. Like the ancient myths of Greece, Rome, and Scandinavia, it will merge into a pantheon. To protest that Shelley or Wells would turn in their graves at seeing their tales mutate and hybridize in this way is beside the point (and not obviously true anyway).

After all, modern myths typically share the same evolutionary arc. They begin with a source text that is often offbeat, scrappy, or half-formed: the diaristic, epistolary, episodic, or Chinese-box assemblages that characterize each of the examples I've discussed. This semi-coherent nature renders boundaries fuzzy and porous from the outset and positively invites later repackaging.

Soon after the myth is precipitated in this amorphous manner, it is pulled apart, sifted, and bowdlerized for the mass market, as in the stage plays that followed hard on the heels of *Frankenstein, Jekyll and Hyde*, and *Dracula*, or the radio and cinema adaptations of *The War of the Worlds* and the Sherlock Holmes stories. In the modern era these secondary texts supply the basis for further elaboration, simplification, and icon-generation. Stock characters are superimposed on the text: the young woman who provides a "love interest," the foolish or protective father, the comic foil, the mad scientist.

Once these popularized versions have all but eclipsed the original text, there comes a period of reevaluation in which a "return to the source" is advertised as a virtue, though what results may be a rather desultory, literal slog through that pristine material with scant thought for the mythical themes that popular culture has begun to distill from it.

Finally comes synthesis and diversification. Freer adaptations offer the opportunity to mine particular aspects of the myth, and to play them against each other, as we come to accept that the subject has long outgrown its origins. Along the way, parodies and spoofs—the Addams Family, the Munsters, *Abbott and Costello Meet . . .*, *Carry On Screaming*, *Hotel Transylvania*—may signal an attempt to neutralize taboo elements in the myth, such as sexual transgression. Don't worry, they say: these creatures are just like you and me really, a big nuclear family of oddballs.

What are the next modern myths going to be? We can't be sure yet which other twentieth-century narratives will make the grade. It takes many decades before a story's mythopoeic potential becomes apparent, not least because time is needed for it to escape from the single dominant reading that each age typically offers. I suspect that a half century is around the minimum time needed for such a story to reveal its prismatic power, its capacity to refract cultural anxieties into many hues. As Gertude Stein wrote, "It just does take about a hundred years for things to cease to have the same meaning that they had before."

Still, there is no lack of candidates, among which we might include Franz Kafka's *Metamorphosis* (1915), Aldous Huxley's *Brave New World* (1932), George Orwell's *1984* (1948), and Ian Fleming's James Bond novels (1953–1966). Edgar Rice Burroughs's Tarzan too is, according to Ursula Le Guin, a "true myth figure"—although she adds that he is "not a particularly relevant one to modern ethical/emotional dilemmas as Frankenstein's monster is," so we shall see.

One thing is for sure: modern myths no longer need to have a predominantly literary genesis. New media are not (as often prophesized) eclipsing the printed word, but they are changing the way our cultural stories evolve, just as print itself once did. The coming of the printing press in Renaissance Europe meant that a largely oral tradition of storytelling began to cede ground to a primarily literary one. Narrative was now all about the text—which enabled it to do new things. Old myths, such as the Faust story, became best-selling books among the literate classes, and in the process their form became fixed and their function more complex. What was once a cautionary tale of hubris and impiety became, in the hands of Christopher Marlowe and later Johann Wolfgang Goethe, a many-layered commentary on the changing relationships between God and humankind, nature and art.

Until the twentieth century, literature was the only medium in which new myths could arise and find purchase in a diversifying, diffused, and secular society. But mass media have changed all that. Not only do radio, cinema, television—and now the internet, social media, and gaming— create new and increasingly dominant ways for modern myths to spread and evolve, but they seem certain to incubate new myths of their own, without any initial expression as words on pages. *Batman* is arguably the earliest example: there is no inceptive Batman novel, and the language of the tale is primarily visual.

It seems very likely that new modern myths will be born from the most visual medium of all: cinema. It's commonplace already, of course, to call the movie industry—Hollywood in particular—a myth machine. It has certainly *shaped* some modern myths, not least because those stories—ill-formed, undisciplined, malleable—lend themselves to cinematic adaptation much more readily than does the conventional canon of literary classics. But cinema is now ready to generate myths de novo. And the first of these has pretty much arrived: the modern myth of the zombie.

◆

Like the vampire and the werewolf, the zombie is a creature of folklore. But a monster is not a myth. What was needed to admit these creatures into modernity was a narrative that articulated, lubricated, and liberated their potential meanings in today's world. I am not sure that story has fully crystallized yet, but we are very close. It goes something like this:

> Undead beings appear and begin to take over society, transforming the living to their own state by contagion, typically spread by body fluids. The zombies are mindless, shambolic, and savage: they have no apparent inner world, no goal or motive beyond attacking the living. The zombie assault grows to apocalyptic proportions as humanity dwindles.

Clearly there is considerable overlap with the Dracula myth—not only was the notion of "the undead" coined in Stoker's book, it was his original title. There are aspects too of Frankenstein's shambling creature (as reformulated for stage and screen), with its precarious physical integrity: you never quite know if bits might fall off. The zombie is also allied to legends of the mummy: according to cultural critic James Twitchell, it is really just "a mummy in street clothes with no love life and a big appetite." And there's a presentiment of the zombie's mental vacancy, its lack of real animation and consciousness, in Jekyll's view of Hyde:

> He thought of Hyde, for all his energy and life, as of something not only hellish but inorganic. This was the shocking thing: that the slime of the pit seemed to utter cries and voices; that the amorphous dust gesticulated and sinned; that what was dead, and had no shape, could usurp the offices of life.

The zombie itself comes from Haitian vodou and folklore, in which context it is freighted with the legacies of slavery, servitude, and colonialism.* That's apparent in one of the first reports of the zombie in the English language, which comes from the traveler William Seabrook's 1929 account of Haiti, *The Magic Island*:

> Obediently, like an animal, he slowly stood erect—and what I saw then, coupled with what I heard previously, or despite it, came as a rather sickening shock. The eyes were the worst. It was not my imagination. They were in truth like the eyes of a dead man, not blind, but staring unfocused, unseeing. The whole face, for that matter, was bad enough. It was vacant, as if there were nothing behind it.

White Zombie, the 1932 movie starring Bela Lugosi, is heavily indebted to such imagery and colonialist rhetoric: Lugosi's villainous "Murder" Legendre hypnotizes the Haitian workers on a sugar plantation to get more labor out of them. In early zombie films these creatures were little more than menacing but ultimately pathetic ghouls summoned through arcane ritual. *King of the Zombies* (1941) was a vodou-inflected spy comedy intended initially as another Lugosi vehicle; *Revenge of the*

---

* There's more than a little of the zombie in Bertha Rochester from *Jane Eyre*, the mad woman in the attic; Jane compares her to a "vampire," but she is from the Caribbean. Val Lewton's 1943 film *I Walked with a Zombie*, set on a Caribbean sugar plantation, includes plot elements from *Jane Eyre*.

Bela Lugosi and friend in White Zombie *(1932).*

Zombies (1943) was a piece of war propaganda featuring a German mad scientist who is trying to make a race of undead for the Third Reich.

The mythical zombie, like the mythical vampire, needed to have some agreed-upon rules of engagement, refined and codified by gradual cultural consensus. We all know now that zombies are mindless, frenzied, frustratingly hard to kill. They like to gnaw on flesh. They are utterly asexual.* And crucially, they hunt in hordes—it's this aspect, perhaps more than any other, that gave them more traction than the mummy, not least through the myth's now vital denouement: the zombie apocalypse.

That image too is a legacy of the vampire, albeit of rather modern origin. It was presented in Richard Matheson's pulp vampire-epidemic

---

* Such rules exist to be subverted, and *Warm Bodies* (2013) offers the bittersweet spectacle of an undead lover struggling with his predicament.

novel *I Am Legend* (1954), which inspired Stephen King's *Salem's Lot*. Here Dracula has, to speak figuratively, been victorious, and the vampire has eclipsed civilization. The hero Robert Neville is an isolated survivor of the "old race" surrounded only by inarticulate vampiric savages: as he puts it himself, "a weird Robinson Crusoe, imprisoned on an island of night surrounded by oceans of death." What makes his situation both poignant and terrifying is that these "dull, robot-like creatures" are the hollowed-out husks of Neville's former friends and neighbors. He faces existential isolation in a world that used to be his own but which has degenerated—as Max Nordau and the social Darwinists feared it might—into abominable squalor.

Neville lives a dreary existence in a sound-proofed house barricaded against the undead who come out at night. His existence is one of "monotonous horror"—a perfect description of the zombie narrative. Every day he goes out to find sleeping vampires and kills them with stakes mass-produced on his lathe, the routine as grindingly relentless as an office job. It is not only boring but banal and pointless, yet what else is there for him to do?

Trying to understand what caused the plague, he reads *Dracula* but finds there only "melodramatics" and "driveling extrapolations." Matheson dispenses with the supernatural and introduces in its place the feature now characteristic of the zombie story: a biological plague, in this case a "vampiris" bacterium that activates the nerves in dead bodies. This reanimation causes behavior resembling that of the traditional vampire, especially bloodlust; garlic acts as an allergen, while an aversion to crucifixes is triggered by a dim memory of the old vampire cliché within the undead mind. The bacterial spores were spread by dust storms and plagues of mosquitoes that followed in the wake of unspecified bombings. "All without bats fluttering against estate windows," Neville records in his journal, "all without the supernatural. The vampire was real." Likewise, almost no zombie story now relies on the magical agency of vodou: it has become a rational affair, explicable—and perhaps even redeemable—by science.

Not all the vampires of *I Am Legend* are a lost cause, however. One day a woman called Ruth turns up, apparently terrified and needing protection, but she has in fact been sent to spy on Neville as one of the "living vampires." These are people who are infected by the bacterium but can keep its influence at bay with drugs. They are trying to forge a new society, but Neville has been unwittingly killing them along with the true vampires: to them, he is simply a monstrous mass murderer. The living vampires capture him and sentence him to public execution, which Neville escapes only by taking a lethal poison.

The book introduces a key ingredient of the zombie myth: apocalyptic civilizational collapse, at that time reflecting fears of Cold War nuclear holocaust. Matheson, himself a successful Hollywood screenwriter, saw his book adapted for the screen in 1964, as *The Last Man on Earth*, starring Vincent Price; 1971, as *Omega Man*, with Charlton Heston; and most recently, 2007, with the original title revived and Will Smith in the lead role.

I Am Legend and *The Last Man on Earth* were acknowledged by George Romero as influences for his 1968 zombie movie *Night of the Living Dead*—which can now claim to be the primary source for the evolving zombie myth. Working on a tiny budget, Romero found the magic blend of lowbrow entertainment, cultural and social criticism, and psychological resonance that marks the true myth. It was here too that the plaguelike contagion of the vampire became written into the zombie narrative, imbuing it with the vital ingredient of infection paranoia.

Night of the Living Dead has the brutal simplicity of the best horror films: a small group of citizens become trapped in an isolated house besieged by an unexplained horde of zombies. Yet the black-and-white, low-budget mayhem doesn't detract from the genuinely subversive political messages of the film; in some ways they are enhanced by the arthouse patina that results. To make the hero a Black man (simply called Ben and played by unknown actor Duane Jones) was shocking enough in 1968. But the manner in which Ben is killed at the end of the film still creates a frisson today. His death—he is shot by the posse of rednecks who have come to dispose of the zombies and mistake him for one of them—has the chillingly casual air of a lynching as his body is dragged by meat-hook, dumped onto the pile of zombie corpses, and set ablaze.

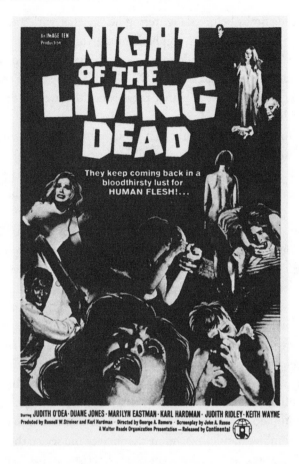

The zombie myth begins to cohere in George Romero's Night of the Living Dead *(1968).*

At the height of countercultural social rebellion and civil-rights unrest, *Night of the Living Dead* exploded a charge under the complacent veneer of American life, challenging norms of family and race and inviting allegorical readings about the ongoing brutality of Vietnam, the anxieties of the Cold War, and the stifling pressures of capitalism. "It came out of the anger of the times," Romero has said. The nuclear family, represented by the Coopers, who take refuge with Ben, is literally pulled apart: the father is an ineffectual coward, and the zombified daughter chews on his limbs before slaughtering her mother with a trowel. The movie's nihilistic ending is an emphatic break from the mythopoeic tales of the late Victorian era, and makes it a harbinger of the horror narratives of today: as poet and critic R. W. H. Dillard says, "the real horror of *Night of the Living Dead* is that there is nothing we can do that will make any difference at all."

If we should be in any doubt that Romero had motives other than to simply scare and disturb audiences with scenes of gory pandemonium, the sequel, *Dawn of the Dead* (1978), dispels it.* With its images of suburbanite zombies lumbering mindlessly through the bright, bland shopping malls of Philadelphia, the film gives us a new metaphor: in thrall to empty consumerism, we are all becoming undead. It is comedic even in its horror, and the gag was later played explicitly for laughs in "zom-rom-com" *Shaun of the Dead* (2004), where Simon Pegg's slacker hero fails even to notice that his neighbors have become zombies as he slouches to the corner shop to buy a pint of milk.

The mall sequences of *Dawn of the Dead* are themselves staged less with intent to scare than with ironic glee. Even one of the survivors, the policeman Stephen, announces as they raid the shops for supplies, "This place is terrific. It really is. It's perfect." Fran, the sole woman in the group, replies, "You're hypnotized by this place, it's so bright and neatly wrapped. You don't see that it's a prison too." Yet once they have slaughtered the zombies wandering among the shelves and sealed the entrances to mall, the group can't resist a joyous consumerist spree, even grabbing a television set although there is nothing being broadcast any more. In the face of the zombie apocalypse, it's the only way left to fill what remains of their lives. Are they then so different from the real zombies who we see pathetically mimicking what used to be normal life, stumbling across the skating rink and milling in the supermarket aisles?

With another Black man in the lead role (Peter, played by Ken Foree) and its opening sequence of racist National Guardsmen surrounding a

---

* In a legal tussle, Romero lost his claim to the phrase "living dead."

building filled with Puerto Ricans resisting the zombie quarantine, it's clear that *Dawn of the Dead* has the class and racial tensions of middle America in its sights. It's an obvious progenitor for Jordan Peele's vastly more sophisticated *Us* (2019), in which a Black family is menaced by zombielike doppelgängers dressed in the red overalls of office cleaning workers. Peele turns the old horror trope into a searing examination of class and privilege in modern America. "Those who suffer and those who prosper are two sides of the same coin," he has said of the film's message. "You can never forget that. We need to fight for the less fortunate." *Us* is the perfect illustration of how, as media scholar Gregory Waller says, zombie stories "explore the nature of the status quo and the relationship between the public and the private, civic duty and personal desire."

In short, zombies have become mythic because they *mean something*— whether they represent mindless consumers, victims of biotechnology or natural disease run amok, the violent mob, or the underclass on which our superficially prosperous society relies. Most importantly—and this is what sets the zombie apart from the beasts of classical myth— everyone is at risk of becoming a zombie. The zombie represents, among other things, a terror of loss of personhood and of bodily autonomy in the age of the commodification of identity. The fear whispers in your ear: have I become a zombie already? Have you?

And unlike vampirism, zombiism is no fun. Zombies are *more dead* than vampires, which makes them more prone to what Freud called the "unconscious hostility" the living project onto the dead. To Freud, the taboos and elaborate rituals surrounding the burial or disposal of the deceased are not so much demonstrations of respect and grief, as attempts to erect a clear boundary between life and death: to make sure that the dead do not come back. "The primitive fear of the dead," he wrote,

> is still so strong within us and always ready to come to the surface at
> any opportunity. Most likely our fear still contains the old belief that

the deceased becomes the enemy of his survivor and wants to carry
him off to share his new life with him.

Here it is surely its very banality that makes the modern zombie so
threatening. So long as raising the dead demands elaborate ritual on
a remote Caribbean island, there is not much to fear.* And there's no
transgressive frisson when Snow White is awoken from what seems to
be a sleep of death, says Freud, because we suspend our expectations in
fairy tales. It's when the undead show up at the supermarket that we
need to worry.

Zombies are a metaphor for life without mind: the word has become
an adjective that connotes an automaton-like, routine action. Philosophers of mind still debate Descartes's question of whether we can ever
know for sure if anyone but ourselves has consciousness, or just the appearance of it. Today's computers and artificial intelligence are zombielike, increasingly acting as though they have minds and goals when we
know they do not; the notion of a "Turing test" to distinguish human
from machine, while given short shrift by AI researchers, remains popular
in our culture because it toys with our anxieties about the increasingly
hazy boundary between the two. Zombies arouse some of the same paranoia and fear we feel when surrounded by others who are not like ourselves, who speak and behave and even seem to think differently. It's no
coincidence that the invasion paranoia of *The War of the Worlds* and *Dracula* converged with the zombie narrative in Don Siegel's film *Invasion of
the Body Snatchers* (1956) at the height of McCarthyite fears of "Reds
in the Bed": Communist agents masquerading as patriotic Americans.

As ever, when the myth shifts—Frankenstein's creature becoming
mute and dull-witted, say—we have to ask what cultural need has occasioned it. The "fast zombie" of *28 Days Later* (2002) and *World War Z*

---

* Even Hammer's *Plague of the Zombies* (1966), set in Cornwall, still required a zombification
ritual (the local squire used to live in Haiti), making it a throwback to *White Zombie* rather
than a contribution to the developing zombie myth.

(2013), as well as Jack Snyder's 2004 remake of *Dawn of the Dead*, signals the declining appeal of Romero's zombie as a satire of the modern consumer wandering gormlessly through a trash culture.* As writer John Levin argues, while the zombies of the new *Dawn of the Dead* may be mindless, "they're less a consumerist mob than a bunch of high-strung car chasers." They're no longer on opioids but on speed. We no longer have time to slaughter them at leisure or, as in *Shaun of the Dead*, to take a selfie before we do so. The object of social satire is then not the dimwitted and dazed consumer but the hyperstimulated gamer frantically chasing the next thrill.

The zombie apocalypse, meanwhile, is an example of the thoroughly modern notion of permanent social rupture, with technology or modern life itself the usual culprit. Romero's *Night of the Living Dead* hints, via a half-seen television broadcast, that the zombie epidemic was caused by radioactive contamination of a space probe sent on a mission to Venus, which exploded on reentering Earth's atmosphere.† Viruses developed in biological research are the explanation in several recent zombie escapades, such as *28 Days Later*. Then the zombie pandemic introduced by Romero becomes itself a metaphor for fears about infectious diseases, especially those spread by physical contact, by body fluids or exhalations: AIDS, SARS, Ebola, bird flu, and the Covid-19 coronavirus that brought to cities across the globe that state of eerie abandonment seen so often in zombie movies.

The most alert of zombie narratives recognize that these associations have a geopolitical dimension. Epidemiology always does. The language

---

* The slow pace of the traditional zombie supplied one of the killer deadpan lines in *Night of the Living Dead*: when a reporter asks a police officer of the zombies "Are they slow-moving, chief?" he replies, "Yeah—they're dead."

† This was apparently just one of three proffered "explanations" for the zombie plague in the film, but Romero says the others were edited out. Had they been retained, it would only have strengthened what is already the implication of these obscure broadcasts: that the authorities are clueless, ineffectual, and no longer of help to the ordinary citizen.

of the zombie epidemic in *World War Z** is that of infectious-disease research: there is a "Patient Zero," the World Health Organization gets involved, and the zombifying plague is at first called "African rabies," an allusion to the belief that HIV first arose in the African jungles. The virus gets spread in the illicit ways they do—or are feared to—today: by people-smuggling of refugees, or a black market for harvested organs in China. In such ways, the film explores anxieties about the decline of the United States as a global superpower and its vulnerability to disruptive forces from the developing world. Protection from zombiism is conferred by status and wealth, as the rich retreat into sealed enclaves while some survivors side with the underclass and become zombie sympathizers. The same discussion of privileged "self-quarantine," while low-paid but essential manual workers are compelled to toil and risk viral exposure, arose during the Covid-19 pandemic.

That crisis has brought an unnerving concreteness to the medicalization of the zombie epidemic, blurring the boundaries of fact and fiction rather as the 9/11 attack on the World Trade Center did with respect to the disaster movie. You could be forgiven for wondering if a real zombie epidemic might be on the way: scientists have used computer models of such an imaginary crisis to explore scenarios for real-world infectious diseases, while survivalist groups in the US have prepared in earnest for a zombie apocalypse. (When Steve Schlozman, a Harvard physician and author of the novel *The Zombie Autopsies: Secret Notebooks from the Apocalypse*, appeared on US radio in 2011 to chat about a fictitious zombie outbreak, he began to receive emails asking how one could protect body and home from the impending onslaught.) Now that a viral pandemic has paralyzed the world for real, we find ourselves living among images every bit as jarring as those in a Romero movie: horrific scenes

---

* The 2013 movie was based on a 2006 novel by Max Brooks, who wrote *The Zombie Survival Guide* three years earlier. His 2020 novel *Devolution*, describing the attempts of a self-isolating family to survive a deadly threat from outside, has secured his reputation as something of a prophet of the latest pandemic.

of emergency hospitals in Central Park, of mass graves and burials attended only by personnel in protective suits, of tourist hotspots fallen eerily silent, and packs of wild animals roaming empty suburban streets. But at the same time we face the banality of shortages of toilet paper and pasta, and negotiate the social niceties of work meetings on Zoom. The scientists look more like heroes and saviors than ever they did in *The Andromeda Strain*, while some leaders of developed nations have shown a level of incompetence, callousness, and denialism beyond anything a scriptwriter would have dared imagine. The Covid-19 pandemic will change many things, and it seems likely that our zombie myth will be one of them.

The zombie myth confronts us with horror at its most visceral, and forces us to ask what these tales are appealing to when they indulge its most recognizable feature: violence. With the zombie, no holds are barred. You can do to it whatever you want, not just with impunity but with justification and necessity: chop off its limbs, squash its head, blast it to pieces. Because after all, the zombie would treat you just as badly, most probably by chomping on whatever portion of your flesh it can lay its decaying hands on. Romero's movies set the tone for zombie ultraviolence, which is a trademark feature of *The Walking Dead* series and more or less the *raison d'être* of Peter Jackson's zombie splatter movie *Braindead* (1992). What is this about?

The ghastly rending of flesh has been a part of Gothic horror ever since *Frankenstein*, never more so than when Victor "tore to pieces" the mate he had made for his creature. The killing of Lucy Westenra in *Dracula* is even more graphic: the stake is hammered messily through her heart and her head is cut off. Horror forces us to face the unsettling materiality, the vulnerable fleshiness, of our existence. It opens up the body, takes it apart, confronts us with the puzzle of how conscious experience is housed in globby viscera.

In the zombie myth all extremes of violence are sanctioned: the zombie must be hacked and pulverized, and what does that matter anyhow when these creatures are already dead? In *Dawn of the Dead*, the bikers who invade the mall don't distinguish between thrusting a cream pie into a zombie's face and blowing its head off. Such actions raise the questions of whether there are more or less "legitimate" forms of killing—an issue that became uncomfortably relevant during the Vietnam War and has haunted foreign policy in conflict zones ever since.

Cultural critic Richard Slotkin argues that killing is an integral part of the "mythology of the American frontier," where it takes on a regenerative, even redemptive quality. That is of course also a feature of Judeo-Christian mythology, not just in Christ's crucifixion but in God's command to Abraham that he sacrifice his son Isaac. These are rituals of purification through violence and death—which, in the form of the death penalty, the United States (among other nations) has never quite relinquished. The zombie myth is in some respects *all about killing*. The zombie is an unclean being onto which taboos like cannibalism and familial homicide can be projected—and which may then be justly slaughtered to exorcise them. In some respects the zombie is a scapegoat, in the original sense of the word.

But it is not as simple as that. Killing never is. For if we too become zombies, then we too are fair game for such brutality. The stark final frames of *Night of the Living Dead* insist that encounters with zombies are inherently brutalizing: we shoot first and ask questions later. "Whoever battles with monsters," Nietzsche wrote in *Beyond Good and Evil*, "had better see that it does not turn him into a monster."

Part of the function of all this violence is surely, too, the function of horror in general: it serves up what James Twitchell calls an "exaggerated reality . . . fantastic, ludicrous, crude, and important distortions of real life situations." By amplifying our fears to grotesque extremes, it may serve a cathartic function. The catharsis offered by *Dawn of the Dead* is not so different from that supplied by the self-mutilation of Oedipus or Theseus butchering the Minotaur.

Yet there's a difference contained in that very word: horror, Twitchell argues, is different from the terror invoked by Grendel or Medusa (see page 255). It is this difference that makes horror "quintessentially modern":

> One of the shifts in sensibility that has informed the modern world
> has been the acknowledgement that horror is not just an aspect of
> human experience but a central part of it. . . . Darwin and Marx and
> Einstein and especially Freud made modern horror not just possible,
> but unavoidable.

Horror, as opposed to terror, is a permanent condition. Grendel creates terror, but he can be killed: we can define monstrosity as distinct from normality, and destroy it. With Dracula that's no longer so: you can kill him, but there will always be other vampires. The same with the zombie: it is the creature of nightmare that you have killed but which refuses, with awful injustice, to stay dead.

This is why we need the modern myth of the zombie. Yes, we have scratched the itch of horror already with *Dracula*, *Jekyll and Hyde*, and all the others, but it hasn't gone away and has more dimensions than those earlier tales can accommodate. We will keep coming back to horror tales, says Twitchell, "until we resolve whatever it is in these myths that is unresolved within ourselves."

***

If we want to indulge in what may well be a fool's game and try to anticipate myths yet to come, we might consider what forms of psychic unease new technologies are generating. Contrary to common belief, the advance of science and technology does not compress the boundaries of what is considered superstitious and irrational: rather, it creates new habitats where ghosts can flourish.

We saw already how the Frankenstein myth is perpetuated and modified by the advent of technologies that intervene in life and procreation: assisted conception, genetic editing and human enhancement, cloning, stem-cell experimentation. Scientists themselves often encourage the perception that their efforts will realize age-old dreams once confined to the realms of myth and magic: immortality, the transmigration of self and consciousness, animal-human chimeras, time travel, invisibility, the artificial creation of life. The new dreams and nightmares that may arise from future technical advance include the multiplication of selves (a variant, one might say, of the old doppelgänger narrative; the possibilities mooted here include cloning and "mind-uploads" onto computer systems); telepathy and mind reading, disembodied consciousness and state-governed mind control; and the acquisition of superhuman powers of mind and body. In all of these areas there are already signs of nascent myth-formation.

Where will myths appear? In literature, for sure: the novels of Philip K. Dick and J. G. Ballard are particularly fertile places to look. But if lack of authorial control is a precondition for the polymorphic character of the modern literary myth, there is now the possibility of dispensing with the author altogether. Collective cultural activities may not just shape and guide but actually generate new myths. Take the genre of storytelling called "creepypasta": horror tales that propagate on the internet, and often take that medium itself as their subject matter. In these tales, people may find unbranded laptops that show them awful images of themselves, or play computer games that surreptitiously invade their minds. The genre already has its own mythical monsters, such as Slenderman: preternaturally thin and tall, faceless, and threatening to inflict something very nasty indeed. Can such an anarchic process of creation produce a narrative robust enough to qualify as a modern myth? That's not yet clear, but movies such as Hideo Nakata's *Ring* (1998), the Facebook horror *Unfriended* (2014), and the lockdown Zoom-call horror *Host* (2020) confirm that today's social networks (and their precursors) tap into profound anxieties that demand tales of their own.

I don't by any means take it for granted that the modern myths of the future will, like all of those I have considered here, be gestated in Western culture. Indeed, I think that unlikely. Modern anxieties about (say) sexuality and gender, race and class, authority and surveillance, reproduction and death, identity and autonomy, individualism and collectivity, time and space, will be refracted in different directions by different cultures, and the stories they produce will be all the richer—and more ambiguous and multivalent—because of it. *Ring*, as well as Yeon Sang-ho's 2016 zombie movie *Train to Busan*, Bong Joon-ho's *The Host* (2006) and *Parasite* (2019), and the entire traditions of manga and anime already make it clear that myths coming out of or through east Asia are going to look very different and hit different targets.

All we can say for sure is that mythmaking will continue. We will do it together, and we will not know that we are doing so until it happens. For we do not choose our myths; they choose us.

❧
Chapter 10

THE MYTHIC MODE

Fantasy has always been a problem. Just as "myth" has come to connote something that is believed yet untrue, so "fantasy" suggests in colloquial speech a foolish delusion. Neither type of narrative is quite respectable: they dupe fools and entertain children.

All of the modern myths I have chosen might be considered to fall, in the broadest sense, under the rubric of fantasy. Even *Robinson Crusoe*, although it includes nothing that might not conceivably have taken place in the eighteenth century, relates a romantic and unlikely tale and was widely relegated to the nursery or the schoolroom.

It's no coincidence that many of these new myths date from the late nineteenth century, for among the major social transformations of this time was a sea change in the forms of the public imagination. New media and modes of discourse offered people novel ways to be enchanted: not, as in former times, via tales drawn from folk superstition, but through conscious and willed engagement with texts that were known to be mere invention. Cultural historian Michael Saler identifies the emergence of a "rational enchantment," in which faith in science and reason did not preclude a suspension of disbelief. Such wonders might be found in novels (widely available now as cheap paperbacks), magazines, music-hall performances, and that most marvelous, enchanting, and disorienting of media, the moving picture. It was at this time too that a true children's literature (that was not just thinly disguised moralizing) became established—and plenty of adults, whether reading to their children or for themselves, joined the trip down the rabbit hole. Out of the Gothic novel and the picaresque fiction of the early part of the century emerged the horror and ghost story, science fiction, fantasy that was not allegorical or satirical (like *Gargantua and Pantagruel* or *Gulliver's Travels*) but meant to be taken at face value. Saler suggests that the public began to enjoy the business of "artful deception": to allow themselves to be taken in. This process had roots in the late eighteenth century with the projection shows of phantasmagoria and theatrical dioramas like the Eidophusikon of Philippe-Jacques de Loutherbourg in London's Leicester Square; it continued with nineteenth-century magic shows at the Egyptian Hall in London and the Théâtre Robert-Houdin in Paris, and culminated with the elaborate, illusionistic early movies of Georges Méliès and George Albert Smith. Writers, artists, and directors created "alternative worlds of wonder," and audiences were delighted.

Gothic fiction was the prototype for these voyages into the fantastical. According to literary scholar David Punter, the Gothic novel "enacts psychological and social dilemmas": precisely the function I am attributing to myth, both historically and today. "Gothic takes us on a tour through the labyrinthine corridors of repression, gives us glimpses of the skeletons of dead desires and makes them move again," Punter says. In doing so,

> It both confronts the bourgeoisie with its limitations and offers it modes of imaginary transcendence, which is after all the dialectical role of most art. Gothic fiction demonstrates the *potential* of revolution by daring to speak the socially unspeakable, but the very act of speaking it is an ambiguous gesture.

That ambiguity is at the heart of myth, which dares to show us things that we might desire but should not, and things that should repel us but do not.

This is not just an important function of literature; it is a vital one. And so we must ask why this form of writing—whether Gothic, fantastical, science-fictional, or what we might call imaginative and speculative in the broadest sense, and which I propose to call the *mythic mode*—has been so often been marginalized, mocked, or ignored by critics (while being voraciously consumed by the public).

The sort of work I have in mind here is not so far from Carl Jung's conception, proposed in *Modern Man in Search of a Soul* (1933), of the "visionary novel." In distinction from the "psychological novel," says Jung, this kind of story

> derives its existence from the hinterland of man's mind . . . it is a
> primordial experience which surpasses man's understanding, and
> to which he is therefore in danger of succumbing . . . [it] allow[s] a
> glimpse into the unfathomed abyss of what has not yet become.

There is, as we see, an old tradition of this kind of writing, and it weaves through the fantastical: from *Frankenstein* to the Morlocks, Gregor Samsa awakening as a giant cockroach, citizens bred in hatcheries, replicants hunted through rain-drenched cities. The mythic mode is challenging and unsettling because that is the nature of the questions with which it grapples. The early Gothic novels, says Punter, arose in a culture that was becoming aware of injustice, globally and domestically. It was a society increasingly bewildered by the pace at which technology was displacing ideas of naturalness, and urbanization and industrialization were transforming how we live and how we create. What emerges from this literature are themes of paranoia, manipulation, domination, degeneration, and the loss of identity. Gothic, Punter argues, "is a form of response to the emergence of a middle-class-dominated capitalist economy." No wonder stories like *Jekyll and Hyde*, *Dracula*, and *The War of the Worlds* were needed to deal with it.

Another reason the mythic mode has been sidelined is that it often does not produce conventionally fine writing. Frankly, it is sometimes quite the opposite, as we have seen. Even when the style is competent enough, as in Shelley, Stevenson, or Conan Doyle, these books have a crooked architecture, full of holes and cracks. They don't quite make sense; the characters are either shallow or inconsistent; the proportions are all wrong. As Punter says, they "do not come out right." And the reason, he says, is because they deal in psychological areas which do not come out right, they deal in those structures of the mind which are compounded with repression rather than with the purified material to which realism claims access.

The crudeness of the tools and the execution is easily mistaken, by arbiters of taste, for poverty of ambition or intent. In her 1965 essay on science fiction films "The Imagination of Disaster," Susan Sontag makes the claim that "there is absolutely no social criticism, of even the most implicit kind, in science fiction films." No criticism, she adds,

> of the conditions of our society which create the impersonality and dehumanization which science fiction fantasies displace onto the influence of an alien It. Also, the notion of science as a social activity, interlocking with social and political interests, is unacknowledged.

To reach such a conclusion—even if it were based on a viewing of nothing more than the monster B-movies of the 1950s—a critic as smart as Sontag would surely have to be viewing through a near-impenetrable filter of preconception and prejudice. For fantasy, and the mythic mode more generally, is (at its best) inherently political, precisely because it shows us how things might be other than what they are—or than what we might like to think they are. "The fantastic," writes literary scholar Rosemary Jackson, "traces the unsaid and the unseen of culture: that which has been silenced, made invisible, covered over and made 'absent.'" She reminds us that the word *phantasticus* has as its root meaning "to make visible/manifest." It excavates what was buried and holds it out to our fascinated, horrified gaze.

Fantasy, Jackson adds, "takes the real and breaks it," for example by exaggerating or distorting nature. That's why fantastical fiction, as well as confronting us with inescapable and perhaps uncomfortable truths about the human condition, can offer alternatives not available to literary realism. We face death, but in fantasy we can overcome it (and ask if we're really better off as a result). We can change our sex, or extinguish it, or mutate it into new forms. We can see the far future and the end of time. In those situations, realism is not only unnecessary but positively obstructive or silly.

"The principle value of studying fantasy fiction is to provide us with a 'negative psychology,'" says Punter: "access to the denied hopes and aspirations of a culture." Without that access, we are finished. "Those who refuse to listen to dragons," writes Ursula Le Guin, "are probably doomed to spend their lives acting out the nightmares of politicians."

The idea that there is a mythic mode of storytelling is nothing new. As historian of religion Bruce Lincoln has argued, it can be considered a part of the very definition of myth: "It is possible—and useful—to consider myth not as *content* but a *style* of narration." If so, it was arguably the original mode of all literature—which the realism of the eighteenth-century novel usurped.

I believe it's possible to trace the divergence of the mythic and realist modes of story-making in the literature to the early nineteenth century: with deliberate oversimplification, I suggest that the choice was between Mary Shelley and Jane Austen. Shelley harked back to a tradition in which plot dictated character, whereas the emergence of literary realism in the eighteenth century presented characters as particular people reacting in a plausible and well-motivated manner to particular circumstances in life, happening in well-defined places at specific times. Causation mattered: past events and traits of character, not inexorable fate or the demands of the plot, determined what happens next. The result is a narrative that passes itself off as an authentic account of actual events. Clearly, *Frankenstein* doesn't fit that mold.*

Don't blame me for that simplistic image; Austen implied it herself. In *Northanger Abbey* she satirizes the Gothic novel, suggesting that it

---

* Daniel Defoe is generally considered to be one of the first authors writing in this realist mode, but he does so in a way that is far from fully formed: we saw how loose the causality is, and how tenuous the plausibility, in *Robinson Crusoe*.

serves only to fill the heads of readers like the naive Catherine Morland with foolish delusions about real life. The arrogant John Thorpe in Austen's book dismisses perhaps the most famous Gothic writer of the day, Ann Radcliffe, saying that "her novels are amusing enough; they are worth reading; some fun and nature in them," while the Gothic aspects of Frances Burney's *Camilla* are dismissed as "such unnatural stuff! . . . the horridest nonsense you can imagine."

For sure, Austen is having some fun here at the expense of the smug people who regard novels of any sort as *infra dig*. But then there are novel and novels, she implies—and the right sort are those "in which the greatest powers of the mind are displayed, in which the most thorough knowledge of human nature, the happiest delineation of its varieties, the liveliest effusions of wit and humour, are conveyed to the world in the best-chosen language." That doesn't sound much like Ann Radcliffe or Mary Shelley. Austen does not want to insist that Gothic novels are worthless—but they are, she implies, ultimately juvenile, or at least you need to revert to your juvenile nature to appreciate them. To develop into adulthood and be ready to marry gallant Henry Tilney, the rather silly Catherine must learn to leave behind the wild fantasies encouraged by her poor reading habits.

It would be wonderful to know what Austen would have made of *Frankenstein*, but of course we never shall; she died in 1817. I don't imagine her opinion would have been very favorable. But this is not a question of whether Austen herself looked down on the Gothic; the issue is rather the contrast between critical opinion of her work and of Shelley's. Austen is rightly regarded as one of the finest proponents of the astutely drawn character; Shelley has sometimes barely been admitted as the author of her own book, which was routinely derided for its sloppiness.

If Shelley had written the books of Austen—or vice versa—I like to imagine the result would have been something like *Wuthering Heights*, which historian and writer Kathryn Hughes calls with good justification a "hot mess." The book "makes too much noise and not enough sense," she asserts, and you can see what she means. But that's only a bad thing if you're inclined to measure it against the template set by Austen and not that of Shelley.

Emily Brontë's reviewers seem to have done more or less that—as Hughes points out, they probably saw *Wuthering Heights* as a "willfully retro" Gothic throwback. "The incidents are too coarse and disagreeable to be attractive," said the reviewer of *The Spectator*. "It is wild, confused, disjointed, and improbable," said the *Examiner*, while the *Athenaeum* complained that it dwelt on acts "the contemplation of which true taste rejects." These judgments are reminiscent of those passed on *Frankenstein* (and there are few more explicit descendants of Frankenstein's creature than Heathcliff). Brontë, reared on the likes of E. T. A. Hoffmann, was not (quite) writing a modern myth, but she was writing in the mythic mode, and by midcentury it was deemed unseemly and unserious, and probably uncanny too. Readers still love it, but honestly, you readers will like *anything*. The critics will take George Eliot any day.

Robert Louis Stevenson was stung by this developing literary hierarchy. He defended "romantic" fictions, of the kind associated at the turn of the century with himself, Rudyard Kipling, Rider Haggard, Arthur Conan Doyle, and H. G. Wells,* against the perception that it was all rather childish in contrast to the "grown-up" psychological realism of Henry James. The value of art, he argued, lay not in how closely it resembled real life but in how it *differed*. His friend the Scottish critic and folklorist Andrew Lang felt that James's style of literary realism risked enshrining a morbid over-intellectualization in place of the traditions of *story*.

H. G. Wells would have agreed, and have added that a good story should be full-blooded and purposeful, not the aimless arrangement of elegant words and lofty ideas that he found in James. Wells and James became friends in the 1890s, but Wells tested that relationship sorely in his novel *Boon* (1915), a copy of which he gave to James. For it contained a searing indictment of James's style, which Wells said was populated with characters who "never make lusty love, never go to angry war, never

---

* Yes, this is not just a very male list but an *assertively* male one. I hope I have made it clear that there are serious problems associated with that fact.

shout at an election or perspire at poker." To you, Wells told James in a subsequent letter, "literature like painting is an end, to me literature like architecture is a means, it has a use." But James had the last laugh, says Andrew Glazzard, a specialist on late-Victorian fiction:

> It is James who remains a canonical writer, and it was his philosophy—
> that fiction should be a morally serious representation of conscious-
> ness, privileging treatment and aesthetic form over theme and message,
> and making the reader work to access its subtleties—that shaped mod-
> ernism, as well as the analytical methods of the emerging academic
> discipline of English literature. Wells, by contrast, is usually typecast as
> a pioneer of science fiction, neither more nor less.

And didn't he know it! He smarted under the condescension of the literary set with whom he had an aptitude for falling out. His friendship with Ford Madox Ford soured too, and the latter wrote with cruel wit of how "we intelligentsia snorted with pleasure at the idea of a Genius whom we could read without intellectual effort." Mr. Wells, said Ford, "went snooping about the world, emitting from time to time a prophecy in the form of an entertaining and magnificently machined gem of fiction." And if those prophecies came true, he added—the "million-wise production of everything, universal sterilization, the asphyxiation of tens of thousands, the razing from the earth of whole cities," it was "largely because Mr. Wells had prepared our minds for those horrors."

That, then, was where the mythic mode was apt to lead: to a world of crude and ugly materialism. For James, in contrast, "It is art that makes life, makes interest, makes importance" — "art" here being the sort of refined artifact that he himself produced.

James Twitchell says that even until the 1980s, literary scholarship took the view that the Gothic novel of the late eighteenth and early nineteenth centuries was "a debasement of the great novelistic tradition in Britain," until the likes of Dickens and Trollope revived it. As the canon of English literature became increasingly curated by academics from the late nineteenth century on, popular culture was denied any say in the matter. Theorists simply didn't know what to do with books like *Frankenstein, Dracula,* and *Jekyll and Hyde*—they lacked the critical tools.

These stories defy explanation, even when they pretend to supply it. Yes, Frankenstein tells us the genesis of his creature, and Jekyll (eventually) of his. Yes, Van Helsing expounds at length on the vampire, and the narrator of *The War of the Worlds* gives us a description of Martian physiology almost suited to the pages of *Nature*. Batman's origin myth is repeated in every new telling. But despite all this exposition, *nothing really gets explained*—because modern myth operates beyond the rational. The features necessary for conventional critical appraisal are missing: the plots in themselves are shallow, poorly formed, and badly proportioned, and the characters are flat. You might get away with it if, like Kafka, you were careful to keep the monsters (mostly) offstage. But to the extent that Melville or Conrad employed the mythic mode of writing, they risked being regarded as a kind of upmarket Rider Haggard, Kipling, or even—perish the thought—Edgar Rice Burroughs.

In this view, what distinguished "serious" literature was the beauty of its prose and the psychological sophistication of its characters. One of the reasons *Robinson Crusoe* came to be regarded as a children's book is that Crusoe himself was considered (with justification, it must be said) immature—indeed, one might almost call him schematic, as he seems not to age or mature at all in the twenty-five years he spends on his island. Even Dickens, drawn as he was to characters often at the extremes of age who do not show much character development, was seen in the early twentieth century to be best suited to juvenile tastes.

By the latter part of that century, the pecking order was firmly established. The mythic mode might pay well for big-name writers of fantasy, horror, and science fiction,* but the realist mode was the only route to critical success. The editors of a book of critical essays on *Frankenstein*

---

* There's a lot to be said for philosopher Noel Carroll's assertion of a close kinship between horror and sci-fi, whereby the supernatural forces of the former are substituted by the futuristic technologies of the latter.

in 1979 expressed a fear that the volume might be seen as a self-parody of the academic gravitas conventionally awarded to the finer works of literature. Even now, scholars who specialize in vampire literature or the Batman oeuvre may find that their work elicits amusement or disdain. At the start of the 1990s, wrote media scholars Roberta Pearson and William Uricchio, "discussions of *Detective Comics* and toy batmobiles still seemed to elicit more snickers than serious reflection."

Author Philip Pullman gives a slightly tongue-in-cheek definition of the difference between genre fiction and literary fiction: they are, respectively, works "where the story is more important than the words and those where the words are more important than the story." He admits that he had aspirations to writing the second kind of book as a young writer, but somewhat to his chagrin found himself writing fantasy—taking a place among the "vulgar storytellers . . . among the fluff under the bed of literature." But amid this fluff there are stories that do much more than telling us a gripping yarn (although that is an honorable tradition in itself). Pullman's *His Dark Materials* trilogy is a part of that select group—although as we have seen, even the pulp horror shockers, formulaic fantasy romances, and routine gumshoe thrillers may sometimes impart flavor and nutrition to this stew.

Pullman reminds us of G. K. Chesterton's words: "Literature is a luxury; fiction is a necessity." But the significance of the mythic mode is not just about a universal thirst for story, or at least it does not *start* from there. Rather, writing in this manner can provide us with *tools for living*. Without tying us to specifics of place or person (or even of physics), it offers stories that are *good for thinking with*. These tales are essential because they help us deal with the irresolvable dilemmas of being human.

Ursula Le Guin understood this, and one of the reasons both why she was marginalized as a science-fiction writer in her early career and why she is revered now is that she sought to mobilize the possibilities of that genre for doing more than indulging in entertaining futurism or space opera. In asserting that "science fiction is the mythology of the modern world," she wasn't content to accept this tidy formulation as more than a "half-truth." Most of its attempts at mythmaking, she complained, involved "blond heroes" and stereotypically evil/good aliens (does that remind you of "a galaxy far, far away" . . . ?), whereas "true myth arises only in the process of connecting the conscious and the unconscious realms."

Philip K. Dick was another of the sci-fi writers able to do this, in his case impelled in later life by the kind of transcendent vision that in earlier times fueled religious mythmaking. Through his work, Le Guin wrote, "one can follow an exploration of the ancient themes of identity and alienation, and the fragmentation of the ego":

> The fact that what Dick is entertaining us about is reality and madness, time and death, sin and salvation—this has escaped most readers and critics. Nobody notices; nobody notices that we have our own home-grown Borges.

For most of their writing lives, authors such as Ballard, Angela Carter, and Le Guin herself fared little better. But things are changing.

The great literary critic Harold Bloom fretted that one day "what are now called 'Departments of English' will be renamed departments of 'Cultural Studies,' where Batman comics, Mormon theme parks, television, movies, and rock will replace Chaucer, Shakespeare, Milton, Wordsworth, and Wallace Stevens."* I doubt anyone who cares about literature would want to substitute one for the other, however. What's happening is instead a greater recognition of the broad functions stories can and should serve.

"Genre fiction" is now moribund as a meaningful label, with its attendant implications of status and significance. Writers ranging from

---

* Bloom's 1959 book *Shelley's Mythmaking* might seem to suggest that he had a wider horizon until you realize it is about Percy, not the Shelley who really *did* make a myth.

Michael Moorcock and Ursula Le Guin to Kazuo Ishiguro, Jeanette Winterson, David Mitchell, and Marlon James have argued that there should be no division, let alone hierarchy, between "literary" and "genre" writing, and have imported fantastical, horror, and science-fictional tropes into the literary world as though it was the most natural thing. Others such as Art Spiegelman and Neil Gaiman have helped break down preconceptions about the audience and ambition of the "comic strip"—which, now rebranded as the graphic novel, has finally gained (not without some grumbling) a degree of critical respectability.

In my lifetime I have witnessed the reevaluation of writers such as Ballard, Dick, and Margaret Atwood: once firmly typecast as purveyors of speculative and science fiction, they are now seen as astute commentators on the social and psychic upheavals brought about by technological change. There was nothing "literary" about Ballard: the psychology of his characters was wafer-thin (or perhaps it's better to say, you simply couldn't see beneath the surface), the narrators often seemed anonymous and interchangeable, and his drowned, crystalline, or high-rise dystopias didn't seem much to resemble the real world. Yet it's ironic that critical opinion began to change at the same time that climate change, urbanization, the destructive effects of globalization, and now the crisis of a global pandemic made those visions suddenly turn up on the news channels—because *that was never the point anyway.* Ballard wasn't writing realist fiction about imagined futures; he was writing mythic fiction about the psychic present, and confronting us with the ruptures that modernity has created in our notions of cause and effect, agency, right and wrong. "The world of assassinations and high-speed car crashes which he depicts," writes David Punter, "has ceased to be amenable to interpretation in terms of natural laws. . . . New chains of connection are needed."

Whether this shift in perception of what matters in literature has anything to do with the overdue critical acceptance of female writers beyond a small, conventional canon, I hesitate to say. But I suspect it played a part. Le Guin's quiet yet fiercely intelligent fantasies were long overlooked in the predominantly male field of science fiction, and Doris Lessing's forays in that field were viewed with perplexity bordering on dismay and ridicule. It took a writer who was bold, rude, and uncaring about critical approval—Angela Carter—to undermine the traditional foundation of critical opinion. Using the mythic mode, Carter wrote a kind of fairy story rather than actual myths—and she woke us up to the fact that, like *Crusoe* and *Batman*, these tales were not (just) for children after all. That only one of the modern myths in this book has come from the work of a woman seems a little less troubling now that the likes of Atwood, Winterson, Carmen Maria Machado, Daisy Johnson, Naomi Alderman, and Samanta Schweblin, as well as, in other media, Jane Campion, Lynne Ramsay, and Julie Taymor are at work. I anticipate that the future of modern myth will be dark, disturbing, and predominantly female.

The key to the mythic mode is ambivalence. It's easy enough to conceive of stories where a hero(ine) fights back against fascist robot overlords or tentacled demons. It's quite another to make those monsters Miltonian—to make us question whether it is the hero we should be cheering on. It's one thing to make the demon repulsive, another to lace that repulsion with sexiness. And it works best when this ambivalence is not just illustrated by but actually *felt* by the author—just as people once felt (and I suspect many still do feel) ambivalence toward their gods. Where there is no ambivalence, there is no mythmaking but only another installment of the *Star Wars* franchise.

The decision to embrace the mythic in fiction should not come just from an egalitarian opposition to cultural snobbery, important though that is. There is plenty of terrible writing, and still more awful filmmaking, in horror, adventure, fantasy, and science fiction. But there is no shortage of poor work among novels that use the realist mode either. Instead of being bound to old conventions of genre, we should acknowledge the mythic mode as a vehicle with an ancient heritage that reveals the psyche of a culture, exposing our most profound and often our most disturbing dreams and fears. Today we are baffled by technological change, wired to machines and surrounded by surveillance, threatened by global catastrophe, challenged by artificial intelligence and deep-fake reality shifts, our modes of existence and reproduction and even our concept of self reconfigured by biotechnology. It seems likely that we will be needing our new myths more than ever before.

ACKNOWLEDGMENTS

I now know what a labor of love is. I hope this particular progeny has not emerged too hideous, but certainly the gestation and birth were long and painful. (Of course a man should not really be speaking in such metaphors, but we have delved so deeply into the realm of the fantastical that I hope they will not seem wholly inappropriate.) My agent Clare Alexander showed more patience and perseverance than I can reasonably have expected by allowing me to write the book I wanted to write, and arguing the case for it. And it would not have appeared if it were not for the trust and belief that my editor Karen Merikangas Darling has shown in me. Seldom too have I found such relief in finding someone who understands what you're trying to achieve as I did when I saw the supportive comments of Jeffrey Kripal, who also offered some excellent advice. What kept my belief in the project alive too were the many stimulating (and often challenging) conversations I have had on the subject with many friends, colleagues, and fellow writers: this is, at the very least, evidently a topic on which pretty much everyone can and will have views. Amanda Rees was particularly generous and insightful in her comments on the proposal (there can be no one better to ask about science fiction, in particular). My copyeditor Joel Score at the Press has done that thing authors always hope for but much more rarely get, which is to leave the voice intact while making you sound smarter and more eloquent than you are. I'm deeply grateful to them all, and to friends and family who have offered safe harbor in strange and disturbing times.

Philip Ball
London, January 2021

NOTES

CHAPTER 1

4   **10 times the speed of light.** Tim Adams, "Who's Going to Write the Algorithm for the Little White Lie?" *Guardian*, 14 April 2019, https://www.theguardian.com/books/2019/apr/14/ian-mcewan-interview-machines-like-me-artificial-intelligence.
     simply should not exist. Baldick, 1.

5   **bewildering variety of contexts.** Doty, 11–12.
     state of semantic disarray. Doty, 1.
     a story of the gods. Dundes, 49.

6   **not of the nature of fiction.** Shinagel, 289.

7   **myth is the quickest way.** Baldick, 1.

8   **no real evidence.** Doty, 13.
     cosmic framework. Johnston, 1.

9   **often convey social norms.** Doty, 23.

10   **logical model.** Lévi-Strauss, 229.
     cultural resource. Rees and Morus, 2.
     not an idle rhapsody. Dundes, 37.
     rational forces that resist. Segal, 39.

11   **real purpose of myth.** Segal, 48.

13   everything becomes possible. Lévi-Strauss, 223.
       accomplishes something significant. Segal, 5.
14   unlike the myths of Greece. Watt, *Myths*, 191.
15   Myths are not the creation. Doty, 14.
16   does not lie in its style. Lévi-Strauss, 210.
17   popular imagination. Twitchell, 51.
       myth may appear in a literary text. Twitchell, 84.
       photography, cinema, reporting. Barthes, 109.
       All versions of each text. Preston, 103.
       mythical value of the myth. Lévi-Strauss, 210.
22   the figures of Don Quixote. Cervantes, xi.
23   noble folly and base wisdom. Watt, *Myths*, 75.
       which are you, Don Quixote. Watt, *Myths*, 75.
       If we should ever see a stick. Watt, *Myths*, 73.
       very congenial. Cervantes, x.
24   Plato worried. Doty, 1.

CHAPTER 2

29   Man cloth'd in Goat-Skins. Rogers, 91.
       He had with him his Clothes and Bedding. Rogers, 92.
30   We may perceive by this Story. Rogers, 96.
31   a Seditious man. Defoe, xiii.
32   plunderers. Earle, 10.
       The stream of his writing. Shinagel, 347.
33   emblematick history. Watt, *Myths*, 164.
35   to write down only. Defoe, 89.
38   the English Ulysses. Seidel, 10.
       the true prototype. Seidel, 11.
       Everything appears as it would. Shinagel, 286.
40   shrewd, middle-class. Grapard and Hewitson, 103.
       an universally popular book. Shinagel, 274.
       I have no doubt. Shinagel, 274.
       so far beyond what was natural. Shinagel, 270.
       unable to live in society. Aristotle, *Politics*, book 1, trans. B. Jowett,
       1253a, 27–29.
41   most suited to human happiness. Defoe, 6.

41   especial favourite with sea-faring men. Michals, 55.
      written in a phraseology. Shinagel, 270.
      like reading evidence. Watt, *Novel*, 34.
42   man-child re-invents the world. Barthes, 65.
      the idea of an island Paradise. Skal, *Blood*, 39.
      Nothing upon earth can be. Campe, iv.
43   state of perfect isolation. Shinagel, 271.
      worn out six stout *Robinson Crusoes*. Wilkie Collins, *The Moonstone*
      (New York: Century Co., 1906), 13.
44   My Island was now peopled. Defoe, 203.
45   without any Teacher or Instructor. Defoe, 186.
      Far from punishing the prodigal. Shinagel, 375.
      quest-romances of the post-Enlightenment. Shinagel, 383.
46   an architect, a carpenter. Shinagel, 323.
47   I had done wrong in parting. Defoe, 31.
      I made him know his Name. Defoe, 174
      If Providence were to watch. Coetzee, 23.
49   a most faithful Servant. Defoe, 235.
      the shrewd vigorous character. Grapard and Hewitson, 27.
50   broad-shouldered, beef-eating John Bull. Grapard and Hewitson, 27.
51   every West Indian child. Walcott, 37.
      has become the property. Walcott, 35.
      the figure of Crusoe. Walcott, 36.
      analysing figures from literature. Walcott, 37.
52   470 Moidores of Gold. Defoe, 237.
53   Moderate though he be. Karl Marx, *Capital*, vol. 1, trans. Samuel Moore
      and Edward Aveling (Champaign: Modern Barbarian Press, 2018), 50.
54   isolated economic man. Michals, 49.
      community of free individuals. Michals, 49.
      if the Crusoe model is correct. Grapard and Hewitson, 38.
      mythical description of international trade. Grapard and Hewitson, 42.
      From each according to his ability. Karl Marx, *Critique of the Gotha
      Program* (1875).
55   This plain Man's Story. Shinagel, 238.
      surest way to rise superior. Campe, v.
      unnecessary rubbish. Campe, v.
      The practice of simple manual arts. Shinagel, 263.
56   best furnished toy-shop. Shinagel, 263.

56   I had never handled a Tool. Defoe, 59.
      vision of a happy nightmare. Seidel, 9.
      nothing is done. Shinagel, 268.
      utility ought to be preferred. Shinagel, 266.
      Say "Robinson Crusoe." Grapard and Hewitson, 27.

57   no book in English literary history. Robert McCrum, "The 10 Best
      Novels: No. 2," *Guardian*, 23 September 2013, https://www.theguardian
      .com/books/2013/sep/30/100-best-books-robinson-crusoe.

59   We're English. Golding, 42.

60   I should have thought. Golding, 201–2.
      He above all the others. Lois Lowry, afterword in Golding, 206.

61   Pacific atoll may not be available. Ballard, vii, http://www.ballardian
      .com/introduction-to-concrete-island.
      Already he felt no real need. Ballard, 176.
      Perhaps, secretly, we hoped. Ballard, viii.

63   translated from the lunar language. Daniel Defoe, *Consolidator* (Lon-
      don: Benjamin Bragg, 1705), title page.

65   [his] survival is not purely physical. Walcott, 40.

66   I need 1500 calories every day. Weir, 18.

67   Every human being has a basic instinct. Weir, 368–69.

                              CHAPTER 3

69   China's Frankenstein. Jon Cohen, "The Untold Story of the 'Circle of
      Trust' behind the World's First Gene-Edited Babies," *Science*, 1 August
      2019. https://www.sciencemag.org/news/2019/08/untold-story-circle
      -trust-behind-world-s-first-gene-edited-babies.

70   perfect myth of the secular. Levine and Knoepflmacher, 30.
      All Mrs Shelley did. Cornwell, 7.

73   beginning to seriously educate myself. Shelley (2012), 427.
      My education was neglected. Shelley (2012), 8.
      Find my baby dead. Mellor, 32.

76   young and inexperienced writer. Shelley (2012), 417.

77   I never was much more disgusted. Frayling, 12.
      Poor Polidori. Shelley (2012), 167.
      speak to the mysterious fears. Shelly (2012), 167.
      When I placed my head on my pillow. Shelley (2012), 168

78   an affection for [*Frankenstein*]. Shelley (2012), 169.
      glanc[es] about everywhere. Philmus, 84.
79   awkwardly written. Twitchell, 165.
      creating out of the void. Shelley (2012), 167.
      Whence, I often asked myself. Shelley (2012), 31.
      spend days and nights in vaults. Shelley (2012), 31.
80   cause of generation and life. Shelley (2012), 32.
      jiggery-pokery magic. H. G. Wells, *The Scientific Romances of H. G.
      Wells* (London: Victor Gollancz, 1933), viii.
      workshop of filthy creation. Shelley (2012), 34.
      on a dreary night of November. Shelley (2012), 35.
      dull yellow eye. Shelley (2012), 35.
81   I thought I saw Elizabeth. Shelley (2012), 36
82   The whole series of my life. Shelley (2012), 128.
      absence of motivation. Barthes, 126.
      utmost extent of malice and treachery. Shelley (2012), 119.
      I will be with you on your wedding-night. Shelley (2012), 121.
83   my friend and dearest companion. Shelley (2012), 132.
      My form is a filthy type of yours. Shelley (2012), 91.
      final and wonderful catastrophe. Shelley (2012), 158.
      exclamations of grief and horror. Shelley (2012), 158.
      I was the slave. Shelley (2012), 159.
84   I shall die. Shelley (2012), 161.
      soon borne away by the waves. Shelley (2012), 161.
      I was created. Shelley (2012), 90.
85   Did I request thee, Maker. John Milton (1667), *Paradise Lost*, book 10,
      743-45.
      principled opposition to tyranny. Baldick, 41.
86   Our taste and our judgement. Shelley (2012), 218.
      whether the head or the heart. Shelley (2012), 219.
87   they would make a great improvement. Shelley (2012), 236-37.
      bordering too closely on impiety. Shelley (2012), 231.
      Upon the whole, the work impresses. Shelley (2012), 231.
      Frankenstein, by Godwin's son-in-law. Shelley (2012), 250.
88   overload[s] the novel. Baldick, 56.
      Frankenstein was the archetype. Aldiss, 47.
89   detailed and reverent description. Mellor, 89.
      hubristic manipulation. Mellor, 89.

90   that branch of natural philosophy. Shelley (2012), 29.
      new and almost unlimited powers. Shelley (2012), 29.
      His gentleness was never tinged. Shelley (2012), 30.
      a great deal of sound sense. Shelley (2012), 30.

92   to modify and change the beings. Mellor, 93.
      nature of the principle of life. Shelley (2012), 168.

93   who preserved a piece of vermicelli. Shelley (2012), 168.
      a corpse would be reanimated. Shelley (2012), 168.
      The jaw began to quiver. J. Aldini, *An Account of the Late Improvements
      in Galvanism* (London: Cuthell & Martin / J. Murry, 1803), 193.

94   I collected the instruments of life. Shelley (2012), 55.
      the more obvious laws. Shelley, *Last Man*, 39.
      Whence, I often asked myself. Shelley (2012), 31.
      To examine the causes of life. Shelley (2012),.

95   natural decay and corruption. Shelley (2012),.
      the corruption of death succeed[s]. Shelley (2012),.
      subtile, mobile, invisible substance. Shelley (2012), 406.

96   implicit warning. Mellor, 114.
      the pesticide DDT. Shelley (2017), xxviii.

97   Nature prevents Victor. Shelley (2017), 241.
      Dr. Frankenstein . . . in a wool sweater. Ian Wilmut, Keith Campbell,
      and Colin Tudge, *The Second Creation* (London: Headline, 2000), 246.
      There is no great invention. K. R. Dronamraju, ed., *Haldane's Daedalus
      Revisited* (Oxford: Oxford University Press, 1995), 36.

98   a McFrankenstein creation. Young, 3.
      Whatever today's embryologists may do. R. G. Edwards, *Life be-
      fore Birth: Reflections on the Embryo Debate* (London: Hutchinson,
      1989), 70.

99   failure of empathy. Shelley (2017), 229.
      In this the direct moral. Shelley (2012), 214.
      causes fully adequate to their production. Shelley (2012), 214.

100   Frankenstein is the degenerate offspring. Shelley (2012), 420.

101   despite the consensus. Shelley (2012), 420.
      My attempts to arouse the students' indignation. Shelley (2012), 420.
      if they do not draw the moral. Shelley (2012), 421.
      how many students disagree. Shelley (2012), 421.
      Despite the modern consensus. Shelley (2012), 422.

103   is but an Artificiall Man. Thomas Hobbes, *Leviathan*, ed. C. Macpherson (London: Penguin, 1985), 81.
       should approach to the faults. Baldick, 17.
       Out of the tomb. Baldick, 19.
       engendered that disastrous monster. Smith, 86.
       as a Galvanic Mass. Baldick, 105.
104   A state without religion. Baldick, 60–61.
       actions of the uneducated. Baldick, 86.
       rise up to life; they irritate us. Baldick, 86.
105   In dealing with a negro. Young, 20.
106   a monster created by a white man. Young, 3.
       I heard of the division of property. Shelley (2012), 83.
       just like the mythical Frankenstein. Young, 1.
107   The United States is extraordinarily gifted. Carlos Fuentes, interview, *Freitag 31*, 23 July 2004, in translation at http://portland.indymedia.org /en/2004/07/293671.shtml.
108   What is it that agitates you. Shelley (2012), 140.
       suddenly I reflected how dreadful. Shelley (2012), 140.
109   Frankenstein seems more aroused. Shelley (2012), 413.
       when a man tries to have a baby. Mellor, 40.
       anti-husband. Levine and Knoepflmacher, 43.
       truly fears is female sexuality. Shelley (2012), 360.
       transform the standard Romantic matter. Shelley (2012), 327.
       horror story of teenage motherhood. Levine and Knoepflmacher, 79.
110   often did my human nature. Shelley (2012), 34.
       motif of revulsion. Levine and Knoepflmacher, 81.
       the hideous corpse. Shelley (2012), 168.
       I am thy creature. Shelley (2012), 68.
       Safie related that her mother. Shelley (2012), 86.
       Dream that my little baby. Mellor, 32.
111   affords a point of view. Shelley (2012), 5.
       elementary feelings of the human mind. Shelley (2012), 213.
       It's the Devil. Smith, 177.
112   We can *create* nothing. Baldick, 180.
       mind had vanquished conscience. W. C. Morrow, *The Monster Maker and Other Stories*, http://gutenberg.net.au/ebooks06/0606131h.html.
       huge, awkward, shapeless. Morrow.

114   **This nameless mode of naming.** Bloom, 67.
        Supremely frightful. Shelley (2012), 168.
        living monument of presumption. Shelley, *Last Man*, 77.
        attributes or prerogatives of the deity. Kirby, 60.
        delves into the physically impossible. Kirby, 60.

115   **The imaginary student.** Charles Dickens, *Great Expectations* (Ware:
        Wordsworth Editions, 1992), 278.
        I don't see why we should. D. H. Lawrence, *The Rainbow* (Mineola,
        NY: Dover, 2017), 440.

116   **By the end of the century.** Shelley (2012), 262.

118   **ways, technical and psychological.** Punter, 349.
        No monster could touch. Twitchell, 182.

121   **no patience for subtlety.** Daniel Argent and Erik Bauer, "Frank Dara-
        bont on the Shawshank Redemption," https://creativescreenwriting
        .com/frank-darabont-on-the-shawshank-redemption/.

122   **almost the exit ceremony.** Twitchell, 200.
        we are all chimeras. Donna Haraway, "The Cyborg Manifesto," http://
        users.uoa.gr/~cdokou/HarawayCyborgManifesto.pdf.

123   **This creature given life.** Smith, 118.

128   **All our inventions and progress.** Karl Marx, speech at the anniversary
        of the *People's Paper*, 14 April 1856. In *Marx/Engels: Selected Works*
        (Moscow: Progress Publishers, 1969), 1:500.
        myth of Frankenstein has become. Baldick, 140.
        the machines are gaining ground. Samuel Butler, "Darwin among the
        Machines," in *The Notebooks of Samuel Butler* ed. Henry Festing Jones
        (Floating Press, 2014), 23.

129   **rule us with a rod of iron.** Samuel Butler, *Erewhon* (Harmondsworth:
        Penguin, 1970), 221.
        In all the worlds there was not one. E. M. Forster, "The Machine Stops,"
        Oxford and Cambridge Review, November 1909, https://www.ele.uri
        .edu/faculty/vetter/Other-stuff/The-Machine-Stops.pdf.

CHAPTER 4

131   **Horror in general attracts us.** Freeland, chap. 4.

132   **What myths are to the race.** Tropp, 1.

134   a fine bogey tale. Twitchell, 234.
      pursued for some crime. Stevenson (ed. Luckhurst, 2006), 160.
      some Brownie. Stevenson (ed. Luckhurst, 2006), 159.
      do one-half my work. Stevenson (ed. Luckhurst, 2006), 159.
      was conceived, written. Wolf, *Jekyll*, 272.
      I cannot but suppose. Stevenson (ed. Luckhurst, 2006), 160-1.

135   created a living text. Wolf, *Jekyll*, 3.
      yet somehow lovable. Stevenson, 5.

136   trampled calmly over. Stevenson, 7.
      There is something wrong. Stevenson, 9.
      If he be Mr Hyde. Stevenson, 14.
      God bless me. Stevenson, 16.

137   This is a private matter. Stevenson, 19.
      I do sincerely take. Stevenson, 19.

138   With ape-like fury. Stevenson, 21.
      odd, upright hand. Stevenson, 25.
      O God, Utterson. Stevenson, 26.
      I am quite done with that person. Stevenson, 29.
      I have brought on myself. Stevenson, 29–30.
      an expression of such abject terror. Stevenson, 33.

139   crying night and day. Stevenson, 37.
      jumped from among the chemicals. Stevenson, 39.
      stagger the unbelief of Satan. Stevenson, 50.
      dangerously like an experiment. Oscar Wilde, "The Decay of Lying," in
      The Works of Oscar Wilde, ed. G. F. Maine (London: Collins, 1948), 912.

140   a certain impatient gaiety. Stevenson, 52.
      a profound duplicity of life. Stevenson, 52.
      that man is not truly one. Stevenson, 53.
      Certain agents I found. Stevenson, 53–54.
      The most racking pangs. Stevenson, 54.
      my new power tempted me. Stevenson, 56.

141   Between these two [selves]. Stevenson, 59.
      came out roaring. Stevenson, 60.

142   I've got my shilling shocker. Reid, 75.
      pouring forth a penny dreadful. Reid, 92.
      not merely strange. Wolf, *Jekyll*, 259–60.
      We would welcome a spectre. Wolf, *Jekyll*, 258.

143   Everything is true. Wolf, *Jekyll*, 147.
144   The mask and mantle. Washington Irving, "An Unwritten Drama of Lord Byron," *The Knickerbocker* 6 (August 1835): 143.
       how utterly thou hast murdered thyself. Poe, 21.
145   Crime belongs exclusively. Oscar Wilde, *The Picture of Dorian Gray*, in *The Works of Oscar Wilde*, ed. G. F. Maine (London: Collins, 1948), 160.
146   a species of the familiar. Freud, 147–48.
147   boundless self-love. Freud, 142.
       in my own hand blasphemies. Stevenson, 65.
       love of life is wonderfu. Stevenson, 65, 62.
       my husband was deeply impressed. Stiles, 879.
148   who live two lives. Stiles, 894.
       morbid disintegration. Ian Hacking, *Rewriting the Soul: Multiple Personality and the Sciences of Memory* (Princeton, NJ: Princeton University Press, 1995), 175.
151   Even it seemed that I. Wells, 177.
153   The harm was in Jekyll. Twitchell, 319–20.
       drinking pleasure with bestial avidity. Stevenson, 57.
       into the details of the infamy. Stevenson, 57.
154   certain gaiety of disposition. Stevenson, 52.
       I felt younger, lighter. Stevenson, 54.
       almost morbid sense of shame. Stevenson, 52.
       abhorrent in themselves. Reid, 99.
       disordered sensual images. Stevenson, 54.
       It was Hyde, after all. Stevenson, 57.
157   with a very pretty manner. Stevenson, 20.
       For us, more than a hundred years. Wolf, *Jekyll*, 142.
       it turns me cold. Stevenson, 17.
       the more it looks like Queer Street. Stevenson, 9.
158   Effeminate men. Smith and Haas, 75.
       can most persuasively be read. Smith and Haas, 69.
       more than any man I ever met. Smith and Haas, 69.
       more sympathy with our frailty. Wolf, *Jekyll*, 143.
       something not quite human. Reid, 104.
161   his jaws retain the grin. Ovid, *Metamorphoses* Book 1, 199–243.
       From this story one learns. Alan Dundes, ed., *Little Red Riding Hood: A Casebook* (Madison: University of Wisconsin Press, 1989), 73.

161 **she'd never seen.** Angela Carter, *Burning Your Boats: The Collected Short Stories* (Harmondsworth: Penguin, 1995), 216.

CHAPTER 5

168 **Supposing there be any reality.** Frayling, 29.
men of low birth. Frayling, 30.
There was no talk of vampires. Frayling, 31.

169 **Capital is dead labour.** Baldick, 130.
For nearly a century. Groom, 97.
It appears that one evening. [Polidori], "The Vampyre," *New Monthly Magazine* 11, no. 63 (1 April 1819), preface.

170 **probably one or two dead bodies.** Frayling, 7.
In spite of the deadly hue. [Polidori], available at https://en.wikisource .org/wiki/The_Vampyre/The_Vampyre.

171 **But first, on earth.** Filmer, 28.
sought for the centres. [Polidori].
you will not impart your knowledge. [Polidori].
Remember your oath! [Polidori].

172 **my bride to day.** [Polidori].
glutted the thirst of a VAMPYRE! [Polidori].
He hastened to the mansion. [Polidori].

173 **Oh, do not touch him.** [Polidori].
In the wake of Polidori's tale. Groom, 123.

174 **He is rather a troublesome acquaintance.** Auerbach, 33.

176 **the foul German spectre.** Charlotte Bronte, *Jane Eyre* (London: Heron Books, 1979), 345.
The monthly activity of the ovaries. Groom, 150.

177 **long, thin, pointed.** Lefanu, 296.
sense of exhaustion. Lefanu, 307–8.
as if warm lips kissed me. Lefanu, 307–8.
Behold! Her bosom. Auerbach, 51.

178 **less the culmination of a tradition.** Auerbach, 64.
paying extravagant lip service. Auerbach, 69.

179 **one of the most obsessional texts.** Skal, *Gothic*, 7.
probably transgressing *something*. Stoker, viii.
a kind of incestuous, necrophilious. Stoker, xi–xii.

179   a violent fit of hysterics. Skal, *Blood*, 159.

180   Soul had looked into soul! Skal, *Blood*, 161.

181   I have never seen. Skal, *Blood*, 64.
       the strongest love. Skal, *Blood*, 64.
       How sweet a thing it is. Skal, *Blood*, 98
       His heart when he was sad. Skal, *Blood*, 103

182   Poor Oscar was *not* as bad. Skal, *Blood*, 355.

183   inflated with literary pretensions. Leatherdale, 0000.

184   a semi-heroic, Everyman quality. Frayling, 79.
       We've told yarns. Stoker, 69.
       good bad book. George Orwell, "Good bad books," *Tribune*, 2 Novem-
       ber 1945, https://www.orwellfoundation.com/the-orwell-foundation
       /orwell/essays-and-other-works/good-bad-books/.
       Like *Dr Jekyll* and *Frankenstein*. Twitchell, 127.
       In the final analysis. Skal, *Blood*, 377.

185   a too-generous helping. Skal, *Blood*, 305.

186   one of the most advanced scientists. Skal, *Blood*, 122.

187   a tall old man, clean shaven. Stoker, 17.
       Listen to them. Stoker, 25.
       crawl down the castle wall. Stoker, 37.
       There are kisses for us all. Stoker, 41.

189   either dead or asleep. Stoker, 56.
       with all their blaze. Stoker, 60.
       a great number of wooden boxes. Stoker, 90.
       its throat torn away. Stoker, 91.

190   something, long and black. Stoker, 101.
       white face and red, gleaming eyes. Stoker, 101.
       in long heavy gasps. Stoker, 102.
       roses in her cheeks are fading. Stoker, 106.
       old friend and master. Stoker, 122.
       she will die for sheer want. Stoker, 131.

191   in a soft, voluptuous voice. Stoker, 172.
       There is peace for her. Stoker, 173.
       I want to cut off her head. Stoker, 176.
       more radiantly beautiful than ever. Stoker, 213.
       There in the coffin. Stoker, 231.

192   This vampire which is amongst us. Stoker, 252.
       into the jaws of Hell. Stoker, 334.

192   **it was butcher work.** Stoker, 394–95.
        the eyes saw the sinking sun. Stoker, 400.
193   **the vampire is an erotic creation.** Frayling, 387–88.
        infantile incestuous wishes. Frayling, 401.
        unaware of the sexuality. Musa, 6.
        sexual fever-dream. Skal, *Gothic*, 28.
194   **playing with a lancet.** Stoker, 65.
        let a girl marry three men. Stoker, 67.
        My soul seemed to go out. Stoker, 109.
195   **The body shook and quivered.** Stoker, 230.
        The sexual implications. Stoker, xxii.
        works of shameful lubricity. Stoker, xxiii.
        The fool was stripped. Rudyard Kipling, "The Vampire," 1897, http://www.kiplingsociety.co.uk/poems_vamp.htm.
196   **Once you give a woman.** Carol Dyhouse, *Heartthrobs: A History of Women and Desire* (Oxford: Oxford University Press, 2017), 26.
197   **The fair girl went on her knees.** Stoker, 45–46.
        A close analysis. Skal, *Blood*, 465.
        a horrible repulsive slime. Bram Stoker, *Lair of the White Worm*, chap. 28.
        almost hallucinatory misogyny. Skal, *Gothic*, 40.
198   **No one asks for probability.** Skal, *Blood*, 485.
        subliminal surprises. Auerbach, 126.
        a ferocity alien. Auerbach, 125–26.
        fed by Wilde's fall. Auerbach, 85.
199   **overwrought protestations of friendship.** Skal, *Blood*, 357.
        I have grown to love you. Stoker, 181.
        He was rather tall. Eric Stenbock, "The True Story of a Vampire," in *Classic Vampire Tales* (Mumbai: Jaico, 2016).
200   **Vardalek always returned.** Stenbock.
        Gabriel stretched out his arms. Stenbock.
        "Your vampires suck blood. Auerbach, 104.
201   **Many years later.** Auerbach, 103.
        My vampire is the Overman. Skal, *Blood*, 398.
        great nineteenth-century syphilis novel. Brian Aldiss, *Dracula Unbound* (New York: Harper, 1992), 227.
202   **I happened to encounter.** Rice, 2.
        You're spoiled. Rice, 121.
203   **I am for the moment.** Arthur Conan Doyle, "The Parasite," *Harper's*

Weekly, 17 November 1894, p. 1086, https://www.arthur-conan-doyle
.com/index.php/The_Parasite.

203   more mesmerist than biter. Auerbach, 118.

204   must feed on the emanations. Eaves, 14.
       It is only possible. Eaves, 29.
       a strain of fourth-race blood. Eaves, 19.
       There are vampires and vampires. Auerbach, 105.

205   like the multitude. Stoker, 340.
       So he came to London. Stoker, 363.
       a new and ever widening circle. Stoker, 60.
       the Adelphi Theatre type. Stoker, 371.

206   to what extent can one be. Punter, 240.

207   our next door neighbor. Luckhurst, *Companion*, 158.
       their moving dark bodies. Stoker, 269.
       The Count is a criminal. Stoker, 363.

208   a negroid of the lowest type. Skal, *Blood*, 478–80.

209   wanting . . . in the higher literary sense. Luckhurst, *Companion*, 46.
       A vampire with a cheque-book. Luckhurst, *Companion*, 46.

210   an agglomeration of points. Skal, *Gothic*, 54.

211   no momentary horror. Skal, *Gothic*, 101.

212   Valentino gone slightly rancid. Skal, *Gothic*, 85.
       dependent on commanding attitudes. Auerbach, 0000.

215   It is terrifying to see. Abbott, 49.

216   prurient debasement. Leatherdale, 9.
       The thing that really scared me. Tim Underwood and Chuck Miller,
       eds., *Bare Bones: Conversations on Terror with Stephen King* (New York:
       Warner, 1988), 5.

220   It's very clear that [Meyer is] writing. https://www.today.com
       /popculture/stephen-king-says-twilight-author-cant-write
       -wbna29001524.

221   Van Helsing is dead. Newman, 20.
       One of the things that's terrifying. Bloom, interview with the author.

CHAPTER 6

223   encased in an elaborate space suit. "Visions of Mars," Planetary Society,
       http://www.planetary.org/explore/projects/vom/.

226   a certain collective death-wish. Bergonzi, 13.

To some extent. Philmus, 22.

one of those books. Will Self, "Death on three legs," *The Times* 23 January 2010.

227   hold the reader to the end. Philmus, 32.

For the writer of fantastic stories. H. G. Wells, *The Scientific Romances of H. G. Wells* (London: Victor Gollancz, 1933), viii.

228   His work dealt almost always. Bergonzi, 17–18.

They belong to a class of writing. Bergonzi, 17–18.

They tell a story. Jorge Luis Borges, *Other Inquisitions* (Austin: University of Texas Press, 1964), 86–88.

230   as some sort of gullible upstart. Bryan Appleyard, "The plot to hide H. G. Well's genius," *Sunday Times*, 26 June 2005, https://www.thetimes.co.uk/article/the-plot-to-hide-hg-wellss-genius-jss9jgjcpjw.

Work that endures. Borges, 87.

must be ambiguous. Borges, 87.

231   the size of a football perhaps. Wells, 65.

more likely to make a reputation. Beck, 100.

232   I know this is an illusion. Wells, 177.

To carry warfare sunward. Wells, 186.

Yet across the gulf of space. Wells, 185.

revolves about the sun. Wells, 185.

235   create sympathy with the aims. "R. A. G." [Richard Gregory], *Nature* 57 (1898): 340.

236   a monstrous tripod. Wells, 218.

238   Directly these invaders arrived. Wells, 311.

It is one of the books. Beck, 144.

239   too much of the young man. Beck, 145.

knew that the future. Beck, 8.

an assault on human self-satisfaction. Parrinder and Philmus, 243.

240   lives insured and a bit invested. Wells, 301.

incredibly dismal. Parrinder and Philmus, 246–51.

241   I'm doing the dearest little serial. Bergonzi, 125.

When I had last seen. Wells, 291.

I found about me the landscape. Wells, 293.

242   We were walking together. Beck, 128.

The Tasmanians. Wells, 186, 297.

bombardment of some village. Beck, 128.

242   How would it be with us. Beck, 128.

243   as remote from. Parrinder and Philmus, 149.

244   the absolute necessity. Beck, 43–44.
       In those days I was writing. Beck, 129.

245   Whereas an allegory. Bergonzi, 42.

246   precision in the unessential. Bergonzi, 45.
       the more impossible the story. Beck, 161.

248   As fiction it is hackwork. Beck, 201.

249   was in fact nationwide! Beck, 211.

250   Throughout New York. Beck, 214–15.
       at least a million Americans. Beck, 229.
       Some ran to rescue loved ones. Beck, 229.
       I ran out into the street. Beck, 215.
       Mr. and Mrs. America. Gallop, 77.

251   a sign of the decadence. Gallop, 118.
       terror-psychosis. Beck, 216.

252   demonstrated beyond question. Gallop, 117.
       We can only suppose. Gallop, 82.
       We were fed up. Beck, 232.

254   my little namesake here. Gallop, 132.
       Such a thing as that. Beck, 224.
       No one would have believed. Wells, 185.

255   They were, I now saw. Wells, 277

258   a crime which is worse than murder. Sontag, 47.
       they took the fresh, living blood. Wells, 277.
       submarine without a conning tower. John W. Campbell, "Who Goes
       There?," https://wp.nyu.edu/darknessspeaks/wp-content/uploads/sites
       /3674/2016/09/who_goes_there.pdf.
       like a saucer if you skip it. Megan Garber, "The Man Who Intro-
       duced the World to Flying Saucers," *Atlantic*, 15 June 2014, https://
       www.theatlantic.com/technology/archive/2014/06/the-man-who
       -introduced-the-world-to-flying-saucers/372732/.

264   "Independence Day" is not just an inheritor. Roger Ebert, https://
       www.rogerebert.com/reviews/independence-day-1996.

265   After 9/11 it began. V. Ecker, "Lights! Camera! Martians!," *Fortean*, Au-
       gust 2005, http://www.forteantimes.com/features/articles/121/lights
       _camera_martians.html.
       the genius of the story. Beck, 345–46.

265   **we will not see anything.** Beck, 263.

       the déjà vu of crowds. Hoberman.

266   **The movies set the pattern.** Hoberman.

       There are just some movies. Hoberman.

       Manhattan getting the crap kicked out. Beck, 263.

       fantasy of living through one's death. Sontag, 44.

267   **The red beach.** Wells, 64.

268   **Cities, nations, civilisation.** Wells, 300.

       I go to London and see. Wells, 319.

269   **Film producers.** Beck, 258.

       dispassionate, aesthetic view. Sontag, 45.

270   **One by one.** Shelley (1998), 326.

       Surely death is not death. Shelley (1998), 413.

272   **Finished, it's finished.** Samuel Beckett, *Endgame* (London: Faber & Faber, 1964), 12.

273   **living through one's death.** Beck, 272.

       science fiction is the mythology. Le Guin, 61.

       an attempt to explain. Le Guin, 62.

CHAPTER 7

279   **Gaboriau had rather attracted me.** Hodgson, 4.

       Cobbler, I see. Sims, 10.

       From close observation. Sims, 10, 5.

280   **a fellow who is working.** Conan Doyle, *Complete Facsimile*, 12. Conan Doyle citations with no title refer to this Wordsworth Editions volume.

       a little too scientific. Conan Doyle, 12.

       You have been in Afghanistan. Conan Doyle, 13.

       The train of reasoning ran. Conan Doyle, 18.

281   **He is a very little fellow.** Poe, 383.

       In my opinion, Dupin. Conan Doyle, 18.

       a miserable bungler. Conan Doyle, 18.

       may be very clever. Conan Doyle, 18.

282   **Broad, low tables.** Conan Doyle, 13.

284   **I have a trade of my own.** Conan Doyle, 17.

285   **Sherlock Holmes took his bottle.** Conan Doyle, 64.

286   **so transcendently stimulating.** Conan Doyle, 64.
       I cannot live without brain-work. Conan Doyle, 67.
       You have done all the work. Conan Doyle, 113.

287   **You really are an automaton.** Conan Doyle, 69.
       love is an emotional thing. Conan Doyle, 113.
       a great hawks-bill of a nose. Klinger, xxxii.
       the middle of the last Holmes story. Lycett, 185.

288   **kill Sherlock Holmes.** Lycett, 191.
       He is the organizer. Conan Doyle, 437.

289   **a flimsy plot device.** Farghaly, 22.

291   **I could not but think.** Conan Doyle, 80.
       had hereditary tendencies. Conan Doyle, 436.
       I must confess to a start. Conan Doyle, 438.
       I have been much blamed. Klinger, xxxiii.

293   **Professional critics.** Klinger, xiv.
       the class of literature. Lycett, 188.
       From a drop of water. Conan Doyle, 17.

296   **interested but very skeptical.** Lycett, 125.

297   **Science will tell you.** Arthur Conan Doyle, *The Mystery of Cloomber* (Mineola, NY: Dover, 2009), 140.

298   **as solid and real as the Eskimos.** Arthur Conan Doyle, *The Coming of the Fairies* (New York: G. H. & Co., 1922), 55.
       recognition of [the fairies'] existence will. Conan Doyle, *Fairies*, 58.

299   **He tried hard to make.** Lycett, 434.

300   **Honestly, I cannot congratulate you.** Conan Doyle, *Complete Facsimile*, 65.

302   **I am a brain, Watson.** Conan Doyle, 970.
       sometimes I found myself. Conan Doyle, 399.

304   **It's the gayest story.** Farghaly, 2.

305   **sort of in love.** Farghaly, 13.

306   **always himself.** Putney et al., ix.

307   **lives in the global imagination.** Edward Docx, "The curious case of the Sherlock pilgrims," *Prospect*, November 2012, 48.
       He represents for us. Putney et al., xii.
       stands before us as a symbol. Klinger, xviii.
       neither Moriarty nor Moran. Putney et al., xii.

308   **Like all superheroes.** Docx, 48.
       Male supremacist society. Putney et al., 276.

308 assuaged again, and again. Hodgson, 378.
309 compensation for an ego. Putney et al., 321.
     The fantasy of detective fiction. Putney et al., 305–6.

CHAPTER 8

312 All these years. Pearson and Uricchio, 23.
     if you get into that. Brooker, 41.
     Batman came out of the darkness. Daniels, 17.
313 You really have to keep in mind. Pearson and Uricchio, 40.
323 His father was Nietzsche. Le Guin, 64.
324 assimilated alien. Picariello, 10.
     Never let us have Batman. Brooker, 60.
326 mothers would object. Brooker, 60.
327 forced to go on a diet. Brooker, 84.
328 a fight between two disfigured people. Brooker, 292.
     A peculiarly American form. Daniels, 41.
     a seething nuclear stockpile. Daniels, 33.
329 All I want is a little kiss! Wertham, 95.
     grow up faster'n anythin'. Wertham, 200.
     The muscular male supertype. Wertham, 188.
     Only someone ignorant. Wertham, 190.
330 I found my liking. Pearson and Uricchio, 153.
     one of the most irresponsible slurs. Brooker, 105.
332 instill the idea that girls. Wertham, 188.
     woman with graceful, appealing figure lines. Wertham, 200.
     dark-skinned people are depicted. Wertham, 103.
     flamboyant melodrama. Hatfield et al., 238.
     semiotic systems. Hatfield et al., 240.
     reading it as a gay man today. Hatfield et al., 241.
333 The homophobic nightmare. Pearson and Uricchio, 37.
334 corruption of the art of drawing. Wertham, 381.
     sex perversion. Brooker, 144.
335 If a comic book could. Daniels, 94.
337 very inferior piece of Camp. Sontag, "Notes on Camp," *Partisan Review*
     31, no. 4 (1964): 515–30, https://monoskop.org/images/5/59/Sontag
     _Susan_1964_Notes_on_Camp.pdf.

338   pure examples of Camp. Sontag, "Camp."
        I reveled in the power. Brooker, 232.
        If you'll notice, my Batman was. Daniels, 113.
340   feel secure and happy. Brooker, 230.
        original conception of Batman. Brooker, 175.
        Pop art and the cult of Camp. Bruce E. Drushel and Brian M. Peters,
        eds., *Sontag and the Camp Aesthetic: Advancing New Perspectives* (Lan-
        ham, MD: Lexington Books, 2017), 179.
341   good and bad are art's only poles. Pearson and Uricchio, 124.
342   Maybe every ten years. Daniels, 17.
343   I think we're at our best. Pearson and Uricchio, 43.
344   an extremely violent character. Pearson and Uricchio, 38.
        Heroes have to work in the society. Pearson and Uricchio, 44.
        He is the American answer. Reichstein, 347.
        the Batman mythos. Durand and Leigh, 31.
        make his devils work. Pearson and Uricchio, 44.
        castrates every hero or villain. Pearson and Uricchio, 45.
        he would be a tyrant. Pearson and Uricchio, 44
        has a lot to tell us. Durand and Leigh, 26.

CHAPTER 9

353   It just does take. Gertrude Stein, *Narration* (Chicago: University of
        Chicago Press, 2010), 1.
        true myth figure. Le Guin, 67.
354   a mummy in street clothes. Twitchell, 261.
355   He thought of Hyde. Stevenson, 65.
        Obediently, like an animal. Seabrook, 101.
357   a weird Robinson Crusoe. Matheson, 76–77.
        dull, robot-like creatures. Matheson.
        monotonous horror. Matheson, 104.
        All without bats fluttering. Matheson, 81.
359   It came out of the anger. Paul Wells, *The Horror Genre from Beelzebub to
        Blair Witch* (London: Wallflower Press, 2000), 80.
360   the real horror of *Night of the Living Dead*. Waller, 289.
361   Those who suffer. Jordan Peele, interview at https://news.avclub.com
        /jordan-peele-explains-how-privilege-influenced-us-in-an-1835225625.

361   **explore the nature of the status quo.** Waller, 15.
        primitive fear of the dead. Freud, 149.

366   **Whoever battles with monsters.** Waller, 352.
        exaggerated reality. Twitchell, 4, 7.

367   **One of the shifts in sensibility.** Twitchell, 22.
        until we resolve whatever it is. Twitchell, 20.

CHAPTER 10

371   **Gothic takes us on a tour.** Punter, 409.
        It both confronts the bourgeoisie. Punter, 417.

372   **derives its existence from the hinterland.** Carl Jung, *Modern Man in Search of a Soul*, trans. W. S. Dell and Cary F. Baynes (Abingdon: Routledge, 2001), 160.
        is a form of response. Punter, 128.
        do not come out right. Punter, 409.

373   **the unsaid and the unseen.** Jackson, 4.
        takes the real and breaks it. Jackson, 4.

374   **The principle value of studying.** Punter, 409.
        Those who refuse to listen. Filmer, 58.
        It is possible—and useful. Doty, 13.

375   **her novels are amusing enough.** Austen, 31.
        such unnatural stuff! Austen, 31.
        hot mess. Hughes.
        makes too much noise. Hughes.

376   **The incidents are too coarse.** Anon., *Spectator*, 18 December 1847.
        It is wild, confused. Anon., *Examiner*, 8 January 1848.
        true taste rejects. H. F. Chorley, *Athenaeum* 25 December 1847.
        never make lusty love. Andrew Glazzard, *New Statesman*, 15 Jan 2020, https://www.newstatesman.com/culture/books/2020/01/how-hg-wells-invented-modern-world.

377   **It is James who remains.** Glazzard.
        we intelligentsia snorted with pleasure. Ford Madox Ford, *Memories . . .* , 415.
        went snooping about the world. Ford, 415.
        millionwise production of everything. Ford, 422.
        It is art that makes life. Glazzard.

379   **where the story is more important.** Philip Pullman, *Daemon Voices*
       (Oxford: David Fickling Books, 2017), 24.
       science fiction is the mythology. Le Guin, 61.
       true myth arises only. Le Guin, 66.
380   **one can follow an exploration.** Le Guin, 68.
       what Dick is entertaining us about. Le Guin, 152.
       what are now called. Harold Bloom, *The Western Canon* (New York:
       Harcourt Brace, 1994), 519.
381   **The world of assassinations.** Punter, 393.

BIBLIOGRAPHY

Abbott, Stacey. *Celluloid Vampires: Life after Death in the Modern World*. Austin: University of Texas Press, 2007.

Aldiss, Brian. *Frankenstein Unbound*. London: Jonathan Cape, 1973.

Atwood, Margaret. *The Penelopiad*. Edinburgh: Canongate, 2005.

Auerbach, Nina. *Our Vampires, Ourselves*. Chicago: University of Chicago Press, 1995.

Austen, Jane. *Northanger Abbey*. Ware: Wordsworth Classics, 1993.

Baldick, Chris. *In Frankenstein's Shadow: Myth, Monstrosity, and Nineteenth-Century Writing*. Oxford: Clarendon Press, 1987.

Ball, Philip. *Unnatural: The Heretical Idea of Making People*. London: Bodley Head, 2011.

Bann, Stephen, ed. *Frankenstein, Creation and Monstrosity*. London: Reaktion, 1994.

Barker, Pat. *The Silence of the Girls*. Edinburgh: Canongate, 2005.

Beck, Peter J. *The War of the Worlds*. London: Bloomsbury, 2016.

Bergonzi, Bernard. *The Early H. G. Wells: A Study of the Scientific Romances*. Toronto: University of Toronto Press, 1961.

Boluk, Stephanie, and Wylie Lenz, eds. *Generation Zombie: Essays on the Living Dead in Modern Culture*. Jefferson, NC: McFarland, 2011.

Bondurant, Tom. "Becoming the Bat: The History of Batman's Many Origins," 10 December 2016. https://www.cbr.com/history-of-batmans-origins/

Brooker, Will. *Batman Unmasked: Analyzing a Cultural Icon*. New York: Continuum, 2005.

Buzwell, Greg. "Dracula: Vampires, Perversity and Victorian Anxieties." British Library, 15 May 2014. https://www.bl.uk/romantics-and-victorians/articles/dracula

Buzwell, Greg. "'Man is not truly one, but truly two': Duality in Robert Louis Stevenson's Strange Case of Dr Jekyll and Mr Hyde." British Library, 15 May 2014. https://www.bl.uk/romantics-and-victorians/articles/duality-in-robert-louis-stevensons-strange-case-of-dr-jekyll-and-mr-hyde

Cantril, Hadley. *The Invasion from Mars: A Study in the Psychology of Panic*. Princeton, NJ: Princeton University Press, 1940.

Carroll, Noel. *The Philosophy of Horror*. New York: Routledge, 1990.

Cervantes, Miguel. *Don Quixote*. Oxford: Oxford University Press, 1992.

Coetzee, J. M. *Foe*. London: Penguin, 1987.

Conan Doyle, Arthur. *Sherlock Holmes: The Complete Facsimile Edition*. Ware: Wordsworth Editions, 1996.

Conan Doyle, Arthur. *Sherlock Holmes: The Major Stories with Contemporary Critical Essays*, ed. John A. Hodgson. Boston: Bedford/St. Martin's, 1994.

Cornwell, Neil. *The Literary Fantastic: From Gothic to Postmodernism*. Hemel Hempstead: Harvester Wheatsheaf, 1990.

Craft, Christopher. *Another Kind of Love*. Berkeley: University of California Press, 1994.

Craft, Christopher. "Kiss Me with Those Red Lips." *Representations* 8 (1984): 107–33.

Daniels, Les. *Batman: The Complete History*. London: Titan, 1999.

Davidson, Brett. "*The War of the Worlds* Considered as a Modern Myth." *The Wellsian* 28 (2005): 39–50.

Doty, William G. *Myth: A Handbook*. Westport, CT: Greenwood, 2004.

Dundes, Alan, ed. *Sacred Narrative: Readings in the Theory of Myth*. Berkeley: University of California Press, 1984.

Durand, Kevin K., and Mary K. Leigh. *Riddle Me This, Batman! Essays on the Universe of the Dark Knight*. Jefferson, NC: McFarland, 2011.

Earle, Peter. *The World of Defoe*. London: Weidenfeld & Nicolson, 1976.

Eaves, A. Osborne. *Modern Vampirism: Its Dangers and How to Avoid Them*. Harrogate: Talisman, 1904.

Ellis, Frank H., ed. *Twentieth Century Interpretations of Robinson Crusoe*. Englewood Cliffs, NJ: Prentice-Hall, 1969.

Farghaly, Nadine, ed. *Gender and the Modern Sherlock Holmes*. Jefferson, NC: McFarland, 2015.

Feldman, Burton, and Robert D. Richardson. *The Rise of Modern Mythology 1680–1860*. Bloomington: Indiana University Press, 1972.

Filmer, Kath, ed. *The Twentieth Century Fantasists*. Basingstoke: Macmillan, 1991.

Foot, Michael. *The History of Mr. Wells*. Washington, DC: Counterpoint, 1995.

Forry, Steven Earl. *Hideous Progenies: Dramatizations of Frankenstein from Mary Shelley to the Present*. Philadelphia: University of Pennsylvania Press, 1990.

Frayling, Christopher. *Vampyres*. London: Faber & Faber, 1991.

Freeland, Cynthia A. *The Naked and the Undead: Evil and the Appeal of Horror*. Boulder, CO: Westview, 2000.

Freeman Loftis, Sonya. "The Autistic Detective." *Disability Studies Quarterly* 34, no. 4 (2014). https://dsq-sds.org/article/view/3728

Freud, Sigmund. *The Uncanny*. London: Penguin, 2003.

Gallop, Alan. *The Martians Are Coming!* Stroud: Amberley, 2011.

Geguld, Harry M., ed. *The Definitive Dr. Jekyll and Mr. Hyde Companion*. New York: Garland, 1983.

Gelder, Ken. *Reading the Vampire*. London: Routledge, 1994.

Grapard, Ulla, and Gillian Hewitson, eds. *Robinson Crusoe's Economic Man: A Construction and Deconstruction*. London: Routledge, 2011.

Groom, Nick. *The Vampire: A New History*. London: Yale University Press, 2018.

Halperin, David J. *Intimate Alien: The Hidden Story of the UFO*. Stanford, CA: Stanford University Press, 2020.

Halperin, David J. "The Moment the Myth of Alien Abduction Was Born." Literary Hub, 26 March 2020. https://lithub.com/the-moment-the-myth-of-alien-abduction-was-born/

Harkup, Kathryn. *Making the Monster: The Science behind Mary Shelley's Frankenstein*. London: Bloomsbury, 2018.

Hitchcock, Susan Tyler. *Frankenstein: A Cultural History*. New York: Norton, 2007.

Hoberman, J. "All As It Had Been." *Village Voice*, 4 December 2001. https://www.villagevoice.com/2001/12/04/all-as-it-had-been/

Huxley, Aldous. *Island*. London: Chatto & Windus, 1962.

Jackson, Rosemary. *Fantasy: The Literature of Subversion*. London: Routledge, 1988.

Johnston, Sarah Iles. *The Story of Myth*. Cambridge, MA: Harvard University Press, 2018.

Jones, Ernest. *On the Nightmare*. London: Hogarth Press, 1931.

Kane, Bob, and Tom Andrae. *Batman and Me*. Forestville, CA: Eclipse Books, 1989.

King, Stephen. *Salem's Lot*. London: Hodder & Stoughton, 2006.

Kirby, David A. "Darwin on the Cutting-Room Floor: Evolution, Religion, and Film Censorship." *Osiris* 34 (2019): 55–80.

Kirk, G. S. *Myth: Its Meaning and Functions in Ancient and Other Cultures*. Cambridge: Cambridge University Press, 1970.

Klinger, L. S., ed. *The New Annotated Sherlock Holmes*. New York: Norton, 2005.

Knight, Stephen. *Towards Sherlock Holmes: A Thematic History of Crime Fiction in the Nineteenth Century World*. Jefferson, NC: McFarland, 2016.

Lawrence, John Shelton, and Robert Jewett. *The Myth of the American Superhero*. Grand Rapids, MI: William Eerdmans, 2002.

Le Guin, Ursula. *The Language of the Night: Essays on Fantasy and Science Fiction*, ed. S. Wood. London: Women's Press, 1989.

Leatherdale, Clive. *Dracula: The Novel and the Legend*. 3rd ed. Desert Island Books, 2001.

Levine, George, and U. C. Knoepflmacher, eds. *The Endurance of Frankenstein: Essays on Mary Shelley's Novel*. Berkeley: University of California Press, 1979.

Luckhurst, Roger, ed. *The Cambridge Companion to Dracula*. Cambridge: Cambridge University Press, 2017.

Luckhurst, Roger, ed. *Science Fiction: A Literary History*. London: British Library, 2017.

Lycett, Andrew. *Conan Doyle: The Man Who Created Sherlock Holmes*. London: Weidenfeld & Nicolson, 2007.

Maixner, Paul, ed. *Robert Louis Stevenson: The Critical Heritage*. London: Routledge & Kegan Paul, 1981.

Maranda, Pierre, ed. *Mythology*. Harmondsworth: Penguin, 1972.

Matheson, Richard. *I Am Legend*. London: Millennium, 1999.

McLean, Steven. *The Early Fiction of H. G. Wells*. Basingstoke: Palgrave Macmillan, 2009.

McLean, Steven, ed. *H. G. Wells: Interdisciplinary Essays*. Newcastle: Cambridge Scholars Publishers, 2008.

McNally, Raymond, and Radu Florescu. *In Search of Dracula*. New York: Hawthorn, 1973.

Mellor, Anne K. *Mary Shelley: Her Life, Her Fiction, Her Monsters*. London: Routledge, 1988.

Michals, Teresa. *Books for Children, Books for Adults: Age and the Novel from Defoe to James*. Cambridge: Cambridge University Press, 2014.

Miller, Elizabeth. *Dracula: Sense and Nonsense*. Westcliff-on-Sea: Desert Island Books, 2000.

Miller, Karl. *Doubles: Studies in Literary History*. Oxford: Oxford University Press, 1985.

Milner, H. N. *Frankenstein; or, The Man and the Monster!* London: John Cumberland & Duncombe, 1826.

Musa, M. T. "A Freudian Analysis of Hysteria in Bram Stoker's Dracula." PhD thesis, Florida Atlantic University, 1994.

Nabokov, Vladimir. *Lectures on Literature*, ed. F. Bowers. London: Weidenfeld & Nicolson, 1980.

Newman, Kim. *Anno Dracula*. London: Titan, 2011.

Novak, M. E. *Realism, Myth and History in Defoe's Fiction*. Lincoln: University of Nebraska Press, 1963.

Page, Michael A. *The Literary Imagination from Erasmus Darwin to H. G. Wells*. London: Routledge, 2012.

Parrinder, P., and R. M. Philmus, eds. *H. G. Wells' Literary Criticism*. Brighton: Harvester Press, 1980.

Pearson, Roberta E., and William Uricchio. *The Many Lives of the Batman*. New York: Routledge, 1991.

Philmus, Robert M. *Into the Unknown*. Berkeley: University of California Press, 1970.

Picariello, Damien K., ed. *Politics in Gotham*. Basingstoke: Palgrave Macmillan, 2019.

Pirie, David. *The Vampire Cinema*. London: Hamlyn, 1977.

[Polidori, John William] (published anonymously). "The Vampyre," *New Monthly Magazine*, 1 April 1819. https://en.wikisource.org/wiki/The_Vampyre/The_Vampyre

Pooley, Jefferson, and Michale J. Socolow. "The Myth of the War of the Worlds Panic." *Slate*, 28 October 2013. https://slate.com/culture/2013/10/orson

-welles-war-of-the-worlds-panic-myth-the-infamous-radio-broadcast-did
-not-cause-a-nationwide-hysteria.html

Prawer, S. S. *Caligari's Children: The Film as Tale of Terror*. Oxford: Oxford University Press, 1980.

Preston, Michael J. "Robinson Crusoe, Gulliver's Travels, and Alice's Adventures in Wonderland: Wonderful Texts." *Merveilles & Contes* 2, no. 2 (1988): 87–105.

Proctor, Richard. "Have We Two Brains?" *Cornhill Magazine* 31 (February 1875): 149–66.

Punter, David. *The Literature of Terror: A History of Gothic Fictions From 1765 to the Present Day*. New York: Longman, 1980.

Putney, Charles R., Joseph A. Cutshall King, and Sally Sugarman, eds. *Sherlock Holmes: Victorian Sleuth to Modern Superhero*. Lanham, MD: Scarecrow Press, 1996.

Rees, Amanda, and Iwan Rhys Morus. "Presenting Futures Past: Science Fiction and the History of Science." *Osiris* 34 (2019): 1–15.

Reichstein, Andrew. "Batman—An American Mr. Hyde?" *American Studies* 43, no. 2 (1998): 329–50.

Reid, Julia. *Robert Louis Stevenson, Science and the Fin de Siècle*. Basingstoke: Palgrave Macmillan, 2006.

Reynolds, Richard. *Super Heroes: A Modern Mythology*. Jackson: University Press of Mississippi, 1992.

Rieger, James. "Dr. Polidori and the Genesis of *Frankenstein*." *Studies in English Literature* 3 (1963): 461.

Robinson, Jack. *Good Morning Mr Crusoe*. London: CB Editions, 2019.

Saler, Michael. "Modernity, Disenchantment and the Ironic Imagination." *Philosophy & Literature* 28 (2004): 137–49.

Segal, Robert. *Myth: A Very Short Introduction*. Oxford: Oxford University Press, 2004.

Seidel, Michael. *Robinson Crusoe: Island Myths and the Novel*. Boston: Twayne, 1991.

Serviss, Garrett P. *Edison's Conquest of Mars*. Los Angeles: Carcosa House, 1947.

Shelley, Mary. *Frankenstein*. Norton Critical Edition, 2nd ed., ed. J. Paul Hunter. New York: Norton, 2012.

Shelley, Mary. *Frankenstein. Annotated for Scientists, Engineers, and Creators of All Kinds*, ed. David H. Guston, Ed Finn, and Jason Scott Robert. Cambridge, MA: MIT Press, 2017.

Shelley, Mary. *The Last Man*. Oxford: Oxford University Press, 1998.

Shinagel, Michael, ed. *Robinson Crusoe*. Norton Critical Edition, 2nd ed. New York: Norton, 1994.

Rice, Anne. *Interview with the Vampire*. London: Sphere, 2008.

Richardson, Ruth. "*Frankenstein*: Graveyards, Scientific Experiments and Bodysnatchers." British Library, 15 May 2014. https://www.bl.uk/romantics -and-victorians/articles/frankenstein-graveyards-scientific-experiments-and -bodysnatchers

Sims, Michael. *Arthur and Sherlock: Conan Doyle and the Creation of Holmes*. London: Bloomsbury, 2017.

Skal, David J. *Hollywood Gothic*. New York: Norton, 1990.

Skal, David J. *Something in the Blood: The Untold Story of Bram Stoker, the Man Who Wrote* Dracula. New York: Norton, 2016.

Smith, Andrew, ed. *The Cambridge Companion to Frankenstein*. Cambridge: Cambridge University Press, 2016.

Smith, E. E., and R. Haas, eds. *The Haunted Mind: The Supernatural in Victorian Literature*. Lanham, MD: Scarecrow Press, 1999.

Sontag, Susan. "The Imagination of Disaster." In *Against Interpretation and Other Essays*. New York: Picador, 1966.

Spadoni, Robert. *Uncanny Bodies: The Coming of Sound Film and the Origins of the Horror Genre*. Berkeley: University of California Press, 2007.

Stashower, Daniel. *Teller of Tales: The Life of Arthur Conan Doyle*. New York: Henry Holt, 1999.

Stevenson, Robert Louis. *Strange Case of Dr Jekyll and Mr Hyde, and Other Tales*. Ed. Roger Luckhurst. Oxford: Oxford University Press, 2006.

Stoker, Bram. *Dracula*. Ed. Maurice Hindle. London: Penguin, 2003.

Sutherland, John. "Sherlock Holmes, the World's Most Famous Literary Detective." British Library, 15 May 2014. https://www.bl.uk/romantics-and -victorians/articles/arthur-conan-doyle-the-creator-of-sherlock-holmes-the -worlds-most-famous-literary-detective

Todorov, Tsvetan. *The Fantastic: A Structural Approach to a Literary Genre*. Ithaca, NY: Cornell University Press, 1975.

Tropp, Martin. *Mary Shelley's Monster: The Story of Frankenstein*. Boston: Houghton Mifflin, 1977.

Twitchell, James B. *Dreadful Pleasures: An Anatomy of Modern Horror*. Oxford: Oxford University Press, 1985.

Turney, Jon. *Frankenstein's Footsteps: Science, Genetics and Popular Culture*. New Haven, CT: Yale University Press, 1998.

Veeder, W., and G. Hirsch. *Dr. Jekyll and Mr. Hyde after One Hundred Years.* Chicago: University of Chicago Press, 1988.

Waller, Gregory A. *The Living and the Undead: Slaying Vampires, Exterminating Zombies.* Urbana: University of Illinois Press, 2010.

Warshak, Richard A. "Batman's Traumatic Origins." *Atlantic*, 6 May 2014. https://www.theatlantic.com/entertainment/archive/2014/05/batmans -traumatic-origins/361638/

Watt, Ian. *Myths of Modern Individualism.* Cambridge: Cambridge University Press, 1996.

Watt, Ian. *The Rise of the Novel.* London: Chatto & Windus, 1967.

Weir, Andy. *The Martian.* London: Penguin, 2015.

Wells, H. G. *The Scientific Fiction.* Vol. 1. London: J. M. Dent, 1995.

Wertham, Fredric. *Seduction of the Innocent.* London: Museum Press, 1955.

Willis, Martin. *Mesmerists, Monsters, and Machines: Science Fiction and the Cultures of Science in the Nineteenth Century.* Kent, OH: Kent State University Press, 2006.

Winterson, Jeanette. *Weight.* Edinburgh: Canongate, 2005.

Wolf, Leonard. *Dracula: The Connoisseur's Guide.* New York: Broadway, 1997.

Wolf, Leonard. *The Essential Dr. Jekyll and Mr. Hyde.* New York: ibooks, 2005.

Young, Elizabeth. *Black Frankenstein: The Making of an American Metaphor.* New York: New York University Press, 2008.

INDEX